Principles of Brain Evolution

PRINCIPLES OF
Brain Evolution

Georg F. Striedter

University of California, Irvine

SINAUER ASSOCIATES, INC. *Publishers*
Sunderland, Massachusetts U.S.A.

About the cover:

Vertebrate brains come in many different shapes and sizes, but our human brains are "special" because they allow us to discover how those diverse brains are interrelated. This montage depicts a human brain superimposed on the center of a "starburst" amacrine cell from the retina of a tree shrew (after Sandmann et al., 1997). The neuron's periphery is lined with lateral views of various vertebrate brains (clockwise from top: mormyrid electric fish, lungfish, lamprey, spiny dogfish, smooth dogfish, opossum, raven, salamander, turtle, hagfish, trout). The background image shows avian neurons that were labeled at their time of "birth" with tritiated thymidine and/or bromodeoxyuridine.

Address editorial correspondence and orders to:
Sinauer Associates, 23 Plumtree Road, Sunderland, MA 01375 U.S.A.
FAX: 413-549-1118
Email: publish@sinauer.com
Internet: www.sinauer.com

Library of Congress Cataloging-in-Publication Data

Striedter, Georg F., 1962-
  Principles of brain evolution / Georg F. Striedter.
      p. cm.
  Includes bibliographical references and index.
  ISBN 0-87893-820-6 (hardcover)
  1. Brain—Evolution. I. Title.

  QP376.S825 2005
  573.8'616—dc22   2004021721

Printed in U.S.A.

6 5 4 3 2 1

For my son, Ian

Contents

Acknowledgments

Writing a book is like washing an elephant: there's no good place to begin or end, and it's hard to keep track of what you've already covered.

—Author unknown

I stumbled across this anonymous quotation when I began writing this book, and it resonated in my mind whenever I was struggling to express and organize the thoughts I wanted to communicate. After two years of intensive work, the book is finished but the elephant continues to have some dirty spots. Still, much of what I wanted to accomplish has, somehow, materialized. For that, I owe a debt of gratitude to several special institutions and people.

First, I thank Berlin's Institute for Advanced Studies (or "Wissenschaftskolleg") for giving me a one-year fellowship to focus solely on writing. Berlin's "Wiko," as it is affectionately known, is one of those rare places where pure scholarship is nurtured without compromise. I still feel guilty for abandoning my laboratory and displacing my family as we moved to Berlin, but I am certain that, without that year of quiet time, this book project would have withered on the vine. While in Berlin, I benefited greatly from discussions with the other Wiko fellows and guests, notably Stephen Simpson, David Raubenheimer, Barbara Finlay, Alex Kacelnik, Luis Puelles, Tecumseh Fitch, Stefano Nolfi, Raghavendra Gadagkar, Gerhard Neuweiler, and Rüdiger Wehner. Thank you, Wiko, for bringing us together and facilitating the pursuit of our various, idiosyncratic dreams. I also thank UC Irvine for exempting me from teaching and committee obligations for that year, and the National Science Foundation for supporting my laboratory research.

Second, I thank those precious friends who gave me valuable feedback on preliminary drafts. David Raubenheimer, Barbara Finlay, Martin Wild and Fabiana Kubke read almost the entire book, corrected numerous mistakes and made extremely helpful suggestions. Michael Arbib, Roland Bender, Richard Darlington, Terrence Deacon, Jon Kaas, Loreta Medina, Glenn Northcutt,

Todd Preuss, Michael Rugg, Charles Stevens, and Norman Weinberger offered similarly useful commentary on individual chapters. Without their insights and attention to detail, this book would contain many more flaws and be considerably less intelligible. Thanks are due as well to Graig Donini, Kerry Falvey, Chelsea Holabird, Christopher Small, Janice Holabird, and Jefferson Johnson of Sinauer Associates for their editorial feedback and support. I always thought Sinauer's books had class; now I know why.

Third, I thank my various mentors. Paul Sherman supervised my undergraduate research and sparked my interest in evolutionary theory. Glenn Northcutt was my Ph.D. advisor; his impact on my thinking has been enormous. Mark Konishi, finally, advised me during the postdoctoral years; he demonstrated how to integrate neuroanatomy with physiology and animal behavior. Without these three mentors, this book would never have happened.

Finally, I thank my family. They keep me balanced and remind me frequently that washing elephants is merely part of what I want to do in life.

GEORG F. STRIEDTER
IRVINE, CALIFORNIA
SEPTEMBER, 2004

1 Evolutionary Neuroscience: This Book's Scope and Ambition

It is interesting to contemplate a tangled bank, clothed with many plants of many kinds, with birds singing on the bushes, with various insects flitting about, and with worms crawling through the damp earth, and to reflect that these elaborately constructed forms, so different from each other, and dependent upon each other in so complex a manner, have all been produced by laws acting around us.

—Charles R. Darwin, 1859

*L*ate in January of 1898, roughly 40 years after Darwin published his *Origin of Species,* Professor Hermon Bumpus grasped a unique opportunity to test Darwin's "law" of natural selection (Bumpus, 1899). Walking to work after a particularly severe winter storm, he found 136 English sparrows lying freezing in the snow. He collected them and brought them to his lab, where he was able to save about half of them. After weighing all the sparrows and measuring their principal body parts, Bumpus observed that "there are fundamental differences between the birds which survived and those which perished. While the former are shorter and weigh less (i.e., are of smaller body), they have longer wing bones, longer legs, longer sternums, and greater brain capacity." (Bumpus, 1899, p. 213). As Bumpus saw it, he had found direct physical evidence of how Nature "eliminates the unfit."

We now have far more extensive evidence for natural selection (Endler, 1986; Grant and Grant, 1989), but evolutionary biologists continue to be intrigued by Bumpus's old data on sparrows and have subjected them to increasingly sophisticated analyses (Grant, 1972; Pugesek and Tomer, 1996). By and large, these later studies have confirmed that the sparrows that died in Bumpus's hands differed from the survivors in at least some anatomical measures. However, none of the later studies has addressed Bumpus's assertion that the surviving sparrows had a larger brain capacity. In part, this omission is due to the fact that Bumpus measured not brain size but skull width, which is at best a highly indirect measure of brain size. However, it is also

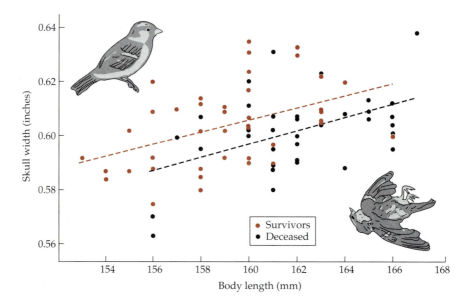

Figure 1.1 Natural Selection in Bumpus's Sparrows   In the winter of 1898, Hermon Bumpus collected more than 100 English sparrows that had died, or nearly died, during a severe snowstorm. After making detailed measurements of both the survivors (red data points) and the deceased (black data points), Bumpus made inferences about the selective forces that had acted on this population. Shown here is a plot of skull width versus body length for the males in Bumpus's data set. If we assume that skull width is a reasonable indicator of brain size (which it probably is in small songbirds) and that body length is a good indicator of body size, this plot reveals that, at any given body size, the survivors tended to have larger brains than the deceased. If one fits a single line to all the data and then compares the residual values for the two groups, the difference is statistically significant ($p = 0.01$; author's analysis). (After Bumpus, 1899.)

likely that subsequent investigators simply found the difference in mean skull width between the two groups of birds to be statistically insignificant (Bumpus himself had noted that the difference was small, but he did not test for statistical significance). But is mean skull width really an appropriate measure if we want to know whether individuals with larger brains were more likely to have survived the storm?

To answer this question, we must consider that, both across and within species, larger individuals tend to have larger brains. This general principle is best seen in graphs of brain size versus body size (Figure 1.1). When one fits a line to the data points on such a graph, the line invariably has a positive slope (see Chapter 4 for more details on such allometric analyses). Now, if we draw such a graph for all of the male sparrows in the Bumpus data set, letting skull width and body length serve as proxies for brain and body size, respectively, we can see that the best-fit line for the surviving sparrows lies well above the best-fit line for the sparrows that died. Therefore, the data suggest that, at any given body size, the males with larger brains were more likely than their smaller-brained associates to have survived the winter storm. Since it is also true that the survivors were slightly smaller, on average, than the deceased, mean absolute brain size (i.e., mean skull width) turns out to be quite similar in the two groups. However, relative brain size (i.e., skull width relative to what one would expect in a bird of a given body length) was significantly larger in the sparrows that survived than in those that died.

Interpreting these results is difficult, of course. For example, we do not know whether Bumpus measured skull width and body length identically in living and dead birds, or how accurately skull width predicts brain size in sparrows. More important, even if skull width is an accurate predictor of brain size, how would an increase in relative brain size help a sparrow survive winter storms? Perhaps survivors were socially dominant and therefore better able to compete for food and shelter, but Bumpus did not collect the behavioral data relevant to that hypothesis. Simply put, we know that brainier sparrows were favored under conditions of extreme climatic stress but we do not know why.

Much more comprehensive studies are needed to answer such evolutionary "why questions" (Mayr, 1961) and, given the time and labor required to demonstrate selective pressures in the wild (Endler, 1986), such studies are inevitably rare. Indeed, I know of no empirical work, other than that of Bumpus, that explores how natural selection affects brain size or structure. (Some work has been done on the effects of selective breeding on brain size, but natural and artificial selection are different beasts [e.g., Hill and Mbaga, 1998]). Therefore, this book has relatively little to say about the role of natural (or sexual) selection in brain evolution. I will, at various points in the book, hypothesize about the behavioral benefits that might accrue from specific changes in brain structure, but only rarely will I try to specify why natural selection might have favored the emergence of particular features. For the most part, such questions are simply too difficult to answer at this point (to get an appreciation for how difficult such studies are, consider the work of Airey et al., 2000).

A second topic that is fascinating but intentionally neglected in this book is the study of "fossil brains" (Edinger, 1929). It is often said that brains do not fossilize, but that is only partially correct. Under particularly favorable conditions, sediment can infiltrate a dead animal's skull and fill its endocranium, the cavity housing its brain. Such fossil endocasts can be examined either by removing the surrounding skull or, in a technique developed recently, by scanning the entire specimen in sophisticated X-ray machines (Figure 1.2). Careful examination of these endocasts can provide important clues about the

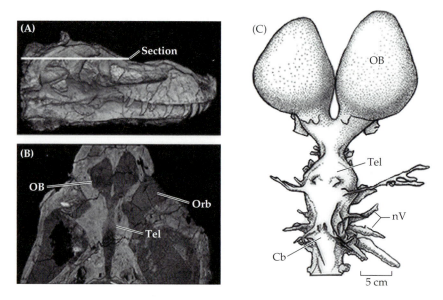

Figure 1.2 Head and Endocast of *Tyrannosaurus rex*   The head of this *Tyrannosaurus rex* specimen, from Chicago's Field Museum of Natural History, was recently subjected to high-resolution computed tomographic (CT) analysis, which allowed the endocast to be reconstructed without removing the overlying skull. (A) Lateral view of the entire head, indicating the plane of section for the scan shown in B. (B) Horizontal scan through the head. (C) Dorsal view of the entire endocast, as reconstructed from multiple sections. Several cranial nerves and some major brain regions can be identified. Given the large size of the olfactory bulbs in this specimen, it is likely that *T. rex* had a well-developed sense of smell. Abbreviations: Cb = approximate site of the cerebellum; nV = branches of the trigeminal nerve; OB = olfactory bulb; Orb = orbit; Tel = approximate location of the telencephalon. (From Brochu, 2000 with minor modifications.)

size and gross structure of an animal's brain (Jerison, 1973; Hopson, 1979; Buchholtz and Seyfarth, 1999; Rogers, 1999; Conroy et al., 2000; Tobias, 2001), but paleoneurology, as the study of endocasts is called, has major limitations. For one thing, in many fishes, amphibians, and reptiles, the brain does not fill more than about half of the endocranial space, which means that estimates of brain size from endocasts are perforce inaccurate. Even in birds and mammals, the extent to which the brain fills the endocranial cavity varies with body size, which is difficult to estimate from fossil data (see Smith, 2002). Moreover, fossil endocasts reveal nothing of the intricate microstructure that fascinates neurobiologists (Figure 1.3). Therefore, this book refers to fossil data only occasionally (in the chapters on mammalian and human brain evolution). Detailed reviews of paleoneurology, and references to much of that literature, can be found in the papers or monographs listed above.

Compared to the scarcity of information that can be gleaned from fossil brains, the amount of information available for living brains is vast. Accord-

(A) The brain of a gymnotoid electric fish (*Apteronotus albifrons*)

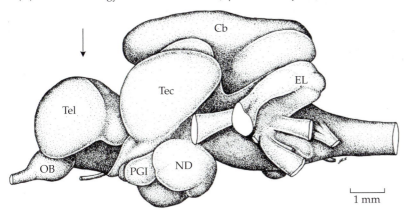

(B) Cross section through the telencephalon

(C) Labeled axons

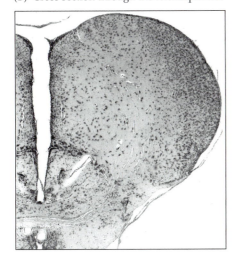
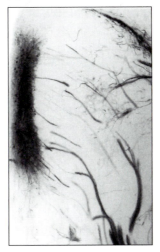

Figure 1.3 Views of a Teleost Fish Brain   One surefire way to demonstrate that species differences in brain anatomy are real is to ask students of mammalian brains to find their way around the brains of teleost fishes, such as the gymnotoid electric fish shown here. (A) A lateral view of this creature's brain reveals some familiar features, such as an olfactory bulb (OB), but also many that seem odd, including an "electrosensory lateral line lobe" (EL; see Chapter 5); a hypothalamic "nucleus diffusus" (ND); a bulging "lateral preglomerular nucleus" (PGl; see Chapter 6) and several large cranial nerves. (B) A Nissl-stained cross section through the right telencephalon (at roughly the level indicated by the arrow in A) reveals cytoarchitectural details that likewise seem alien if you have studied only mammalian brains. Where is the neocortex? Where is the hippocampal formation (see Figure 2.16)? (C) Studies of neuronal connections can reveal further details (shown here are axons projecting from a diencephalic cell group to the dorsomedial telencephalon), but they do not resolve all species differences. In fact, even among teleosts, neuronal connections differ considerably (e.g., see Striedter, 1992). Additional abbreviations: Cb = cerebellum; Tec = optic tectum; Tel = telencephalon. (Panel A after Striedter, 1992.)

ing to most estimates, there are roughly 50,000 species of extant vertebrates (about 50% are bony fishes, 20% birds, and 10% mammals). Within each of the seven vertebrate classes, only a handful of species has been studied in detail, but these model species have already yielded a treasure trove of information. This becomes evident when one examines the major treatises that have tried to summarize what is known about vertebrate brains. The most influential classic text, published by Ariëns Kappers et al. in 1936, contains 1,845 pages and about 5,900 citations to the literature. The most recent comprehensive text, published by Nieuwenhuys et al. in 1998, contains more than 2,200 pages and 10,000 references. Reading such massive books (see also Kuhlenbeck, 1967–1978) is a daunting task, and I suspect that few readers have ever managed more than a few pages at a time. Even some of the authors have, at times, been overwhelmed. As Rudolf Nieuwenhuys remarked near the end of his own book:

> *Looking back on this whole endeavor, spanning as it does more than two decades of work, we are struck by a combination of frustration and wonder. We are frustrated because it would be satisfying to conclude by highlighting some clever and subtle principle that made sense of all that has gone before, to reveal the secret of brain structure and its organization. Instead, we are left with a sense of awe at the myriad complexity of it all. Looking at the brains of many animals, one has the feeling of being in a room, on the floor of which are closed boxes of many shapes and sizes. Opening any box, large or small, reveals something like an immensely intricate and exquisite assembly of components and circuits. The contents of each box resemble the contents of the others, yet each has a distinct identity and the assemblies are obviously functioning, but not according to any known paradigm. (Nieuwenhuys et al., 1998, pp. 2189–2190)*

This statement is exaggerated, for Nieuwenhuys himself has helped to identify some fairly general and clever principles of brain evolution (Nieuwenhuys, 1994), but the complexity of vertebrate brains surely inspires awe, and their diversity challenges the order-loving mind.

One way to make sense of this diverse complexity is to reconstruct the history of how it came about. Because past attempts at brain historiography have often been fanciful (Sagan, 1977), it is tempting to condemn all such efforts as mere speculation, but that position is no longer justified. Methodological advances, particularly the development of cladistics (Chapter 2), have largely removed the subjective element from phylogenetic reconstruction and replaced it with rigorous hypothesis testing. Therefore, we can now, at least in principle, make sense of Nieuwenhuys's "myriad complexity" by scientifically reconstructing the process that gave rise to it.

Unfortunately, there are not yet enough data to allow a full historical account of how brains became so complex and diverse (see Butler and Hodos, 1996). Nor, I suspect, would such an account be much fun to read, for it would have to contain innumerable statements about what is homologous to what (or not!) in the various vertebrate lineages, followed by an endless list of which features appeared when. It would be useful to specialists, much as any historical record is useful to professional historians, and nonspecialists could use it as an encyclopedia to look up the homologies that interest them, but it would recruit few newcomers into the field of evolutionary neuroscience. It would likely be a "tombstone for the field" (Northcutt, 1998). That is one reason why this book will not emphasize the purely historical approach to brain evolution.

My second major reason for eschewing brain historiography is that the organizational structure required of such historical narratives tends to interfere with the recognition of general principles. Given the great complexity of brains, any systematic discussion of their history must proceed one region (or functional system) at a time; it must also treat one lineage at a time because evolutionary history has generally proceeded differently in different lineages. The book by Nieuwenhuys et al. (1998), for example, contains separate chapters for hagfishes, lampreys, cartilaginous fishes, various bony fishes, amphibians, reptiles, birds, and mammals; each of these chapters deals separately with gross brain structure, spinal cord, medulla, cerebellum, midbrain, diencephalon, and telencephalon. The main advantage of such an organization is that readers can easily locate most of the information that is of interest to them (e.g., what does the lateral geniculate nucleus look like in a bird?). Its main disadvantage is that logical connections between phylogenetic changes in different taxonomic groups, and/or in different brain regions or systems, remain difficult to grasp. For example, the many similarities in how bird and mammal brains evolved (Chapter 8) remain for careful readers to extract. In order to avoid this problem and still adopt the historical approach, one may write additional, more integrative sections or chapters (see Nieuwenhuys et al., 1998), but such discussions easily become disconnected from their more descriptive counterparts. As you shall see, this book adopts a rather different organization, focused specifically on principles and general ideas rather than historical and factual particulars. The following section provides a simplified preview.

The Book's Scope and Major Themes

I have written this book for advanced undergraduate and graduate students, as well as some more senior scientists, who already know something about the brain but want a deeper understanding of how diverse brains evolved. In order to make the book broadly accessible, I have avoided neuroanatomical jargon and other terms that only experts tend to use. My emphasis throughout is on the subjects that I find most intriguing. I have, for instance, virtually ignored the central nervous systems of invertebrates, for I know little about them. Even within vertebrates, I cover some species far more extensively than others, giving *Homo sapiens* a chapter all its own. Generally, my aim has been to pique your interest, not to tell you everything. As I reviewed above, encyclopedic works are already published (Ariëns Kappers et al., 1936; Kuhlenbeck, 1967–1978; Nieuwenhuys et al., 1998), and you can use the approximately 1,200 cited references to help you track down more details. Hopefully, this book will help you deal constructively with that intimidating literature.

As I conceive of it, evolutionary neuroscience is a vast multilayered enterprise that draws on diverse disciplines that range from molecular genetics to comparative psychology. It involves scrutinizing brain anatomy as well as physiology, and it attempts to link those realms to animal behavior and ecology. It is grounded in history but benefits as well from purely theoretical ideas about how complex systems generally vary with time. I here draw on all of these approaches, but neuroanatomy provides the starting point for most of the discussions in this book. Partly this is due to my own expertise, for I was trained as a comparative neuroanatomist (Striedter and Northcutt, 1989), and partly it reflects the field's own history, since the first scientists to seriously investigate how brains evolved were all comparative neuroanatomists (Chapter 2). Additionally, comparative neuroanatomists have gathered most of the available data on species differences in brains. By comparison, molecular neurobiologists and neurophysiologists tend to report mainly species similarities, which makes it difficult to say much more than "conservation has occurred." Comparative psychologists have shown more interest in the differences between species, generally adopting Darwin's view that evolution is "descent with modification," but their analyses are often tenuous and/or divorced from neuroscience. For all of these reasons, this book is anchored firmly in comparative neuroanatomy, fanning out to other levels of analysis only when and where good data are available.

Overall, my purpose in this book is, as the title promises, to explicate some major rules or principles of how brains generally evolve. In doing so, I highlight not one law that solely governs how all brains evolve, but a variety of principles that coexist and interact. This pluralistic perspective seems radical if one thinks of evolution as proceeding steadily from simple to complex, following a universal "law of progression," but that once-popular, monistic view of evolution is no longer tenable. Still, there is no doubt that having a multiplicity of principles makes comprehension more difficult: Of all the rules and principles, which ones are most significant? How do they all cohere?

These questions are profound, and this whole book is my attempt to answer them. Although I cannot offer a complete theory of how brains generally evolve, I do review some major steps toward that goal. Some principles do seem to be central, and some of their interrelationships can be discerned. Because these principles, relationships, and take-home messages are sometimes difficult to discern, I introduce them now in an abbreviated, stream-lined form. That way you will be primed for them.

The most important principle of brain evolution is that many aspects of brain structure and function are conserved across species, with closely related species tending to have brains that are more similar than those of distant relatives. Generally speaking, the degree of conservation is highest at the lowest levels of organization (genes and other molecules), and embryonic brains tend to be far more similar than adult brains. I review those most conserved aspects of brains in Chapter 3. In that same chapter, however, I also stress that species differences are real and should not be dismissed. They may be more prominent at some levels than at others, but they are important and deserve to be explained. We should ask, for instance, how conserved genes can direct brain development to different ends, and how adult structures can be so conserved when their genetic and developmental foundations have continued to evolve. In other words, how do conservation and divergence at the various levels of organization relate to one another? I raise this question most explicitly in Chapter 3, but it recurs throughout the book. Additionally, I often ask, what kinds of changes in development might explain divergences in adult brain anatomy? Most of the time, we must plead ignorance. However, some aspects of brain development are known, and that allows us to hypothesize how evolutionary changes in development might effect changes in adult anatomy. In other words, we can try to determine whether the rules of *brain development* have consequences for how brains *evolve*. As you shall see, some principles of brain evolution do, indeed, seem rooted in the rules of brain development.

In terms of understanding how brains vary across species (instead of how they are the same), the most important general principle is that brains tend to change in internal organization as they vary in size. Across all vertebrates, absolute brain size varies by 5 orders of magnitude, from less than 20 mg to more than 2 kg; that variation in brain size correlates with variation in diverse structural respects, including neuron number, size, and density, as well as neuronal connectivity and the size of various brain regions relative to one another. More important, these attributes all scale at different rates (relative to absolute brain size) and tend to be interrelated causally. This means that evolutionary changes in absolute brain size by necessity entail a slew of structural changes, which collectively are likely to have serious effects on an animal's behavior. Curiously, many neuroscientists tend to dismiss absolute brain size as a relatively boring and old-fashioned variable (I, at least, used to do this). Even evolutionary neuroscientists who are intrigued by measures of brain size tend to neglect absolute brain size in favor of relative brain size, which controls for the effect of body size on absolute brain size (see Figure 1.1 and Chapter 4). This emphasis on relative brain size does make some sense, since it is only by this measure that we humans have the largest brains (in terms of absolute brain size, whales and elephants have larger brains). How-ever, absolute brain size is more concrete and easier to think about than rela-tive brain size, and it is starting to attract more attention among evolutionary neuroscientists. I will now briefly spell out three of the most central principles of brain scaling in vertebrates.

The best-studied principle of brain scaling is that a brain region's size rela-tive to other regions (i.e., its proportional size) tends to change predictably with absolute brain size (Chapter 5). The dorsal telencephalon, for instance, becomes disproportionately large as absolute brain size increases, both in mammals and in birds. Indeed, each major brain region tends to scale against brain size with a characteristic slope, and different regions typically have dif-ferent slopes. Consequently, the proportions of the various brain regions change predictably with absolute brain size. In aggregate, this means that individual brain regions tend to evolve not independently of one another, but in concert. This initially counterintuitive notion remains controversial, for exceptions to the rule clearly exist, especially when we compare major taxo-nomic groups such as primates and rodents. Overall, however, I consider those exceptions to be relatively rare. More important, we now know why changing absolute brain size should have predictable effects on the propor-tions of the major brain regions: as long as evolution creates larger brains by "stretching" brain development (which it generally seems to do), brain regions that mature relatively late should become disproportionately large. This rule of "late equals large" (Finlay and Darlington, 1995), is a prime example of an evolutionary principle that springs from underlying principles of brain development. It may not explain all the variation among brains, par-ticularly if the comparisons involve some distant relatives, but its explanatory power is significant.

A second major scaling principle is that, as individual brain regions change in size, they tend to change in internal structure (Chapter 6). Specifically, brain regions tend to fractionate into more subdivisions, nuclei, or areas as they enlarge phylogenetically. This size-related proliferation of brain subdi-visions may be due to the addition of some truly "new" brain areas or to the segregation of components that are "old." Either way, complexity increases. That, in turn, is likely to allow different areas to specialize for different func-tions, leading to improved performance on at least some tasks. Closely related is the principle that enlarged brain regions tend to become laminar (their neu-rons form sheets rather than nuclei). The mammalian neocortex is the best-known laminar structure, but lamination evolved independently in many dif-ferent brain regions, all of which are large relative to their ancestral and unlaminated state. Why does lamination correlate with changes in brain region size? The finding that in laminar brain regions, axons and dendrites tend to course parallel or at right angles to the cellular layers in laminar brain regions suggests that lamination tends to minimize connection lengths. This, in turn, suggests that lamination conserves space and energy, which become more limited as neuron number is increased. In addition, lamination may allow for temporally precise forms of information processing that are difficult to carry out in nonlaminar structures. Thus, both the tendency of enlarged brain regions to become more subdivided and their tendency to become laminar can be explained in terms of improved functionality. In addition, both phylogenetic subdivision and lamination may be understood in terms of brain development. Specifically, I suggest that regions subdivide as they enlarge because the distance over which developmentally important molecules can diffuse or interact is physically limited, and that lamination evolves frequently because it is quite "easy" to develop, using mechanisms already employed to construct maps within the brain. These developmental explanations remain vague and speculative, but they are amenable to more research.

The third major aspect of brain organization that changes predictably with absolute brain size is neuronal connectivity (Chapter 7). Specifically, connection density (the proportion of a brain's neurons that are interconnected directly) tends to decrease as neuron number increases. This rule arises naturally from the fact that individual neurons are limited in how many other neurons they can innervate (or receive input from); it is functionally significant because without it, neuronal wiring costs would increase exponentially with increasing brain size. Decreasing connection density also forces brains to become more modular, by which I mean that distant brain regions become functionally more independent and diverse. Up to a point, this increased modularity is beneficial but too much modularity impedes coordination. Apparently, over the course of evolution brains have minimized this problem by adopting a "small world" architecture that allows distant neurons to communicate efficiently even when connection density is low. Superimposed upon the rule of decreasing connection density is Deacon's (1990) rule of "large equals well-connected," which links changes in a region's connectivity to changes in its size relative to other areas. It states, for instance, that as regions become disproportionately large, they tend to "invade" regions that they did not innervate ancestrally. This increases the ability of the enlarged, invading region to influence other brain regions and thereby makes the enlarged region more important for normal brain function and behavior. Empirically, Deacon's rule is not yet firmly established, but it follows logically from what we know about how axons generally grow during development. Combined with the aforementioned rule of late equals large, it is likely to have played a major role in brain evolution.

Overall, the message is that evolving brains are subject to a tangled web of rules and principles, containing many constants and some crucial variables, chief among them absolute brain size. Of course, brain size is not everything! If it were, then same-sized bird and mammal brains should be identical, which they are definitely not (Chapter 8). As I mentioned earlier, there are exceptions to the rule: sometimes brain regions change in size independently of overall brain size, some regions become simpler even though they have increased in size (e.g., the neocortex of cetaceans; see Chapter 10), and some neuronal connections vary independently of origin or target size. Nonetheless, as long as we restrict our comparisons to species in a single taxonomic class (e.g., to mammals or birds), absolute brain size is a fascinating and informatory variable. In comparisons within orders (e.g., within primates) it becomes downright invaluable. This conclusion may seem odd in our modern age, where most research is aimed at neuroanatomical or physiological details, but it is incorrect to think of size as being independent of such details. As I take pains to emphasize, evolutionary changes in brain size frequently go hand in hand with major changes in both structural and functional details. Indeed, they often demand them! This insight is not new (e.g., see Deacon, 1990; Finlay and Darlington, 1995), but it has never been presented as extensively as in this book. Therefore, it is fair to ask how well this size-driven view of vertebrate brain evolution "works" in the real world. It is fine to talk of abstract rules and principles, but can they help us understand specific brains? In an effort to answer this question, I devote Chapters 8 and 9 to the brains of mammals and, within mammals, humans. How do the brains of these two taxonomic groups differ from the brains of other vertebrates? What, if anything, makes them special? How well do our scaling principles apply?

Although mammals comprise only about one-tenth of all the vertebrates, they are a reasonably successful class of animals, particularly if we include humans. This success is due to various factors, notably the ability to generate internal body heat and an extended hearing range. Neurobiologically, mammals are distinguished mainly by their neocortex, which has nonmammalian precursors but is highly modified and genuinely new (Chapter 8). Although mammal brains are larger than the brains of reptiles or amphibians at any given body size, the brains of early mammals were extremely small in absolute terms. Therefore, the origin of the neocortex cannot be explained as a simple, automatic consequence of increased absolute brain size. Instead, it probably involved size-independent (and still largely mysterious) changes in neurogenesis, migration, and axon guidance. On the other hand, absolute brain size did increase enormously after mammals first evolved, and those increases in brain size were accompanied by major changes in brain organization. Moreover, several of those changes are explicable in terms of size-related principles. Specifically, the fact that larger mammals tend to have proportionately larger neocortices than their smaller relatives is consistent with the rule of late equals large. Likewise, the observation that larger mammals tend to have more areas within their neocortex is consistent with the principle that enlarged regions tend to become subdivided. Therefore, we can conclude that the neocortical expansion and elaboration that is so central to the story of mammalian brain evolution is precisely what you would expect, given the well-documented increases in absolute brain size and our scaling principles.

Humans are unique among primates in terms of absolute brain size, which increased in fits and starts during the last 6 million years but essentially plateaued 100,000 years ago. At first glance, modern human brains seem similar to their chimpanzee counterparts in everything but size. However, the comparative data on chimp and human brains suggest that human brains have been considerably "reorganized." The neocortex:medulla ratio, for instance, is roughly twice as large in humans as in chimpanzees. This change in proportions is approximately what you would expect given the difference in absolute brain size, but it is nonetheless significant. Specifically, it probably explains, via the rule of large equals well-connected, why the human neocortex seems to have unusually extensive projections to the medulla. This, in turn, is likely to explain (at least in part) why humans are so remarkably dextrous in their hand, eye, mouth, and vocal-fold movements. Another special feature of the human brain is that its lateral prefrontal cortex is disproportionately large. Again, this is not unexpected, since the lateral prefrontal cortex generally enlarges disproportionately with increasing absolute brain size. Still, it is functionally significant because the enlargement of the lateral prefrontal cortex probably helped to increase the ability of humans to suppress reflexive responses to stimuli. This increased behavioral "freedom" probably helped *Homo sapiens* evolve symbolic language 50,000–100,000 years ago. Once human language had evolved, it caused major and accelerating changes in behavior that were largely independent of brain size. Thus, I conclude that increases in absolute brain size helped to reorganize our brains and brought us to the point where language could evolve. After that, absolute brain size became much less significant.

It is important to note that these arguments are all considerably simplified. For instance, as you might recall from our discussion of Bumpus's sparrows, relative brain size can be a very useful variable to contemplate. In fact, some might say that natural selection in those sparrows acted specifically on relative brain size, ignoring absolute brain size. In response, I would suggest that natural selection might just as well have acted to maintain a constant absolute brain size while body size decreased. As I mentioned above, determining what selection "tried to do" is not simple. Evolutionary neuroscience is not for those who hate complexity! As you shall see, caveats and complications are strewn throughout this book.

Although I stress repeatedly that brains evolve according to some common themes and principles, this book is not exclusively about those core ideas. It will take you on some side paths and detours, causing the aforementioned principles to recede at times. Ultimately, though, they tend to reemerge. For me, that is part of the fun, to see what basic themes and principles reappear consistently from what at first appears to be chaos.

Because evolutionary neuroscience has such a venerable history, dating back to Darwin's days, you may be surprised, perhaps even dismayed, to find it so immature in terms of theoretical development, its rules and principles so vague, uncertain, and debated in the literature. Indeed, I am impatient, too. On the other hand, once the limitations are revealed, we become aware of what still must be done. And if you disagree with statements I have made, I trust that you will work to set the record straight. As Darwin once proclaimed: "False facts are highly injurious to the progress of science, for they often endure long; but false views, if supported by some evidence, do little harm, for every one takes a salutary pleasure in proving their falseness." With that dictum in mind, I have included only data that I deem trustworthy, but have been less constrained in writing about rules and principles that help us weave those data into stories that make sense. In the next section I discuss this attitude in more detail.

My hope is that the ideas I have gathered here will cause you to have ideas of your own. Who knows where inspiration lies? As Stuart Kauffman (1993, p. viii) wrote, "Ideas . . . once loose upon a page harbor their own lives, follow their own unsuspected paths, mature in unforeseen ways, and mingle with their own logic. If useful, they have progeny."

Philosophical Preamble

Given the unorthodox organization of this book, I feel compelled to spell out its philosophical foundation. As far as I am concerned, the main aim of evolutionary neuroscience is not just to document the history of brain evolution, but also to explain it. Reconstructions of what likely happened when and where are interesting and important but, ultimately, they are not enough. Henri Poincaré, the French mathematician and philosopher, clearly expressed this sentiment:"Science is built with facts, as a house is built with stones; but a collection of facts is no more a science than a pile of stones is a house. . . . Above all, the scientist must make predictions." (Poincaré, 1902; author's translation.)

In this passage, Poincaré argues that scientists must focus not on specific facts but on the regularities that tie them together. Only the recognition of those regularities, including so-called laws, enables scientists to predict future events. This much is widely accepted, but the same logic applies also to past events, for historical explanations are really "predictions in hindsight": given what we know, could we have predicted what actually came to pass? If the answer is affirmative, then we have explained the past. This "nomological-deductive" approach to historical explanation has been discussed extensively in the philosophical literature (e.g., Hempel, 1942), but its application in biology presents some problems that are worth discussing in detail (see also Lauder, 1981; Bock, 1999).

One problem is that biological history is governed not by a single almighty Law but by a plethora of different laws. Once upon a time, biologists tended to think that evolution was governed by a single law of progression that caused simple organisms to become complex and induced lower species to ascend "the scale" (Chapter 2; see also Figure 2.4), but this view is no longer tenable. We now know that evolution sometimes drives toward simplicity, and that complexity may take a variety of forms. So, has the old law of progression been replaced with some new universal and almighty law of evolution? According to some authors, the answer is yes: the law of natural selection applies universally (Rosenberg, 2001). That proposition is also dubious, however, for the essence of Darwin's "survival of the fittest" argument is a mathematical truth rather than an empirically determined law (which is why it is sometimes misidentified as a tautology; see Sober, 2000). Moreover, natural selection does not prescribe what should evolve. It yields "jury-rigged compromises" (Conway Morris, 1991) that may be good at what they do but are not predictable in their details. Nor does natural selection act on random variants. There clearly are some developmental and/or physical "constraints" that help to channel the course of evolution (Amundson, 1994). Therefore, natural selection is far from omnipotent; it must work with raw materials that are subject to a variety of other laws.

Because biology has many laws, rather than just one, biologists tend to debate which laws are most pertinent in any given instance or domain. They argue, for instance, about whether natural selection or constraints were more important in explaining a feature (Gould and Lewontin, 1979). Such debates disturb at least some philosophers of science, who view them as an unproductive form of "theoretical pluralism" (Beatty, 1997). Would it not be better to have just a single law in each domain? Should biologists not strive to eliminate, through cleverly designed experiments, one putative law after another until only one remains standing? Such critiques sound plausible because they resemble the kind of calls for falsification and "strong inference" that have long been popular in science (Platt, 1964; Stamos, 1996; O'Donohue and Buchanan, 2001).

But the issue here is different. We are talking about laws rather than hypotheses, and there is no reason to believe that biological systems should be governed by a single law. Even physical systems (e.g., a bullet fired from a gun) generally require the application of multiple laws (e.g., gravitation and friction) before they become entirely predictable. Therefore, what biologists really need is not to whittle down their set of laws, but to devise a unitary theory that accommodates and unifies a lot of different laws. Given such a unifying theory, the coexistence of multiple laws would no longer be troublesome but natural, for the whole point of unifying theories is that they provide a causal framework that explains how and why various laws, forces, and factors interact with one another. Viewed from that perspective, the problem in biology is not that there are too many laws but that the unifying theory is incomplete. That is why most evolutionary explanations are partial explanatory "sketches" (Hempel, 1942) rather than full-fledged theories.

A second problem with the nomological-deductive approach to historical explanations is that biological laws are generally riddled with exceptions. Mendel's Law of Independent Assortment, for example, is far from universal,

Figure 1.4 Allen's Rule in Rabbits and Hares Allen's rule is evident in hares. Shown here are (from left to right): the arctic hare *Lepus arcticus*, the snowshoe hare *L. americanus*, the blacktailed jackrabbit *L. californicus*, and the antelope jackrabbit *L. alleni*. Taken in this sequence, they live in progressively hotter climates (from Greenland to the deserts of southern Arizona) and have progressively longer legs, ears, and tails. This makes sense, since having long extremities should, according to some basic principles of geometry and physiology, facilitate heat loss.

since meiotic drive and a host of other factors can cause genes to be inherited in nonrandom, biased ways (see Oster, 1998). Even the supposedly universal genetic code, which governs how RNA is translated into protein, is not without exception, since some bacteria, some yeasts, and many mitochondria use a slightly different code (Jukes and Osawa, 1993; Ohama et al., 1993). That is why biologists generally talk about "rules" and "principles" rather than laws. They recognize, for example, that arctic animals tend to have shorter extremities than animals that live in hot deserts (Figure 1.4), but they do not think of Allen's rule, as this principle is known, as a law because it applies only to comparisons between closely related species (or races) and is, even then, not without exception.

Similarly, biologists have realized that the size of a bird's eggs scales quite predictably with body size (Rahn et al., 1975), but they also know that this egg-scaling rule varies across taxonomic orders and is subject to some striking exceptions (Figure 1.5). This troubles some philosophers of science, who worry that, if every rule must be augmented with a list of provisos, then the rules cannot be very scientific or useful (Beatty, 1995). If the rules are full of exceptions, how can they be used to predict anything? Such critiques are intriguing but ultimately misguided (see Earman and Roberts, 1999; Cartwright, 2002; Lange, 2002). They fail to reflect accurately how scientists in general (even the physicists!) think about and use their so-called laws.

For one thing, scientists appreciate that not all laws are deterministic; many are true only on average and are, therefore, inherently full of exceptions. Naturally, such probabilistic laws can still be useful, particularly if you are trying to predict the average behavior of many particles or organisms. In addition, most scientists are well aware that even deterministic laws may not

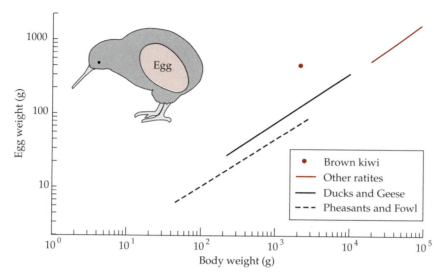

Figure 1.5 The World's Most Exceptional Egg   The brown kiwi's egg:body ratio is one of biology's most remarkable "exceptions to the rule." Within most orders of birds, egg size scales quite predictably with body size, though the allometric lines for different orders often have different *y*-intercepts (Rahn et al., 1975). The eggs of the brown kiwi, however, are more than 3 times heavier than one would expect for a ratite species of its body size (other ratites in this data set are ostriches, rheas, emus, and cassowaries). Most likely, this exceptionally high egg:body ratio is due to a relatively recent but dramatic reduction in absolute body size. The inset depicts a brown kiwi hen with an egg to show their relative sizes. (After Calder, 1984.)

yield certain predictions. As Poincaré himself had pointed out and the chaos theorists have stressed, "prediction becomes impossible [when]. . . small differences in the initial conditions produce very great ones in the final phenomena" (Poincaré 1908, pp. 68–69; see Aubín and Dalmedico, 2002). That is why long-term weather forecasts are notoriously uncertain, why it is difficult to predict when humans will become extinct, and why most evolutionary explanations are indefinite. Again, however, such caveats should not be overemphasized. Limited weather forecasts are possible, and we can predict that pesticide-resistant pests, for instance, are fairly certain to evolve in the long run. Furthermore, if explanations are predictions in hindsight, then at least some aspects of the evolutionary past can be explained scientifically.

Finally, most scientists are used to making simplifying assumptions. That is, they recognize that in order to make predictions they must disregard some negligible factors and hope that no non-negligible factors intrude unexpectedly. Predictions of a spacecraft's flight, for instance, involve numerous assumptions that are generally safe, but occasionally something does go wrong. Similarly with evolution: It may be predictable in rough outline, but when an asteroid hits Earth all bets are off. Accidents do happen frequently in evolution, but it is not entirely haphazard (Gould, 1970; McIntyre, 1997). Therefore, evolutionary explanations may be partial and imprecise, but they are not condemned to being mere speculation or just-so stories (see Gould and Lewontin, 1979).

The third major problem with the nomological-deductive approach to history is that it does not yet work well for the most complex systems in our universe: brains. Much has been learned in the last few decades about how brains evolved, develop, and function, but it is still just a drop in the proverbial bucket. This dearth of information makes it difficult to come up with statistically valid laws of brain evolution, to explain why those laws exist, and to comprehend why they sometimes do not apply. In addition, several previously advocated "laws of brain evolution" (Ariëns Kappers, 1910; Bishop, 1959) have failed to stand the test of time. Nonetheless, evolutionary neuroscience has transcended its speculative roots (Striedter, 1998a) and is now ready to construct the kind of causal framework that leads to unifying theories. This book is my attempt to push it further down that road. It offers some attempts at synthesis but falls short of yielding a "synthetic theory." Hopefully, that failure will spur you to action. Borrowing some words from Hans Spemann (1927, p. 187), a famous embryologist, "we still stand in the presence of riddles, but not without hope of solving them. And riddles with the hope of solution—what more can a man of science desire?" With that apology and entreaty, I welcome you, men and women of science, into the tangled bank that is brain evolution.

2 A History of Comparative Neurobiology

Only when we know what has been done by earlier contributors can we judge the present scene.

—Walter Cannon, 1945

Many well-known giants of neuroscience, such as Galen, Willis, Broca, Ramón y Cajal, Ferrier, Brodmann, Papez, and Adrian, were also comparative neurobiologists. A full historical account of comparative neurobiology is clearly beyond the scope of this book; instead, I will focus on a few key people and episodes that are particularly interesting and remain relevant to the field today. I firmly believe (with Walter Cannon) that many of the conceptual problems facing comparative neurobiologists today are easiest to understand—and perhaps resolve—when they are seen in the context of earlier contributions. Richard Feynman, the famous physicist, was supposedly proud of his ability to rediscover the brilliant revelations of previous thinkers, but for most of us it is more efficient to peer out from atop the shoulders of the giants who have gone before. In particular, I have often found that the scientists who first struggled with a particularly difficult problem thought about it most deeply and freely. Perhaps their struggles can help us see more clearly where we stand today and, more important, where we might go from here.

With that agenda in mind, I recognize six major periods in the history of comparative neurobiology, namely: (1) the birth of comparative neuroanatomy (400 B.C.–1600 A.D.); (2) the debate between Owen and Huxley, which took place during Darwin's time (1850–1870); (3) the heyday of comparative cytoarchitectonics (1870–1936); (4) advances in comparative hodology and histochemistry (beginning in the 1960s); (5) the rise of neurocladistics (beginning in the 1980s); and (6) the rejuvenation of comparative neuroembryology (beginning in the 1990s; Figure 2.1). These divisions are somewhat arbitrary and the last three periods overlap, but each epoch is characterized by a unique combination of technical and theoretical approaches to comparative neurobiology.

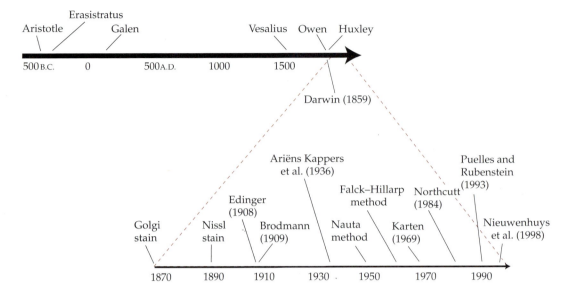

Figure 2.1 A Timeline of Comparative Neurobiology   Comparative neurobiology began with Aristotle but did not flourish until after Darwin published his *Origin of Species* in 1859. The field was boosted by the development of new histological methods— especially the discovery of Golgi and Nissl stains—at the end of the nineteenth century. The field stagnated between 1936 and the 1950s, but was then revived by new techniques for tracing axonal connections (e.g., the Nauta method) and visualizing neurochemical features (e.g., the Falck–Hillarp method). More recently, evolutionary neurobiologists have become excited about using cladistics to reconstruct phylogenetic changes (e.g., Northcutt, 1984) and using molecular genetics to rejuvenate comparative neuroembryology (e.g., Puelles and Rubenstein, 1993). The monumental treatise by Nieuwenhuys et al. (1998) summarizes most of what evolutionary neurobiologists have learned to date.

The Birth of Comparative Neuroanatomy

Scientists have always found it easier to work with animal brains than with human brains, and this was particularly true in ancient Greece, where contact with human corpses was expressly forbidden by sacred laws (von Staden, 1992). Anyone who touched a dead or dying human was thought to be defiled and was required to undergo extensive rites of purification. Even eye contact with a deceased human being was not allowed. In a play by Euripides, for example, Artemis leaves the dying Hippolytus with these words: "Farewell! Sacred law forbids me from looking at the dead or staining my eye with the exhalation of death!" Given this cultural and legal context, the dissection of human brains was out of the question (or at least much less accepted than it was by the time of the early Renaissance). Aspiring neuroanatomists therefore turned to animal brains.

Probably the most prolific ancient dissector was Aristotle, who examined at least 49 different animal species, ranging from sea urchins to elephants (Finger, 2000). Despite Aristotle's enormous accomplishments, his views on the brain seem strange to us today. Particularly unusual was his belief that the heart controls feelings and thoughts, whereas the brain (with its extensive vascularization) functions primarily to cool the blood. This belief, coupled with the observation that humans have unusually large brains, led Aristotle to conclude that humans must produce more heat than animals do. Thus, Aristotle was the first scientist to make an inference about brain function from comparative neuroanatomical data. The long-term impact of his "radiator hypothesis" was minimal, however. Even in ancient Greece, the radiator hypothesis was a minority view, since Hippocrates and his followers had already argued that "from nothing else but the brain come joys, delights, laughter and sports, and sorrows, griefs, despondency, and lamentations" (quoted in Finger, 2000, p. 29). This Hippocratic view of the brain as "central command" ultimately displaced Aristotle's radiator hypothesis.

The idea that the brain is the body's central command was promoted most effectively by Galen, Rome's greatest physician–scientist, who was born around 130 A.D. Galen published extensively on human anatomy and physiology, including several lectures specifically on the brain. Because the prohibitions against human dissection were still in place, Galen studied Barbary apes (a tail-less kind of macaque), barnyard animals (e.g., oxen), cats, dogs, weasels, camels, lions, wolves, stags, bears, mice, and even an elephant. He also dissected some fishes, birds, and reptiles (but apparently drew the line at invertebrates) (Finger, 2000). Galen then used this broad knowledge of animal anatomy to make inferences about numerous details of human anatomy that he had not seen for himself. Thus, Galen furnished fairly detailed accounts of the human brain (among other organs), and his conclusions were accepted as gospel truth for nearly 1,500 years.

Eventually, however, Galen's descriptions were questioned by the Flemish anatomist Andreas Vesalius (1514–1564; Figure 2.2A) (see Vesalius, 2002). Vesalius was free of the ban against contact with human corpses and consequently dissected humans regularly. To his great surprise, Vesalius found nearly 200 mistakes in Galen's descriptions of human anatomy. Most notoriously, Galen had described a fine plexus of blood vessels, a rete mirabile, at the base of the human brain (Figure 2.2B), but Vesalius could find it only in oxen and a few other animals, not in humans (where, instead, he found what we now call the "circle of Willis"). Vesalius had been a great admirer of Galen's and was perplexed by the magnitude of Galen's mistakes until, one day, Vesalius compared the skeletons of an ape and a human side by side. Suddenly, he realized that Galen had described the ape skeleton perfectly, but had incorrectly extrapolated his findings from apes to humans. Thus, Vesal-

(A)   (B)   Rete mirabile

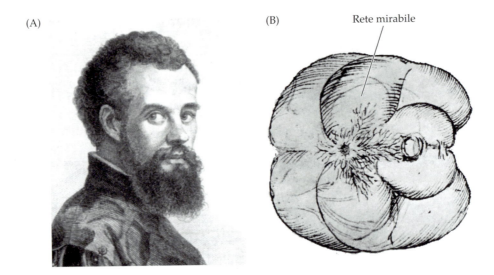

Figure 2.2 Vesalius and the Rete Mirabile   (A) Andreas Vesalius (1514–1564) had noted that Galen made many serious errors in his descriptions of the human brain. For example, Galen had reported the existence of a rete mirabile, a complex network of blood vessels, at the base of the human brain, but Vesalius could find this structure only in oxen and other nonhuman animals. (B) A drawing by Leonardo da Vinci, showing an ox brain with its prominent rete mirabile. (A from Vesalius, 1543; B from da Vinci, 1911–1916; see also Polyak, 1957.)

ius became the first famous scientist to realize how dangerous it is to assume that species differences are negligible.

In addition to Aristotle, Galen, and Vesalius, I must mention Erasistratus, a Greek who practiced comparative neuroanatomy in Egypt, where prohibitions on contact with corpses were less severe, around 300 B.C. None of his writings survive, but we know from Galen that Erasistratus regularly dissected humans and probably performed experiments on human prisoners (Christie, 1987). We also know that Erasistratus tried to correlate behavioral traits to the size and shape of different brain regions. Most notably, Erasistratus observed that "deer, hares, and other fast-running animals have more intricately folded cerebellums than less active animals . . . [and that] . . . we humans are smarter than other animals because our cerebral hemispheres have more convolutions than those of the brutes" (Finger, 2000, p. 35). Although later workers pointed out that sloths also have highly convoluted cerebellums and that donkeys have complexly folded cerebral hemispheres, Erasistratus deserves recognition as one of the founders of comparative neuroanatomy.

Darwin's Time: The Owen–Huxley Debate

Long after its birth, comparative neuroanatomy was transformed by the theory of evolution. Erasmus Darwin (Charles's grandfather), Jean Baptiste Lamarck, and others had theorized about evolution in the late eighteenth and early nineteenth centuries, but Charles Darwin dramatically increased its acceptability when, in 1859, he proposed natural selection as a mechanism for evolution (Darwin, 1859). Although Darwin said little about human evolution in his *Origin of Species*, most of Darwin's contemporaries quickly realized that his theory radically challenged the traditional notions about "man's place in nature" (Huxley, 1863). The burning question was whether apes and humans are separated by an unbridgeable gulf or continuous with one another as ancestor and descendant. For an answer, many looked to comparative neuroanatomy.

One of the first scientists to study ape brains was Richard Owen, a contemporary of Darwin and Britain's most famous anatomist. One of Owen's most important contributions was that he coined the word "homologue" and defined it as "the same organ in different animals under every variety of form and function" (Owen, 1843, p. 379). Owen also coined the word "dinosaur" and published extensively on platypus, giant moa, and gorilla anatomy. In 1857, at the height of his career, Owen proposed a novel classification of mammals, based primarily on neural characters. He divided the Mammalia into four subclasses: (1) the Lyencephala, composed of the "loose-brained" mammals that lack a corpus callosum; (2) the Lissencephala, consisting of mammals that have a corpus callosum and "smooth" cerebral hemispheres; (3) the Gyrencephala, all nonhuman mammals with prominent cerebral convolutions; and (4) the Archencephala, consisting only of humans. Owen justified this taxonomic sequestration of humans by pointing out that human brains are unique:

> In Man, the brain presents an ascensive step in development, higher and more strongly marked than that by which the preceding subclass was distinguished from the one below it. Not only do the cerebral hemispheres overlap the olfactory lobes and cerebellum, but they extend in advance of the one and further back from the other. Their posterior development is so marked that anatomists have assigned to that part the character of a third lobe; it is peculiar to the genus Homo, and equally peculiar is the "posterior horn of the lateral ventricle" and the "hippocampus minor" which characterizes the hind lobe of each hemisphere. The superficial grey matter of the cerebrum, through the number and depth of convolutions, attains its maximum extent in man. Peculiar mental powers are associated with this highest form of the brain, and their consequences wonderfully illustrate the value of the cerebral character; according to my estimate of

which, I am led to regard the genus Homo, as not merely representative of a distinct order, but of a distinct subclass of the Mammalia, for which I propose the name Archencephala. (Owen, 1857, pp. 19–20)

In essence, Owen used comparative neuroanatomy to reinforce his belief that humans and apes are separated by a vast gulf. Most likely, Owen did this because he was nervous about Darwin's impending *Origin of Species* (Owen certainly had some knowledge of what Darwin was about to publish) and wanted to bolster the arguments for *Homo*'s special place in the creation. Owen is not easily pigeonholed, however, for his ideas were complex and nearly evolutionary at times. Nonetheless, Owen clearly opposed Darwin's ideas about the "struggle for existence" because they threatened the moral fabric of Victorian society (and, Owen worried, they might foster racism). By clearly separating humans from other mammals, Owen probably hoped to ward off Darwin's "dangerous idea" (Dennett, 1995).

Owen's strategy backfired, however, and instead provoked a fierce battle with T. H. Huxley, Darwin's most aggressive defender (and the grandfather of Julian and Aldous Huxley). Huxley had been converted to Darwinism in the early 1850s and was ready to use his talents (mainly a razor-sharp tongue and pen) to fight for Darwin's idea. He had already tussled with Owen over other issues and was chafing at Owen's imperious demeanor (Desmond, 1994). He may also have heeded the ancient oracle's advice that the best way to become a great man is to "slay one." In any case, Huxley jumped at the chance to do battle with Owen over the classification of mammals. His rebuttal of Owen's 1857 argument was straightforward and effective (Huxley, 1861). He reviewed the prior literature on ape brains and showed that the features Owen claimed were characteristic of humans were not, in fact, unique to humans. Specifically, Huxley argued that previous authors had already demonstrated the existence of a hippocampus minor (Figure 2.3) in apes and that Owen must have known about this prior literature, implying that Owen willfully exaggerated the neuroanatomical differences between apes and humans.

Owen later toned down his claims, conceding that apes have rudimentary posterior lobes, rudimentary posterior horns, and a rudimentary hippocampus minor. However, he continued to assert that there is a "much greater difference between the highest ape and lowest man, than exists between any two genera of Quadrumana [i.e., nonhuman primates]" (Owen, 1859, p. 269). Consequently, Huxley kept up his attack. To help him in this endeavor, Huxley enlisted several friends who then published detailed descriptions of primate brains in the *Natural History Review,* which Huxley owned. None of these descriptions supported Owen's position and, by 1863, Huxley's arguments had carried the day. According to Huxley and his associates, the neuroanatomical data warranted only that humans be placed in a separate genus. In other words, humans should take their place in the same order as apes and

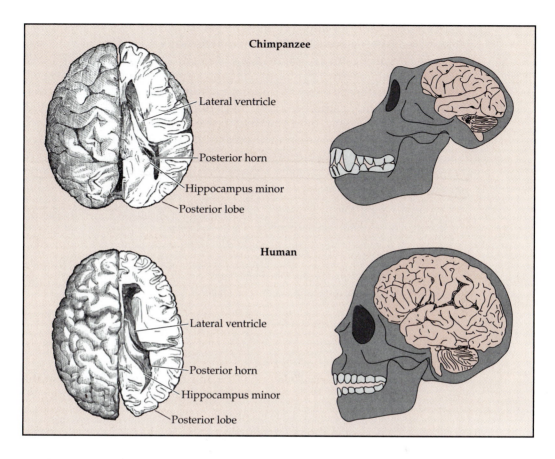

Figure 2.3 Chimpanzee and Human Brains   T. H. Huxley noted that Richard Owen had been wrong to claim that only humans had a posterior lobe, a posterior horn of the lateral ventricle, and a hippocampus minor (a ridge in the floor of the posterior horn of the lateral ventricle). This criticism was appropriate, but Huxley craftily minimized the differences in size between chimpanzee and human brains by drawing both of them to the same length (left). In reality, human brains are almost three times as large as chimpanzee brains, even though adults of the two species are similar in body size (right). (From Huxley, 1863.)

lemurs—that is, among the primates. Huxley conceded that "there is a very striking difference in absolute mass and weight between the lowest human brain and that of the highest ape" (Huxley, 1863, p. 120), but he considered these size differences to be of minor importance.

Although Huxley emerged from this battle as the unambiguous winner, and thus transformed comparative neuroanatomy into evolutionary neurobiology, Huxley's hero status is tainted by several facts (Cosans, 1994). First, Huxley used deceptive illustrations to minimize the differences in size between ape and human brains (see Figure 2.3). Second, Huxley never addressed Owen's primary concern about how one could modify ape development to generate human morphology. From our modern perspective, Owen was certainly correct to claim that apes could never be transformed into humans by simply prolonging their morphological development, which is how most evolutionists then thought about the transmutation of species. Finally, Huxley did a great disservice to evolutionary neurobiology by promoting the idea that evolutionary change is unilinear and unrelentingly progressive. For example, Huxley wrote:

> The brain of a fish is very small . . . In Reptiles, the mass of the brain, relatively to the spinal cord, increases and the cerebral hemispheres begin to dominate over the older parts; while in Birds this predominance is still more marked. The brain of the lowest Mammals, such as the duck-billed Platypus and the Opossums and Kangaroos, exhibits a still more definitive advance in the same direction. The cerebral hemispheres have now so much increased in size as, more or less, to hide the representatives of the optic lobes, which remain comparatively small . . . A step higher in the scale, among the placental Mammals, the structure of the brain acquires a vast modification . . . The appearance of the 'corpus callosum' in the placental Mammals is the greatest and most sudden modification exhibited by the brain in the whole series of vertebrated animals . . . In the lower and smaller forms of placental Mammals, the surface of the cerebral hemispheres is either smooth or evenly rounded, or exhibits a very few grooves . . . But in the higher orders, the grooves, or sulci, become extremely numerous, and the intermediate convolutions proportionately more complicated in their meanderings, until, in the Elephant, the Porpoise, the higher Apes, and Man, the cerebral surface appears a perfect labyrinth of tortuous foldings. (Huxley, 1863, pp. 112–114)

Thus, Huxley may have been the first scientist to describe brain evolution as proceeding linearly from fish to man. Of course, Huxley was not the only early Darwinian to view evolution as unilinear and progressive. Most of Darwin's contemporaries were quite ready to recast Aristotle's *scala naturae* (Figure 2.4) as a phylogenetic scale (Figure 2.5).

This willingness to view evolution as a struggle for progress may have been socially motivated. Huxley, for example, knew poverty well and strove hard to advance his social position in what he probably perceived as a perpetual struggle for existence. In that context, it is interesting to note that Darwin had always been wealthy and was less inclined to see evolution as unfailingly progressive. Moreover, Darwin realized far more clearly than most of his contemporaries that speciation by means of natural selection produces not linear scales but family trees (Figure 2.6). As he put it in one of his early note-

Figure 2.4 A Classic "Scale of Nature"   At least since Aristotle, people have
sought to order all beings, objects, and elements into some sort of gigantic natural
scale, or *scala naturae*. Shown here is a medieval depiction of the *scala naturae*, which
ranks humans above rocks, fire, plants, and animals, but below the heavens, angels,
and God. (From Ramon Llull's *Ladder of Ascent and Descent of Man*, 1305; see Ruse,
1996.)

books: "The tree of life should perhaps be called the coral of life, base of
branches dead; so that passages cannot be seen. – this again offers contradic-
tion to constant succession of germs in progress" (Notebook B, pp. 25–26;
quoted in Barret, 1987).

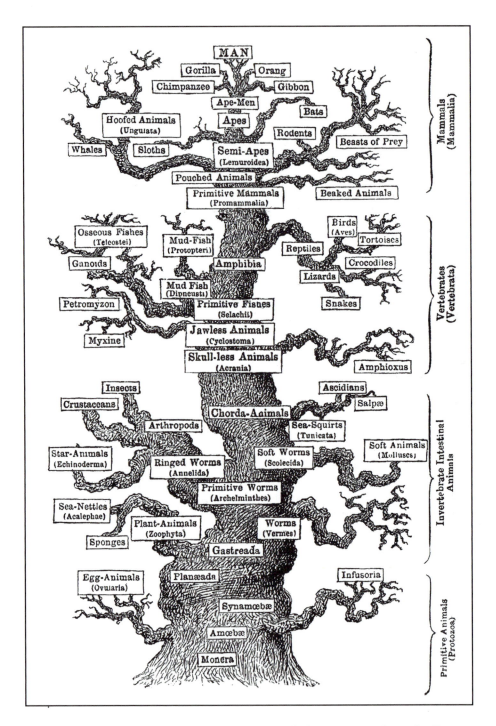

Figure 2.5 A Classic "Phylogenetic Tree"   This phylogenetic tree, drawn by Ernst Haeckel, preserves many elements of the *scala naturae*, since man is at the top and the "lower" animals near the bottom of the tree. Although the tree has numerous branches, these are depicted as diversions off the principal lineage, which led to man. (From Haeckel, 1893.)

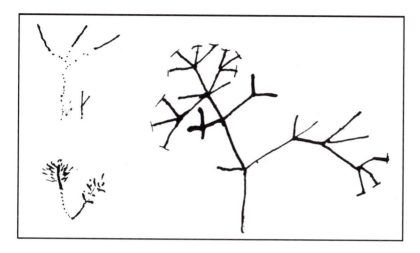

Figure 2.6 Darwin's "Corals of Life"   In one of his early notebooks, Charles Darwin sketched these "corals of life" to illustrate the genealogy of different species. Compared to Haeckel's phylogenetic trees, Darwin's "corals" are relatively free of *scala naturae* thinking. (From Barret, 1987.)

However, even Darwin rarely stressed this "excessively complicated" aspect of his theory, and it was lost on most of his followers. Instead, most Darwinists followed Huxley's lead and viewed evolution as proceeding steadily up the phylogenetic scale (Bowler, 1988). Since then, many scientists have attempted to purge *scala naturae* thinking from evolutionary theory, but traces of this association remain.

The Era of Comparative Cytoarchitectonics

Neuroanatomy in Darwin's day was based almost exclusively on gross dissections. Usually, entire brains were hardened by immersion in alcohol and cut with dissecting knives. This method revealed the brain's macroscopic structure, including the distribution of gray and white matter, but did not divulge the brain's histological detail—its cytoarchitecture. In the latter half of the nineteenth century, however, the neuroanatomist's armamentarium expanded dramatically (Clarke and O'Malley, 1986; Shepherd, 1991; Northcutt, 2001). Microscopes with achromatic and spherically corrected lenses became widely available, and more effective fixatives (e.g., chromic acid and formaldehyde) were introduced. Microtomes suitable for cutting thin serial sections were also developed, hand in hand with new tissue-embedding media such as celloidin and paraffin wax (Figure 2.7). Finally, many new dyes were introduced at this time, and some of these stained neurons selectively.

Figure 2.7 Section through an Embryonic Brain This frontal section through the celloidin-embedded head of a 3-month-old human embryo was produced by Wilhelm His in 1904. At this stage of development, the cerebral ventricles are very large and most neurons are still undifferentiated. The ventrolateral portions of the telencephalon bulge into the ventricles, forming the so-called "ganglionic eminences." The neocortex is relatively undifferentiated. (From His, 1904.)

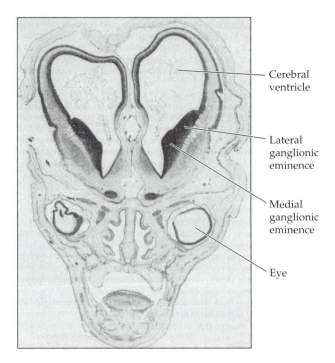

Cerebral ventricle

Lateral ganglionic eminence

Medial ganglionic eminence

Eye

Most influential were von Gerlach's carmine stain (invented in 1858), Golgi's silver impregnation method (reported in 1875), and Nissl's methylene blue stain (described in 1894). Collectively, these methods enabled neuroanatomists to obtain their first glimpses of brain histology.

An excellent example of the comparative cytoarchitectonic studies spawned by these technical innovations was Korbinian Brodmann's work on the cerebral cortex. Between 1900 and 1909, when he worked in Berlin, Brodmann generated and examined numerous sections through the cerebral hemispheres of all mammalian orders (except cetaceans—dolphins and whales). He then used these comparative neuroanatomical data to conclude, in opposition to earlier workers, "that the original, primitive pattern of cortical layering in the whole mammalian class is six-layered, and that this six-layered pattern is visible in all orders, either permanently or at least as a temporary ontogenetic stage in the embryo" (Brodmann, 1909, p. 17). Brodmann also showed that, while many cortical areas can be homologized across mammals, others are unique to particular lineages. For example, Brodmann noted that primates have more cortical areas than other mammals (see Chapter 9). This phylogenetic increase in the number of cortical areas intrigued Brodmann, but he was quick to point out that "the struggle for existence through natural selection not only produces progressive changes, but also regressive ones" (Brodmann, 1909, p. 205). Because Brodmann's observations on phylogenetic regression did not fit with the idea of progression along a phylogenetic scale, they were largely ignored at the time.

In addition to Brodmann's excellent work on mammals, the first few decades of the twentieth century saw an enormous increase in the amount of cytoarchitectural data on nonmammalian brains. For example, G. Elliot Smith, J. B. Johnston, and others showed that the telencephalon in reptiles and birds harbors huge intraventricular ridges that are not readily comparable to anything seen in mammals (see Chapter 8). Santiago Ramón y Cajal, C. Judson Herrick, Cornelius Ariëns Kappers, Ludwig Edinger, and Hartwig Kuhlenbeck (to name just the most prominent contributors) also worked during this time. Collectively, these neuroanatomists generated a veritable mountain of data on the cytoarchitecture of various different vertebrate brains. This rapid data influx generated enormous confusion, however, as different authors often used different names for similar structures and the same names for different structures. The resultant "tower of Babel" dilemma created an urgent need for synthesis.

Ludwig Edinger was the first to attempt such a synthesis of the comparative cytoarchitectonic data in his *Bau der nervösen Zentralorgane*, an enormously influential textbook that went through nine editions (Edinger, 1908b). Edinger's most important contribution was the recognition that forebrain organization differs dramatically between the major vertebrate groups, whereas midbrain and hindbrain organization is highly conserved (barring minor adaptations to special ways of life). In an effort to systematize the variation in forebrain structure, Edinger distinguished between a palaeencephalon, which is found in all vertebrates, and a neencephalon, which is found (as a clearly distinguishable structure) only in animals "higher" than cartilaginous fishes (Figure 2.8). Within the higher vertebrates, Edinger argued, the neencephalon increases in both size and complexity along the phylogenetic scale, reaching its apex in man. Although the terms "palaeencephalon" and "neencephalon" are no longer used, Edinger's core idea, that the forebrain evolved by means of the sequential addition of parts, continues to have its adherents. For example, the "triune brain" theory that Paul MacLean and Carl Sagan popularized in the 1970s is clearly derived from Edinger's original idea (MacLean, 1973; Sagan, 1977).

After Edinger, a second major synthesis was forged by C. Ariëns Kappers, Carl Huber, and Elizabeth Crosby. Over the course of ten years, starting in 1926, these authors gathered together all of the available information on vertebrate brains and distilled it into a two-volume "bible" of comparative neuroanatomy (Ariëns Kappers et al., 1936; republished as 3 volumes in 1960). Aside from simply compiling the unwieldy data, their major contribution was

Figure 2.8 Edinger's View of Brain Evolution
Around 1908, Ludwig Edinger proposed that the
major event in vertebrate brain evolution was the
appearance of a "neencephalon" (solid black) in
tetrapods. According to him, the neencephalon first
appeared in cartilaginous fishes, where it was small
and poorly differentiated. The neencephalon then
increased steadily in size and complexity as one
ascends the phylogenetic scale, through lizards and
rabbits to man. (From Edinger, 1908b.)

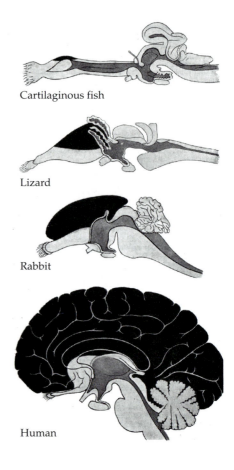

Cartilaginous fish

Lizard

Rabbit

Human

to elaborate Huxley's and Edinger's vision of how brains, through the addition of new components, had progressed along the phylogenetic scale. For
example, they proposed that not only the cortex but also the striatum and
cerebellum were divisible into phylogenetically old and new parts. Accordingly, Ariëns Kappers et al. created many new neuroanatomical terms with
the prefixes "paleo-," "archi-," and "neo-," to denote when a structure
evolved. In an effort to explain why brain evolution had been so unrelentingly progressive, they postulated that all organisms are imbued with an
"inner urge" that results in "the progressive development of the brain in
accordance with a general plan [and] in the progressive differentiation and
adjustment of its constituents" (Ariëns Kappers et al., 1936, pp. xii–xiii).

Although the book by Ariëns Kappers et al. was enormously influential, it
sparked little new research and, in fact, nearly destroyed the field. Perhaps
the huge size and encyclopedic nature of the book suggested to budding com-

parative neuroanatomists that everything had been described already and led them to believe that there was nothing further to investigate (Northcutt, 1998). World War II also played a major role in comparative neuroanatomy's torpor because it generally thwarted science in Europe, where comparative cytoarchitectonic studies had gotten their start. Edinger's institute, for example, lost its leadership due to the emigration of Jewish scientists from Nazi Germany. Even in the United States the war effort interrupted research, and comparative neuroanatomy seemed on the brink of extinction.

Comparative Hodology and Histochemistry

Fortunately, a series of technical innovations in the 1950s and 60s revived comparative neuroanatomy. First, Walle Nauta and others developed new methods for selectively staining degenerating axons, which allowed them to lesion a cell group and then trace its efferent connections by looking for degenerating axon terminals (Nauta, 1950; Nauta and Gygax, 1954; Fink and Heimer, 1967). Shortly thereafter, radioactively labeled amino acids and horseradish peroxidase (HRP) were introduced as effective anterograde and retrograde tracers, respectively (Lasek et al., 1968; LaVail and LaVail, 1972). Concurrently with these advances in hodology, the study of neuronal connections, major breakthroughs were made in neuronal histochemistry. First came the Falck–Hillarp method, which used fluorescence to detect biogenic amines (e.g., dopamine and serotonin) in neural tissue (Falck et al., 1962). Soon after that, enzyme histochemistry and immunohistochemistry were invented (Sternberger et al., 1970). Collectively, these new methods revitalized comparative neuroanatomy because they generated entirely new kinds of data.

The principal insight gained as a result of these technical developments was that brain evolution was a more conservative process than previous authors had envisioned. Ariëns Kappers and C. J. Herrick, for example, had thought that the avian telencephalon consists mostly of striatal areas and lacks a clear homologue of mammalian neocortex (if you are not familiar with terms such as "neocortex," striatum," or "dorsal thalamus, "refer to Chapter 3, where I provide an overview of the major subdivisions in vertebrate brains). Histochemical data obtained in the 1960s, however, suggested that the striatum in birds is confined to the ventral telencephalon, which implied that the more dorsal areas might be homologous to neocortex (Figure 2.9). Harvey Karten and his collaborators then used Nauta's axon degeneration method to show that the dorsal telencephalon in birds (including the dorsal ventricular ridge; see Figure 2.9) receives massive ascending inputs from the dorsal thalamus (Karten, 1969). Since dorsal thalamic input is generally considered to be a hallmark of neocortex, these data suggested that birds actually

(A) Pigeon telencephalon

Dorsal
ventricular
ridge

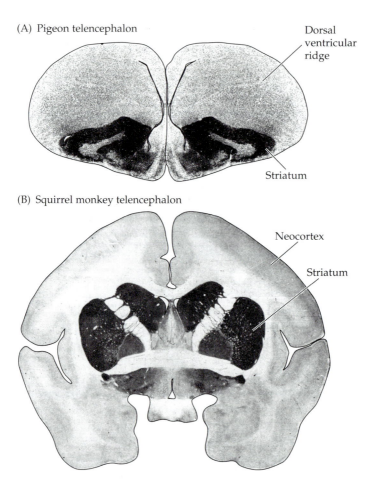

Striatum

(B) Squirrel monkey telencephalon

Neocortex

Striatum

Figure 2.9 The Fruits of Comparative Histochemistry When comparative neuroanatomists realized that the mammalian striatum stains heavily for acetylcholinesterase (AChE), they applied the same stain to nonmammalian vertebrates.
Fairly quickly, they realized that only a relatively small ventral portion of the avian
telencephalon is homologous to the mammalian striatum. Shown here are (A) an
AChE-stained cross section through a pigeon's telencephalon and (B) an equivalent
section through the telencephalon of a squirrel monkey. For a discussion of how the
avian telencephalon compares to that of mammals, see Chapter 8. (A after Karten,
1969; B after MacLean, 1972.)

have a neocortical homologue of substantial size (for more details and complications, see Chapter 8). Because this putative neocortex homologue in birds
is not six-layered (as neocortex is in mammals), Karten and others suggested
that homologous structures need not display similar cytoarchitecture (Karten

and Shimizu, 1989). In other words, connections and histochemistry might be phylogenetically conserved even when cytoarchitecture is not (Chapter 3).

In parallel with this work on birds, hodological studies on fishes and amphibians reinforced the idea that neuronal connections are highly conserved. C. J. Herrick (1948) and others had thought that in all fishes the telencephalon receives only olfactory projections. Herrick also thought that in amphibians thalamic projections reach only the ventral telencephalon (the striatum), and that thalamic projections to the dorsal telencephalon (the pallium; see Chapter 3) first evolved in the last common ancestor of reptiles, birds, and mammals. This "invasion hypothesis" of telencephalic evolution was roundly rejected, however, when Sven Ebbesson, Frank Scalia, Glenn Northcutt, and their collaborators showed that the olfactory bulb in fishes and amphibians projects only to very restricted portions of the telencephalon (Figure 2.10). In addition, several experimental studies showed that the dorsal thalamus in fishes and amphibians does have projections to the dorsal telencephalon (Northcutt, 1981). These findings, in conjunction with similar data on other fiber systems, prompted Northcutt (1984) to conclude that "many, if not most, neural pathways appear to be very stable phylogenetically, and the majority of these pathways appear to have arisen with the origin of vertebrates or, shortly after, with the origin of jawed vertebrates" (p. 710).

Thus, the nineteenth century theory that brains evolved by the sequential addition of novel parts was toppled by the discovery of putative neocortical homologues in "lower" vertebrates and, more generally, by the finding that neuronal connections are more conserved across species than the theory had predicted. Indeed, the new data obtained in the 1970s and 80s suggested that all vertebrate brains consist of the same basic parts, not only in the midbrain and hindbrain (as Edinger had supposed) but in the forebrain as well. Consequently, many old neuroanatomical terms became obsolete. The journal *Brain, Behavior and Evolution,* for example, began in 1992 to warn its authors that "the descriptive prefixes paleo-, archi-, and neo-applied to cerebellum, cortex, or striatum imply an invalid phylogenetic sequence and should . . . be avoided." By the 1990s, most comparative neuroanatomists believed that all vertebrate brains are built according to a common plan that varies only in its details (Chapter 3). According to one recent textbook, "all vertebrate central nervous systems share a common organizational scheme so that someone who is familiar with the brain of any vertebrate will also be on familiar ground when first encountering the brain of any other species" (Butler and Hodos, 1996, p. xv).

This "conservative revolution" in the study of brain evolution was facilitated by the long-awaited demise of *scala naturae* thinking in most of the biological sciences. During the 1940s and 50s, the architects of the "evolutionary synthesis" had revitalized Darwin's idea that phylogenies are branched rather than unilinear. Most important, they replaced the old phylogenetic "scale" with a phylogenetic "tree" and argued that evolutionary change is often not linear but divergent, following different trajectories in different lin-

Figure 2.10 Limited Projections of the Olfactory Bulb The techniques invented by Walle Nauta and his colleagues to stain degenerating axons have been used by numerous neuroanatomists to study axonal projections in many different vertebrates. Shown here is a map of the olfactory bulb projection targets (red) in (A) a bullfrog and (B) a polypterus, one of the ray-finned fishes with an everted telencephalon; see Figure 2.16. In bullfrogs the olfactory projections are bilateral, whereas in polypterus they are mostly ipsilateral, but in both cases they cover only a limited portion of the telencephalon, disproving the hypothesis that "lower" vertebrates have a telencephalon that is dominated by the sense of smell. (A after Northcutt and Royce, 1975; B after Northcutt and Braford, 1980.)

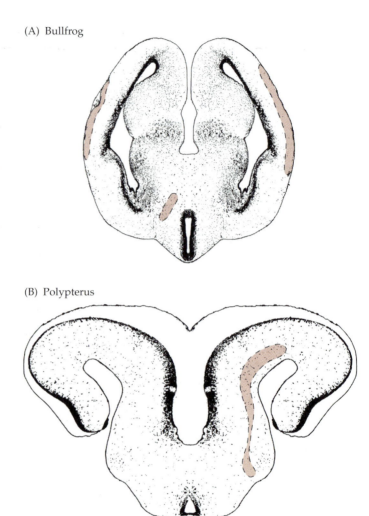

(A) Bullfrog

(B) Polypterus

eages. George Gaylord Simpson, for example, wrote that: "some such sequence as dogfish-frog-cat-man is frequently taught as 'evolutionary,' i.e., historical. In fact the anatomical differences among these organisms are in large part ecologically and behaviorally determined, are divergent and not sequential, and do not in any useful sense form a historical sequence" (Simpson, 1958, p. 11).

This genealogical way of thinking about evolution was slow to infiltrate comparative neuroanatomy but did eventually take hold. William Hodos and Boyd Campbell, for example, pointed out that "rats were never ancestral to cats nor were cats ancestral to primates; rather, each represents a different evolutionary lineage" (Hodos and Campbell, 1969, p. 341). Similarly, Hodos and Campbell argued that it is "practically meaningless" to speak of amphibians as representing a higher degree of evolutionary development than teleost fish, because "they have each followed independent courses of evolution" (Hodos and Campbell, 1969, p. 339). This position was officially adopted by *Brain, Behavior and Evolution* in 1992, when it declared that "vague, subjective descriptors such as 'higher' and 'lower' should be avoided" when referring to animal groups (*Brain, Behavior and Evolution*, January, 1992). With thoughts of a *scala naturae* officially banished, most comparative neurobiologists began to view brains as fundamentally similar across all vertebrates, differing only in the details of their organization.

The Rise of Neurocladistics

Despite the fundamental similarity of all vertebrate brains, someone who knows only the brain of a rat, for example, will not recognize many familiar features in the brain of a teleost fish (see Figure 1.3). Across the major vertebrate groups, even homologous brain regions often differ in size, shape, position, cytoarchitecture, histochemistry, connections, and/or function. So how is one to make sense of this variation? In the days of *scala naturae* thinking, species differences were readily interpreted as representing different stages of evolution. Features in "lower" species were considered primitive, those in higher species derived (i.e., advanced). But in the absence of a *scala naturae*, living species cannot be assumed to exhibit only primitive features, for any species is likely to exhibit at least some features that are derived for its particular branch of the phylogenetic tree. Moreover, without a *scala naturae*, one can no longer be certain that evolution always proceeds from simple to complex. These problems created an urgent need for some kind of procedure that allows species differences to be sorted and understood in terms of historical, phylogenetic processes.

This need was filled by cladistics, which Willi Hennig had developed in 1950 as a formal methodology for the classification of species (Hennig, 1950, 1966). Hennig's method was largely ignored until the late 1960s and 70s, when a small band of avid "cladists" promoted its use (Hull, 1988). Cladistics then spread like wildfire through the field of systematics, largely displacing the competing schools of thought. It did not, however, impact comparative neurobiology until the mid-1980s, when Glenn Northcutt showed how cladistic principles could be used to reconstruct the phylogenetic history of neural traits (Northcutt, 1984, 1985a). In order to show how Northcutt's brand of "neurocladistics" (Nieuwenhuys, 1994a) changed comparative neurobiology, I must first explain at least some of the concepts on which cladistics is based. More detailed expositions of cladistic methodology can be found in several excellent books (e.g., Eldredge and Cracraft, 1980; Ridley, 1986).

Cladistics begins with the proposition that classifications should be based exclusively on the phylogenetic relationships (i.e., the genealogy) of the species involved. Cladists then argue that these phylogenetic relationships can, at least in principle, be inferred from the distribution of features that are found in the species available for study (be they extant or extinct). This is true simply because descendants generally inherit the features that distinguished their last common ancestor (just as children tend to inherit the peculiarities of their parents). Imagine, for example, a long extinct species that "invented" (phylogenetically speaking) hair and mammary glands. Then imagine that this species thrived and gave rise to a large family of descendant species. Most of the descendants probably retained the hair and mammary glands, and these shared features then distinguish these descendants from other species (assuming that hair and mammary glands were not invented numerous times independently). In the language of cladistics, hair and mammary glands are *shared derived* characters that characterize a monophyletic group (a lineage with a unique origin). By examining the distribution of these shared derived features, the cladist can, in a sense, work backward and reconstruct which groups of animals are most closely related to one another.

This much is relatively straightforward, but cladistics (or phylogenetic systematics) is complicated because not all similarities are shared derived characters. Consider, for example, the fact that both monotremes (e.g., platypus) and marsupials (e.g., kangaroos) lack a corpus callosum. This similarity might tempt one to classify monotremes and marsupials together (as Owen had done). However, reptiles and amphibians also lack a corpus callosum, suggesting that the absence of a corpus callosum was not an invention of monotremes and marsupials (they did not "lose" the corpus callosum) but a holdover from premammalian days. As a cladist would put it, the absence of a corpus callosum is probably a *shared primitive* character for monotremes and marsupials, rather than a shared *derived* feature, because it also characterizes their nearest outgroups—their closest nonmammalian relatives.

This distinction is important because, according to cladists, shared primitive characters do not characterize monophyletic groups and should not be used to derive classifications. Indeed, we now know from a vast array of molecular and morphological evidence that monotremes and marsupials do not constitute a monophyletic group. Instead, most systematists have concluded that marsupials and placentals are more closely related to one another than either one is to monotremes (Figure 2.11).

Cladistic classifications are generally based on many different characters, which range from bones and behavior to DNA sequence data. Ideally, all of these characters point to a single classification scheme (since life's history is unique, there can be only one historically accurate family tree and only one classification scheme that is based on that tree). Unfortunately, however, the systematist's world is not ideal, and different characters often support different phylogenetic trees. This problem arises because very similar features may evolve several times independently in different branches of the phylogenetic

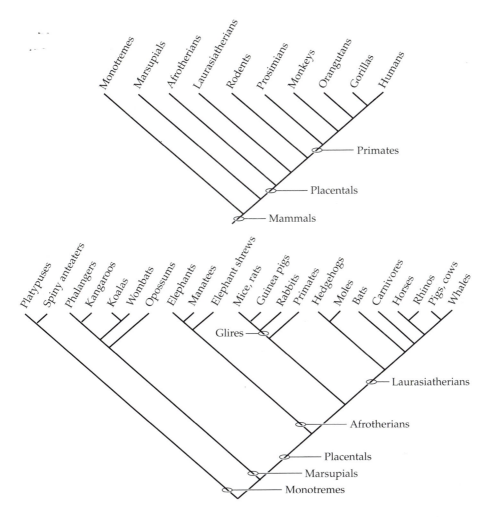

Figure 2.11 The Branching Topology of Cladograms   Cladograms represent the phylogenetic relationships of different animal groups in their branching topology. These two cladograms of mammals appear very different at first glance but are topologically consistent with one another. In (A), the primates are overrepresented and humans are placed in the rightmost position, which gives the impression that humans are the "most advanced" species. (B) This cladogram includes more marsupial, afrotherian, and laurasiatherian species, and some of its branches are rotated (twisted around a nodal point) relative to (A). In (B) it is more apparent that primates are just one of many branches on the mammalian phylogenetic tree. (After Kirsch et al. 1997, Liu et al., 2001; Madsen et al., 2001; Murphy et al., 2001.)

tree. As long as the tree's topology is unknown, these independently evolved characters are indistinguishable from shared derived characters. This means that some of the features that cladists provisionally score as shared derived features are not "shared" at all. They are red herrings in the cladist's character soup. Fortunately, cladists can identify these specious characters by (1) examining as many different characters as possible, (2) identifying which tree topology is most parsimonious (i.e., implying the lowest number of independent evolutionary events), and (3) determining a posteriori which characters are consistent with this most parsimonious tree. Only the characters that are consistent with the most parsimonious tree are bona fide shared derived characters. The others are due to independent evolution. As long as the rate of independent evolution is relatively low, this procedure is logically sound.

Consider a concrete example. As I mentioned above, Owen had classified all mammals with complexly folded hemispheres as Gyrencephala, but they were a motley crew, including spiny anteaters, donkeys, dolphins, elephants, and chimpanzees. Moreover, all of the mammalian orders that we now recognize as being "well established" contain at least some smooth-brained (lissencephalic) species. Among primates, for example, marmosets and other small monkeys have almost perfectly smooth hemispheres (Figure 2.12). Therefore, Owen's classification breaks up what we now consider to be "the primates" and classifies some of them (e.g., marmosets) with nonprimates. But marmosets share many more features with other primates than they do with other smooth-brained placental mammals (e.g., hedgehogs). Under Owen's classification, marmosets would have had to evolve all of these primate-typical features (e.g., large frontal eyes; see Chapter 9) independently of the other primates. This is highly unparsimonious. Far more parsimonious is the hypothesis that complexly folded hemispheres evolved several times among mammals. Most important, this conclusion emerges naturally from any cladistic analysis of mammalian phylogeny that is based on a reasonably large number of different characters.

Most cladistic analyses are based on non-neural data, but in the early 1980s John Kirsch and his collaborators used 15 neural characters (e.g., the presence or absence of "barrels" in the somatosensory cortex) to derive a cladistic classification for 134 species of mammals (Kirsch and Johnson, 1983; Kirsch et al., 1983). The resultant phylogeny was interesting but lacked resolution because it was based on too few characters. Similar problems also plagued other, later attempts to reconstruct phylogenetic relationships from exclusively neural data (Northcutt, 1986; Pettigrew, 1989). Given the brain's complexity, it may seem strange that it is so difficult to find enough neural characters for a successful cladistic analysis, but one must remember that cladists can use only shared derived characters in their analyses. They must ignore the far more numerous primitive similarities. In practical terms, this means that cladists must sift through a mountain of raw similarity data to find a few nuggets of cladistic gold. Since neurohistology is labor intensive, shared derived neural characters (the presence of a novel feature or the absence of an otherwise widespread feature) are difficult to find. Therefore, all well-established phy-

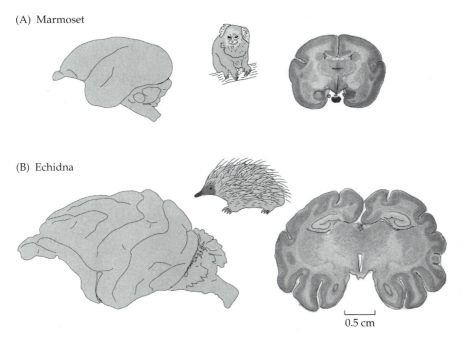

(A) Marmoset

(B) Echidna

0.5 cm

Figure 2.12 Gyrencephalic Brains Evolved Repeatedly   Most mammalian orders contain some representatives with lissencephalic (smooth) brains and some with gyrencephalic (complexly folded) brains. Shown here are (A) the brain of a marmoset monkey and (B) the brain of a spiny anteater, or echidna. Clearly, the echidna's brain is more complexly folded than that of the marmoset, even though the echidna, being a monotreme (see Figure 2.11), is generally regarded as a "more primitive" mammal. Lateral views of each brain are shown at left; Nissl-stained transverse sections at right. The sections and brains are drawn to approximately the same scale. (After images on the Comparative Mammalian Brain Collections website [http://brainmuseum.org], from the University of Wisconsin–Madison Brain Collection.)

logenies are based almost exclusively on non-neural data. Why then did neurocladistics generate any excitement at all?

From my perspective, neurocladistics became exciting when Northcutt (1984, 1986) showed how one could reconstruct the phylogenetic history of individual neural characters by examining their distribution among species whose phylogenetic relationships were already known (from cladistic analyses of non-neural data). In essence, Northcutt started with the idea that many different evolutionary scenarios could account for the phylogenetic distribution of any given character. By counting the number of phylogenetic gains and losses associated with each scenario, he argued, it should be possible to find the most parsimonious scenario—the one requiring the smallest number of gains and losses. Assuming that gains and losses are equally likely and that independent evolution is relatively rare, the most parsimonious scenario is most likely to reflect the character's actual phylogeny. Some cladists had already used this kind of parsimony analysis to study the evolution of non-neural characters (Ridley, 1983; Maddison et al., 1984), but Northcutt was the first to apply it to neural data. For example, he argued convincingly that "relatively large brains have evolved independently in some members of most, if not all, vertebrate radiations" (Northcutt, 1986, p. 362; see Chapter 4). He also deduced that pathways from the spinal cord to the thalamus evolved independently in cartilaginous fishes and amniotes (reptiles, birds and mammals; Figure 2.13).

Northcutt originally applied neurocladistics to overall brain size and to the presence or absence of neuronal aggregates and major axon pathways. Soon, however, neurocladistics was extended to other kinds of neural features, including cytoarchitectural features and neuronal circuits (e.g., Northcutt and Wullimann, 1988; Striedter, 1991). Even neuronal cell types, embryonic struc-

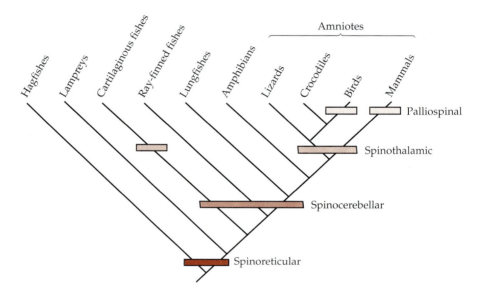

Figure 2.13 Phylogenetic Distribution of Ascending Spinal Pathways Cladistic analyses allow us to infer when and where several major axonal projection systems first evolved. Spinoreticular projections are found in all vertebrates and are therefore likely to have evolved with the origin of vertebrates. Spinocerebellar projections are found in all vertebrates except hagfishes and lampreys, which means that they probably evolved in the last common ancestor of all jawed vertebrates. Spinothalamic projections probably evolved independently in amniotes (reptiles, birds, and mammals) and cartilaginous fishes. Palliospinal projections, finally, are most parsimoniously interpreted as having evolved independently in birds and mammals. (After Northcutt, 1984.)

tures, and developmental mechanisms are now being subjected to cladistic analyses (e.g., Schlosser et al., 1999; Striedter, 1999). Thus, cladistics found fertile ground in the brain. The recent comparative neuroanatomy book by Ann Butler and William Hodos (1996), for example, is couched entirely in a cladistic framework. Although neurocladistics has some limitations, particularly when dealing with features that are "easy" to evolve (which means that independent evolution can no longer be assumed to be rare), it is the most rigorous available method for reconstructing the history of brain evolution. As such, it has effectively filled the void left by the demise of the phylogenetic scale.

Perhaps the most important consequence of cladism's success is that most neurobiologists (indeed most biologists) now think differently about the concept of homology. Since 1859 most biologists had defined homology in terms of common ancestry, but comparative neuroanatomists had often disagreed about which criteria one should use to homologize brain regions (Nieuwenhuys and Bodenheimer, 1966; Campbell and Hodos, 1970). Some authors emphasized that homologous structures must occupy similar relative positions within the brain, while others based their homologies primarily on similarities in cytoarchitecture, connections, histochemistry, and/or embryonic origin. When the different criteria point to different homologies, which homologies should win out? No one could agree. This seemingly interminable debate prompted one scholar to exclaim in frustration that "a neuroscientist today can hardly utter the word homology without finding a half-dozen other neuroscientists at his throat" (Jones, 1985, p. 763). In response, some comparative neuroanatomists tried to replace the term "homologous" with "equivalent" (Karten and Shimizu, 1989), but this hardly settled the argument.

From the perspective of cladistics, however, the problem of homology is relatively straightforward, because homologues can be defined simply as the shared derived features that characterize monophyletic groups. In other words, two or more features are homologous only if they did not evolve independently of one another (Figure 2.14). The corpus callosum, for example, is homologous across all placental mammals because it most likely evolved just once, with the origin of placental mammals, and was then retained with a continuous history in the descendants. Conversely, any similarities that evolved independently of one another must be considered nonhomologous, or as the cladists say, "homoplasous," regardless of their degree of similarity. Cerebral gyri in dolphins, chimps and elephants, for example, cannot be homologous to one another (as gyri) because they are most parsimoniously interpreted as the result of independent evolution. From this perspective, similarities in position, connections, and other attributes do not constitute "criteria of homology" but rather "criteria of character identification" (Striedter, 1999). They help the investigator identify which features should be compared to one another. If two or more features are so dissimilar that they cannot be

Figure 2.14 Cladistics and Homology The cladist's views on homology and nonhomology (homoplasy) are best illustrated schematically. The stars in (A) are all homologous to one another because the character "star" is most parsimoniously interpreted as having evolved just once in the last common ancestor of taxonomic groups d–g. In contrast, the stars in (B) are not all homologous to one another because it is most parsimonious to conclude that the character "star" evolved twice, once in the lineage leading to c and again in the last common ancestor of f and g. The alternative hypothesis, that "star" evolved once in the common ancestor of c–g and was then lost twice, in lineages d and e, is less parsimonious because it involves three separate evolutionary changes instead of two.

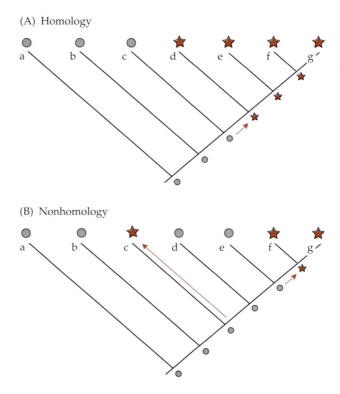

(A) Homology

(B) Nonhomology

identified as the same character, then they cannot enter into a homology comparison. On the other hand, if they can be identified as the same character, then the cladist's criterion of nonindependent evolution (more commonly referred to as the criterion of phylogenetic continuity) becomes the ultimate arbiter of their homology.

In sum, the advent of neurocladistics has clarified the homology concept because it presents a clear methodology for reconstructing the phylogenetic history of individual characters. Naturally, some features remain difficult to homologize, but this is not necessarily due to problems with the homology concept itself. Instead, most of these controversial homologues are simply so different between species that investigators cannot agree how to identify the character. And if features in different species cannot be identified as the same character, then they simply cannot be homologized (Striedter, 1998, 1999). This conclusion does not trouble most cladists, however, for cladistics was never about trying to homologize everything. The cladists only need to find enough homologues (i.e., shared derived characters) to build robust classifications. By pursuing this goal, they have advanced not only systematics but also comparative neurobiology.

The Rejuvenation of Comparative Neuroembryology

Embryology had been an essential component of evolutionary studies in Darwin's days. Ernst Haeckel, in particular, was an ardent admirer of Darwin's and a prolific builder of phylogenetic trees, which were based largely on embryological data (Haeckel, 1889). Haeckel's core idea was that an organism's evolutionary history (its phylogeny) can be read directly from its developmental history (its ontogeny) because "ontogeny recapitulates phylogeny" (except when natural selection specifically creates larval adaptations). Haeckel based this idea on the assumption that evolution generally fashions novel adult forms by appending new developmental stages to the ancestral ontogenies.

Although Haeckel's theory of recapitulation was enormously influential, it was in serious trouble even before Haeckel was born. Back in 1828, Karl Ernst von Baer had already refuted pre-Darwinian versions of the recapitulation theory by showing that "the young stages in the development of an animal are not like the adult stages of other animals lower down on the scale, but are like the young stages of those animals" (quoted in de Beer, 1958, p. 3). Von Baer had also noted that "during its development an animal departs more and more from the form of other animals" (von Baer, 1828). These observations flatly contradicted Haeckel's theory of recapitulation. Interestingly, they did not contradict the idea of evolution itself, although von Baer was opposed to evolution as Darwin envisioned it. In fact, Darwin himself frequently cited von Baer's observations as strong support for evolution. After all, community of descent readily explained embryonic similarities, and the transmutation of species could be caused by developmental divergence just as readily as by the addition of new stages to the end of ancestral ontogenies (von Baer, 1828; de Beer, 1958).

In the 1920s, Nils Holmgren began to apply this comparative embryological approach to the nervous system in vertebrates (Holmgren, 1922, 1925). He and his followers, notably Harry Bergquist and Bengt Källén (Bergquist and Källén, 1953a,b), examined an enormous number of different embryos, in many different species, and noted that all embryonic brains shortly after neurulation exhibit a very similar pattern of organization. According to this embryonic brain Bauplan, or archetype (described at length in Chapter 3), embryonic brains are divided into numerous transversely oriented proliferative zones from which young neurons migrate toward their adult positions (Figure 2.15). By following the fate of these embryonic proliferative zones, Holmgren and his associates were able to determine in many instances how developmental divergence creates adult species differences. This, in turn, allowed these investigators to clarify several homologies that had remained controversial as long as only adult data were considered (see Striedter, 1997). For example, Holmgren's work helped to clarify that the telencephalon in

(A) Neuromeres according to Bergquist, 1952

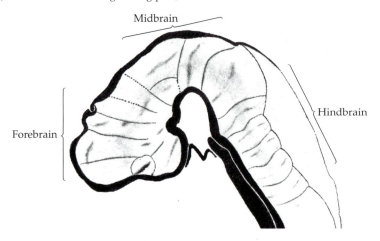

(B) Neuromeres according to Puelles et al., 1987

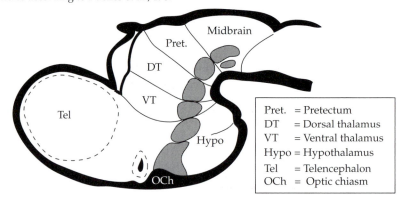

Pret.	= Pretectum
DT	= Dorsal thalamus
VT	= Ventral thalamus
Hypo	= Hypothalamus
Tel	= Telencephalon
OCh	= Optic chiasm

Figure 2.15 Versions of the "Neuromeric Model"   Comparative neuroembryologists in Sweden during the first half of the twentieth century proposed that embryonic vertebrate brains are divisible into numerous transverse segments, or neuromeres, which are separated from one another by external grooves. This neuromeric model of vertebrate brain organization languished until it was revived by Luis Puelles, John Rubenstein, and others, who used histochemical and molecular biological methods to confirm that neuromeres exist. Shown in (A) is an illustration of the neuromeric model produced by Harry Bergquist in 1952; (B) is a diagram of AChE-positive cell clusters (gray) that Puelles et al. (1987) used to illustrate their version of the neuromeric model. (A after Bergquist, 1952; B after Puelles et al., 1987.)

teleost fishes everts rather than evaginates during development. Recognition of this developmental difference, in turn, helped to clarify various telencephalic homologies between teleosts and other vertebrates (Figure 2.16).

Although Holmgren was well known, his comparative neuroembryological approach did not become adopted outside of Sweden. One reason for this lack of success was that embryonic proliferative zones are extremely difficult to see in normally stained brain sections. Many therefore doubted the existence of these proliferative zones, particularly in the diencephalon and telencephalon. In addition, Herrick and others objected that these embryonic divisions were transient and unrelated to the functional divisions of adult brains. Finally, comparative neuroembryology did not take hold because embryology had generally become dissociated from evolutionary studies (Gilbert et al., 1996). Indeed, for most of the twentieth century, most embryologists were not interested in evolutionary questions (perhaps because of Haeckel's speculative excesses), and evolutionary biologists were interested in paleontology and population genetics, not embryology. Therefore, evolutionary biologists generally came to think of phylogeny as a succession of adult forms. As William Bateson pointed out long ago, however, phylogeny is not a sequence of adults but a sequence of entire ontogenies:

> *An illustration will perhaps help to make clear the point at issue. The received view of homology supposes that a varying form is derived from*

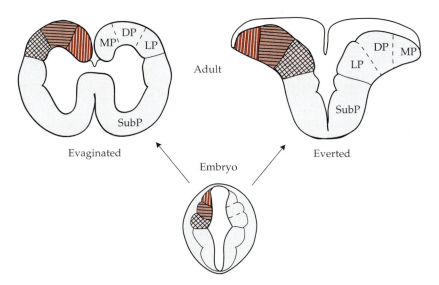

Figure 2.16 Telencephalic Evagination versus Eversion   In most vertebrates, the telencephalon evaginates during development to form the cerebral hemispheres (top left). In ray-finned fishes, however, the telencephalon does not evaginate. Instead, its dorsal portion folds outward in what is called an eversion process. In such an everted telencephalon, the mediolateral positions of the dorsal telencephalic (pallial) areas are reversed relative to what they are in an evaginated telencephalon. Abbreviations: DP = dorsal pallium; LP = lateral pallium; MP = medial pallium; SubP = subpallium. (After Holmgren, 1922.)

*the normal much as a man might make a wax model of the variety from a
wax model of the type, by small additions to, and subtractions from, the
several parts. This may, to our imagination, seem, perhaps, the readiest
way by which to make the varying form if we were asked to do it; but the
natural process differs in one great essential from this. For in nature the
body of the varying form has never been the body of its parent and is not
formed by a plastic operation from it; but in each case the body of the offspring is made again from the beginning, just as if the wax model had
gone back into the melting pot before the new model was begun. (Bateson,
1892, p. 115)*

Bateson's warning was heeded by some, notably Walter Garstang and
Gavin de Beer, but most evolutionary biologists simply ignored embryology.
This began to change after 1977, when Stephen J. Gould eloquently reviewed
the prior literature on ontogeny and phylogeny and proposed that most evolutionary changes are changes in developmental timing (Gould, 1977). Moreover, Gould convinced many scientists that the question of how development
relates to evolution is important. At about the same time, developmental biologists began to think about the molecular mechanisms that underlie evolutionary changes in development (Raff and Kaufman, 1983). Thus, the timing
was right for the rebirth of evolutionary developmental biology, or evo-devo,
as it is often called (Hall, 2000). This field has grown remarkably during the
last twenty years and now sports its own journal, textbook, and program at
the National Science Foundation. Predictably, evo-devo also infected neurobiology.

In the late 1980s, Luis Puelles and his collaborators began to reinvestigate
the embryonic compartmentalization of vertebrate brains. They showed,
using primarily histochemical and gene expression data, that the embryonic
nervous system of chicks and mice is indeed subdivided into a large number
of distinct compartments, much as Bergquist and Källén had suggested
(Puelles et al., 1987; Puelles and Rubenstein, 1993; Puelles et al., 2000) (see Figure 2.15). This compartmentalization is particularly evident in the hindbrain,
where eight distinct neuromeres are revealed by the expression patterns of
several regulatory genes. However, similar neuromeres are also evident,
according to Puelles et al. (1987), at forebrain levels. This "neuromeric model"
of embryonic brain organization (discussed in more detail in Chapter 3) has
undergone significant modifications in recent years but has proven broadly
applicable across different vertebrate groups. In general, this work has created excitement because it revealed that the embryos of different vertebrates
are unexpectedly similar to one another (Gilland and Baker, 1993) and, I suspect, because it provided a golden opportunity for molecular neurobiologists
to address evolutionary problems in novel ways.

During that same period, Northcutt began a cladistic analysis of the
embryological development of electroreceptors and lateral line sense organs,

which are found in the skin of aquatic vertebrates (Northcutt, 1992). His approach consisted of three largely separate steps. First, he identified comparable stages in the development of these receptors across the major vertebrate groups. Then he used these data to reconstruct the most likely ancestral sequence of development for this system. Finally, he used this ancestral sequence to identify phylogenetic changes in the development of these receptor systems. One of Northcutt's most interesting conclusions was that (in this system) the addition of developmental stages to the end of ancestral ontogenies was relatively rare compared to terminal deletions or nonterminal changes in development. As unhappy as Haeckel would have been about that conclusion, von Baer might have been pleased.

In sum, the last 10–15 years have seen a dramatic rejuvenation of comparative neuroembryology. Once again, comparative neuroembryological data are viewed as critical for understanding how brains evolved, but in contrast to the days of Holmgren and his followers, the comparative neuroembryologists of today have at their disposal the powerful tools of both molecular biology and cladistics. Therefore, the opportunities for progress are immense. Most exciting to me is the prospect of using these methods to obtain a mechanistic understanding of how phylogenetic changes in development have led to species differences in adult morphology. For example, Schlosser and his collaborators (1999) have already shown that direct-developing frogs (which lack an aquatic larval phase) probably lost their lateral line organs because their skin lost the ability to respond to some (as yet unknown) inductive signals. From there, it seems but a small step to understanding precisely what molecular alterations have caused the skin of these direct-developing frogs to lose its responsiveness to inductive stimuli.

Conclusions

Even a cursory scan through the history of comparative neurobiology reveals that this field has become progressively more rigorous and sophisticated. For many years brains were compared almost exclusively within the context of a *scala naturae,* or phylogenetic scale, but this changed after Darwin's vision of branching phylogenies was fully realized and cladists used rigorous hypothesis testing to reconstruct the phylogeny of neural characters. The other major advance came with the realization that phylogeny is not a succession of adult forms but a sequence of entire ontogenies, for this perspective suggested new hypotheses about how ontogenetic changes cause species differences in adult morphology. Thus, evolutionary neurobiology has come a long way from its speculative beginnings (remember Aristotle's cerebral radiators!) and now finds itself on the brink of a very exciting time when phylogenetic hypotheses can be phrased and tested in mechanistic terms.

The historical perspective also reveals a protracted tug-of-war between those who emphasize species differences in brain organization and those who dwell on similarities. Owen, for example, was fascinated by species differences among mammalian brains, while Huxley tried to minimize those differences, at least with regard to human brains. Similarly, Edinger and Ariëns Kappers were struck by species differences in forebrain organization, while Holmgren and other comparative neuroembryologists saw all vertebrate forebrains as being built on a common plan. The experimental neuroanatomists in the second half of the twentieth century also focused primarily on species similarities, yet one can hardly review their histochemical and connectional data without stumbling onto numerous species differences. Why else would the most recent textbook of comparative neuroanatomy (Nieuwenhuys et al., 1998) be more than 2,200 pages long? Clearly, the brains of different vertebrates do vary in more than negligible respects.

So, why do some investigators see species differences where others see similarities? Perhaps some investigators see species differences as an obstacle to model systems research, though this perception is flawed, since model systems are often useful precisely because of species differences. Or perhaps species differences are stressed preferentially by those who are most familiar with animal behavior, since behavioral species differences are hard to ignore and must ultimately correlate with species differences in brain organization. Both factors are likely to be important, but I also suspect that it is a general characteristic of the human mind, when confronted with input as complex and multifarious as the data on vertebrate brain organization, to seek order (similarities) amidst confusion. Once order has been detected, the mind can once again latch on to species differences and begin a new round of looking for similarities. If the human mind indeed works this way, then it is only natural that some scientists, at some points in time, should emphasize similarities, while others emphasize differences. According to the quantum physicist David Bohm (1957), it is precisely this tension between similarities and differences, between order and disorder, that leads to the discovery of scientific laws and principles (see Chapter 1). In any case, it is probably prudent for evolutionary neurobiologists to be mindful of both species differences and similarities in brain organization. In that spirit, the next chapter presents a summary of what is relatively constant across vertebrate brains and thus lays a foundation for the subsequent analyses of variation.

3 Conservation in Vertebrate Brains

It is known that Nature works constantly with the same materials. She is ingenious to vary only the forms. As if, in fact, she were restricted to the same primitive ideas, one sees her tend always to cause the same elements to reappear, in the same number, in the same circumstances, and with the same connections.

—E. Geoffroy Saint-Hilaire, 1807

The idea that all creatures are built according to a common plan is very old and powerful. As early as 1753, Georges Buffon recognized that there is a "primitive and general design - which one can follow quite far - according to which everything seems to have been conceived" (Buffon, 1753; cited in Goethe, 1981, p. 392). This insight then flourished in the mind of Etienne Geoffroy Saint-Hilaire, who argued that all animals are composed of the same basic elements and that the spatial arrangement of these elements is conserved even among animals that appear, at first glance, to be highly dissimilar. For instance, Geoffroy Saint-Hilaire proposed that the nervous system of insects and crustaceans is topologically equivalent to the nervous system of vertebrates, even though the former occupies a ventral position while the latter lies dorsal to the gut. He also proposed that the insect exoskeleton is structurally equivalent to the vertebrate skeleton, and that insect legs correspond to vertebrate ribs (see Russel, 1916).

Although Geoffroy Saint-Hilaire's work inspired many naturalists to seek further evidence of Nature's unity, most of his specific homologies (he called them "analogies"; the word "homology" was not coined until 1848) met with considerable skepticism. Indeed, Geoffroy Saint-Hilaire soon found himself embroiled in a vigorous debate with the great French anatomist Georges Cuvier, who protested that animals were created not according to a single plan, but four different plans. Specifically, Cuvier proposed that there are four fundamentally different types of animals, namely vertebrates, mollusks, radiates (coelenterates and echinoderms), and articulates (insects, crustaceans, and annelids). Within each type, Cuvier argued, the major organs and body parts can be homologized with relative ease, but drawing homologies between different types (e.g., between insects and vertebrates) is fraught with difficulty. As Geoffroy Saint-Hilaire's proposals became more and more speculative, Cuvier felt compelled to rebut them publicly, prompting spirited counterattacks by Geoffroy Saint-Hilaire (see Appel, 1987). Thus began one of the most important debates in the history of biology, one that was followed eagerly by scientists all over Europe (e.g., Whewell, 1837; Goethe, 1981).

Looking back on the Cuvier–Geoffroy Saint-Hilaire debate, it is not so easy to say who won. Darwin's theory of evolution clearly supports Geoffroy Saint-Hilaire's contention that all organisms share certain essential features, whereas Cuvier's strict separation between different animal types presents more of an obstacle to evolutionary thought. On the other hand, many of Geoffroy Saint-Hilaire's proposed homologies still strike us as rather fanciful today, and contemporary morphologists recognize not one but at least 35 distinct animal body plans (Raff, 1996). Thus, both men were partially correct and their respective insights had to be reconciled with one another. By now, this rapprochement is largely complete, as most morphologists agree that shared body plans are evidence for evolution, but also acknowledge that new body plans may arise during phylogeny (e.g., Burke et al., 1989). Cuvier correctly cautioned against overeager homologizing between animals of different types, but Geoffroy Saint-Hilaire's speculations presaged the now widely accepted position that all animals are built from highly conserved elements at the cellular and molecular levels of analysis. Thus, the realization that conservation at one level of analysis may go hand in hand with innovation at another level largely resolves the controversy that lay at the heart of the Cuvier–Geoffroy Saint-Hilaire debate.

Since most biologists now agree that there is a basic, highly conserved plan for vertebrate bodies, we can ask whether there is also a basic plan for vertebrate brains. Historically, this question has been approached in two different ways. On the one hand, anatomists such as Herrick (1948) focused on comparisons between adult brains; they derived what is best called an "adult brain archetype." On the other hand, comparative neuroembryologists such as Holmgren (1922) compared primarily embryonic brains and consequently derived an "embryonic brain archetype." These two lines of inquiry clearly influenced one another, but they are logically distinct because embryonic features differ fundamentally from adult features.

Accordingly, I discuss adult and embryonic comparisons separately for most of this chapter: only toward the chapter's end will I explore how the adult and embryonic brain archetypes can be linked to one another. But first I must digress briefly to review just who these vertebrates are and how they are related to one another.

A "Who's Who" of Vertebrates

In this section I will briefly review the vertebrates—all animals with a backbone. Most readers will be familiar with most of the major groups of vertebrates, but not all readers will know all of them, and we need to consider all of them to determine whether vertebrate brains are all built according to a common plan. Moreover, if we do not know how the major groups of vertebrates are related to one another, then we cannot reconstruct how their brains have changed phylogenetically. The relations between the various vertebrate groups are shown in the cladogram in Figure 3.1

To start our examination of the vertebrates, we will first consider their closest invertebrate relatives, the cephalochordates (see Figure 3.1), which possess a cartilaginous notochord but no bony vertebrae. By far the most intensively studied cephalochordate is the lancelet, *Branchiostoma lanceolatum*, more commonly known as amphioxus (Figure 3.2A). This slender animal is a few centimeters long and spends most of its time buried in sand, leaving only its anterior end exposed. It lacks jaws but has a huge mouth and pharynx, which are used for filter feeding. It has an anterior eyespot, but no paired sensory organs. In fact, amphioxus does not have much of a head at all, which is why it is sometimes referred to as an acraniate animal. Although the central nervous system of amphioxus has been studied extensively, many questions remain (Bone, 1959; Lacalli, 1996; Nieuwenhuys et al., 1998). The spinal cord of amphioxus resembles that of vertebrates in many ways, but its brain is decidedly dissimilar, being no thicker than the spinal cord and lacking the flexures and evaginations that are typical of vertebrate brains. Gene expression data indicate that amphioxus has a hindbrain and midbrain of sorts, but the telencephalon seems to be lacking entirely (Lacalli et al., 1994).

The vertebrates themselves are traditionally divided into jawless vertebrates (agnathans) and jawed vertebrates (gnathostomes). The jawless vertebrates comprise two major groups, namely hagfishes and lampreys. By most accounts, lampreys are the closest relatives of jawed vertebrates, and hagfishes are the closest relatives of lampreys and jawed vertebrates. Although lampreys and hagfishes have similar body shapes, they are very different from one another (see Figure 3.2B,C). Lampreys spend much of their life as larvae, buried in mud, but then become active swimmers as adults. They use a large, suckerlike mouth to attach themselves to other fish and suck them dry. Hagfishes, in contrast, are bottom-feeding scavengers that spend most of their time buried in sand. When threatened, they secrete copious amounts of viscous slime, which is why they are sometimes called "slime eels." Although not much is known about hagfish behavior and ecology, it appears that hagfishes lead rather simple, passive lives. Strangely, however, their brain is significantly larger and more complex than that of the more active lampreys (see Figure 3.2C). Both lampreys and hagfishes are represented by only a few

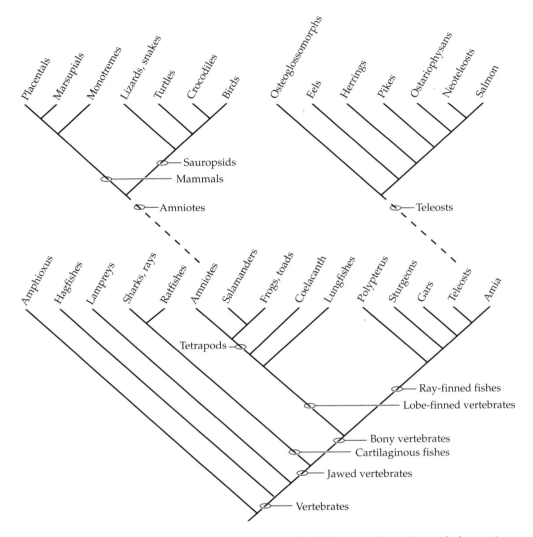

Figure 3.1 Cladogram of Vertebrates   In this cladogram of vertebrate phylogenetic relationships, common names have been substituted for several proper taxonomic names, and some well-known species have been used as stand-ins for larger taxonomic groups (e.g., amphioxus is a cephalochordate). (After multiple sources including, most prominently, Carroll, 1988.)

species today, but the fossil record indicates that our planet was once inhabited by many different kinds of jawless vertebrates.

The jawed vertebrates are generally divided into cartilaginous fishes and bony vertebrates. Since cartilaginous fishes, with approximately 800 extant

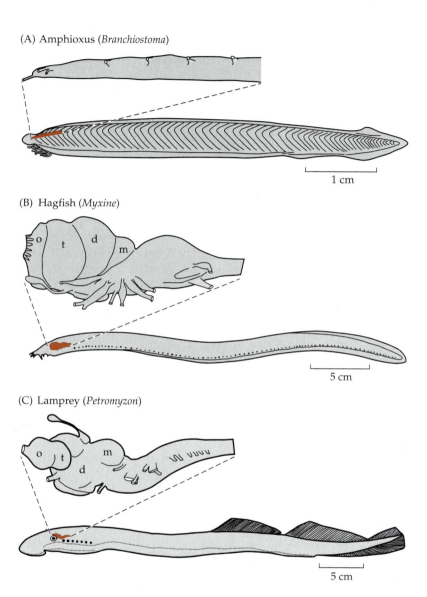

(A) Amphioxus (*Branchiostoma*)

1 cm

(B) Hagfish (*Myxine*)

5 cm

(C) Lamprey (*Petromyzon*)

5 cm

Figure 3.2 A Comparison of a Cephalochordate and Jawless Vertebrates Lateral views of the body and the brain for (A) amphioxus, one of the cephalochordates, the closest invertebrate relatives of vertebrates; (B) a hagfish, one of the jawless vertebrates and the closest relative of lampreys and jawed vertebrates; and (C) a lamprey, another jawless vertebrate and the closest relative of the jawed vertebrates. The scale bars apply to the bodies, not to the magnified brains. Abbreviations: d = diencephalon; m = mesencephalon; o = olfactory bulb; t = telencephalon. (After Nieuwenhuys et al., 1998; hagfish and lamprey bodies after Romer and Parsons, 1977.)

Figure 3.3 Cartilaginous Fishes   (A) The ratfish belongs to a primitive group of car- ▶
tilaginous fishes that has a rather small, elongated telencephalon. (B) The spiny dogfish
is one of the so-called squalomorph sharks, which tend to have relatively small brains.
(C) The smooth dogfish belongs to the large-brained galeomorph sharks. Abbreviations:
c = cerebellum; d = diencephalon; m = mesencephalon; o = olfactory bulb; t = telencephalon. (A,C after Ebbesson, 1980; B after Nieuwenhuys et al. 1998.)

species, are the smaller of these two groups, let us deal with them first. As the
name implies, cartilaginous fishes possess a cartilaginous skeleton and lack
proper bone. Most of us know that sharks, skates, and rays are cartilaginous
fishes, but relatively few people have heard of ratfishes, a rare kind of cartilaginous fish that lives in deep water and has a rather peculiar, small, and
elongate brain (Figure 3.3A). Among sharks, skates, and rays, brain organization varies tremendously. Some sharks (e.g., the spiny dogfish; Figure 3.3B)
have relatively small and simple brains, while others (e.g., the smooth dogfish; Figure 3.3C) have much larger and more complex brains. Similarly,
among skates and rays, some species (e.g., the common skate) are relatively
small-brained, while others (e.g., the manta rays) are equipped with surprisingly large and complex brains (see Chapter 4). Also noteworthy is the Torpedo ray, which sports a powerful electric organ that once provided neurobiologists with a rich source of acetylcholine receptors. Innervating this electric
organ is a pair of medullary "electric lobes" that comprise as much as 60% of
the brain (Roberts and Ryan, 1975).

The bony vertebrates consist of ray-finned fishes (actinopterygians) and
lobe-finned vertebrates (sarcopterygians). True to their name, the ray-finned
fishes have fins that are supported by bony fin rays. Lobe-finned vertebrates,
in contrast, possess finlike appendages that are attached to the body via fleshy
lobes (in tetrapods these appendages form arms and/or legs). Both groups are
extremely diverse, containing more than 20,000 species each. Traditionally, ray-
finned fishes are considered "lower" than lobe-finned vertebrates (of which
we are one), but ray-finned fishes have been no less successful in terms of speciation and populating the planet. It is mainly by convention, then, that I here
discuss ray-finned fishes before the lobe-finned vertebrates.

Most ray-finned fishes are teleosts, an extremely diverse group that contains more than half of all vertebrate species and includes virtually all of the
fishes you would find in a lake, stream, or aquarium store. One "key innovation" of teleosts is their protrusible mouth, which effectively sucks in prey
when it is opened rapidly. Teleosts are also fascinating because they continue
to grow throughout their adult life, which means that their brains likewise
continue to add new cells and grow. Some teleosts, notably the mormyrid and
gymnotoid teleosts (members of the osteoglossomorph and ostariophysan
radiations, respectively; see Figure 3.1) can generate and detect weak electric
fields, an ability that has been well studied by neurobiologists (Bullock and
Heiligenberg, 1986). Mormyrids are particularly interesting from an evolu-

(A) Ratfish (*Hydrolagus*)

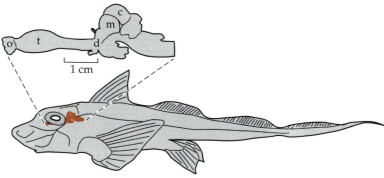

(B) Spiny dogfish (*Squalus*)

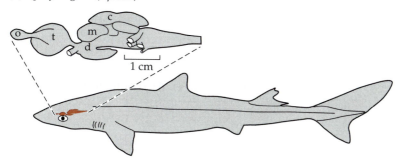

(C) Smooth dogfish (*Mustelis*)

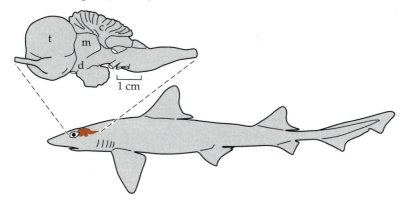

tionary perspective because they have extremely large brains with a brain:body ratio similar to that of humans (Figure 3.4A; see Chapter 4). In addition to teleosts (such as trout; see Figure 3.4B), there are some notable nonteleost ray-finned fishes, such as long-nosed gars, caviar-yielding stur-

(A) Mormyrid electric fish (*Gnathonemus*)

(B) Trout (*Salmo*)

(C) Bichir (*Polypterus*)

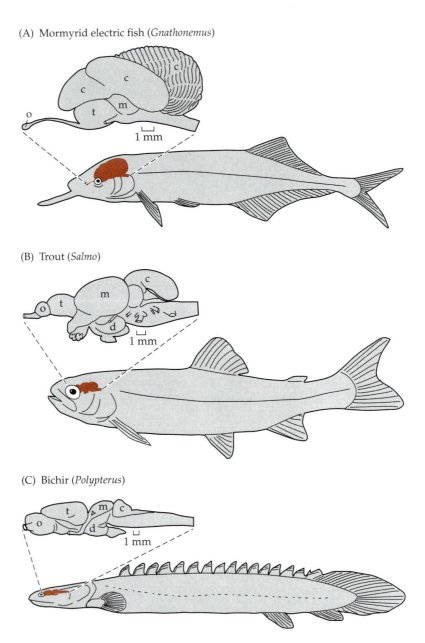

geons, and polypterids, a group of unusual creatures that sport numerous, small dorsal fins, a peculiar nose, and an odd-looking brain (see Figure 3.4C). Although the phylogenetic position of polypterids remains controversial, they are probably the nearest living relatives of all other ray-finned fishes.

◀ **Figure 3.4 Ray-Finned Fishes**   (A) Mormyrid electric fishes (also known as elephant-noses) have an extremely large brain with an enormously enlarged cerebellum (technically, it is mainly the cerebellar valvula that is so enlarged). Part of this enlarged cerebellum is used to detect how objects near the animal distort the electric fields generated by the fish, but it may well have some additional functions. (B) The trout is in many ways a rather typical teleost, which possesses a relatively small telencephalon but a large hypothalamus, optic tectum, and cerebellum. (C) The bichir, also known as *Polypterus*, is a nonteleost ray-finned fish that retains many primitive features. Abbreviations: c = cerebellum; d = diencephalon; m = mesencephalon; o = olfactory bulb; t = telencephalon. (A and B after Nieuwenhuys et al., 1998; C after Papez, 1929.)

   The lobe-finned vertebrates comprise coelacanths, lungfishes, and tetrapods (discussed later in this chapter). The coelacanth (*Latimeria chalmunae*) is a classic living fossil because it was known exclusively from the fossil record until a few specimens were caught off the coast of southern Africa (Figure 3.5A). Although the coelacanth has persisted largely unchanged for more than 300 million years, it exhibits many features that are highly specialized. For example, it gives birth to live young, osmoregulates by means of urea retention, and possesses a unique rostral organ that is probably used to detect weak electric fields generated by potential prey. The coelacanth's brain is extremely small (occupying less than 1% of the total endocranial volume) and appears in some respects as if it were a mosaic of features from teleosts and cartilaginous fishes (Millot and Anthony, 1965).
   Lungfishes, the second major group of lobe-finned vertebrates, live in tropical climates and can, as their name suggests, resort to breathing air during seasonal droughts. Lungfish brains exhibit very little histological differentiation and are among the simplest of all vertebrate brains (Figure 3.5B). Various morphological and molecular data suggest that lungfishes are more closely related to tetrapods than coelacanths are (Rosen et al., 1981; Meyer and Dolven, 1992), but other data suggest that coelacanths are the closest living relatives of tetrapods.
   The tetrapods are divisible into amniotes (discussed later in this chapter) and amphibians. The amphibians, in turn, are generally divided into anurans (frogs and toads), urodeles (newts and salamanders), and caecilians (legless salamanders). Typical of most amphibians is an aquatic larval stage that ultimately metamorphoses into a terrestrial adult form. All amphibians also have relatively small and simple brains. Most neurobiological studies have focused on salamanders of the genus *Ambystoma* (see Figure 3.5C), specifically the tiger salamander and the axolotl. The former was made famous by C. J. Herrick in *The Brain of the Tiger Salamander* (Herrick, 1948); the latter is popular among developmental neurobiologists who appreciate its large, accessible embryos and are fascinated, for example, by its ability to regenerate brain areas that have been experimentally removed (Kirsche and Kirsche, 1964). Other studies have focused on bolitoglossine salamanders, which have

(A) Coelacanth (*Latimeria*)

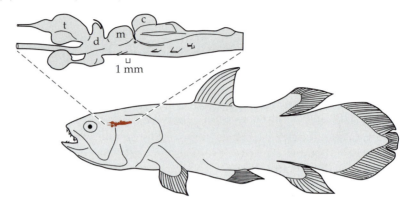

(B) Lungfish (*Lepidosiren*)

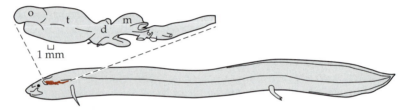

(C) Salamander (*Ambystoma*)

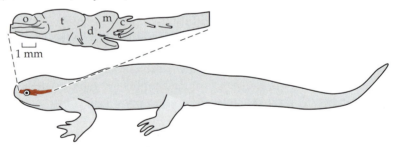

Figure 3.5 Lobe-Finned Vertebrates (A) The coelacanth (also known as *Latimeria*) is one of the most peculiar vertebrates. It is a living fossil and has a rather small and unusual brain. (B) Lungfishes, especially those from Africa (shown here), look like eels with tiny toeless legs. Their brains, as well as their bodies, exhibit many features that are typical of very young animals in other species. (C) Salamanders also exhibit numerous juvenile features, which may explain why their brains look so similar to those of lungfishes. Abbreviations: c = cerebellum; d = diencephalon; m = mesencephalon; o = olfactory bulb; t = telencephalon. (After Nieuwenhuys et al., 1998; body in A after Romer and Parsons, 1977.)

exceedingly small and histologically simple brains. Anurans typically undergo metamorphic changes that are more pronounced than those in urodeles and involve major reorganizations of both sense organs and brain. In addition, anurans are characterized by numerous specializations that facilitate vocal communication, including a tympanic membrane and a well-developed central auditory system (Fritzsch et al., 1988). The anurans that are most commonly used in neurobiological studies are the North American bullfrog *Rana catesbeiana* and the African clawed frog *Xenopus laevis,* which is fully aquatic and ideal for embryological research. The third major group of amphibians, the caecilians, consists of wormlike creatures that generally live underground. They have slender heads, small eyes, and unusually elongate telencephalons (Northcutt and Kicliter, 1980). The phylogenetic relationships among the various amphibian groups remain controversial.

Amniotes derive their name from the fact that they produce eggs with well-developed extraembryonic membranes, one of which is the "amnion." There are two major groups of amniotes, namely mammals and sauropsids (reptiles and birds). Although sauropsids are sometimes classified as "submammalian" vertebrates, extant sauropsids are neither more primitive (in the sense of having evolved earlier) nor less complex (as a group) than mammals. Therefore, one might as well discuss mammals first.

The mammals are divisible into monotremes, marsupials, and placentals. The monotremes include the echidna (spiny anteater) and the duck-billed platypus. Both species lay eggs and incubate them. Their snouts are equipped with numerous electroreceptors, which are probably used to detect prey. The echidna's telencephalon exhibits numerous sulci and gyri (see Figure 2.12), while that of the platypus is fairly smooth. The marsupials are a large radiation that includes opossums (Figure 3.6A) and kangaroos, to mention only two of the better-known groups. They all give birth to altricial young that require further development in the mother's marsupial pouch, where they obtain milk from mammary glands. Because marsupial young are born so prematurely, they are ideally suited for developmental studies that aim to elucidate, for example, the mechanisms of spinal cord regeneration. The brains of marsupials all lack a corpus callosum (see Chapter 2) but are otherwise quite diverse. The placental mammals require little discussion at this point, since they are covered extensively in Chapters 8 and 9. For now, I merely point out that the placental mammals all have a corpus callosum and comprise more than 20 distinct orders that differ greatly in terms of behavior and brain organization. Their phylogenetic interrelationships remain controversial, but most authors agree that the placental mammals are a monophyletic group, that their closest relatives are the marsupials, and that monotremes are the closest relatives of placentals and marsupials (see Figure 2.11).

The sauropsids, finally, include reptiles and birds. Although birds share many features with mammals (e.g., homeothermy and complexly folded cere-

Figure 3.6 Amniotes
(A) Opossums are marsupials with relatively small and simple brains. (B) Turtle brains are interesting because, among other things, they can function with very little oxygen. (C) Ravens are the largest living songbirds; they have remarkably large brains and are surprisingly intelligent. Abbreviations: c = cerebellum; d = diencephalon; m = mesencephalon; o = olfactory bulb; t = telencephalon. (A and B after Nieuwenhuys et al., 1998; C After Madge and Burn, 1994 and Stingelin, 1958.)

(A) Opossum (*Didelphis*)

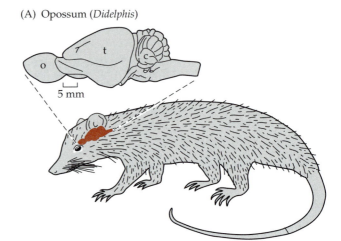

(B) Tortoise (*Testudo*)

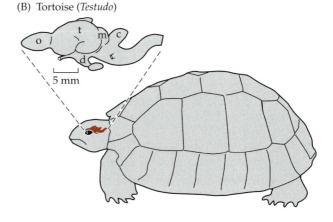

(C) Raven (*Corvus*)

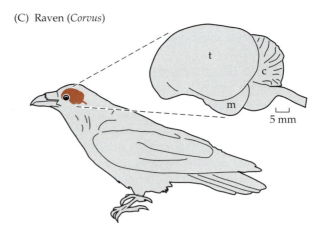

bella), they are more closely related to reptiles than mammals (Gauthier et al., 1988; see Figure 3.1); in fact, as the likely descendants of dinosaurs, birds could be called "flying reptiles." Among earthbound reptiles, the closest relatives of birds are the crocodilians (Hedges, 1994). Turtles are probably the nearest relatives of crocodilians and birds, but this hypothesis remains controversial (Zardoya and Meyer, 2001). The other major group of sauropsids, the lepidosaurs, comprises mainly lizards and snakes. Relative brain size varies greatly among sauropsids, with turtles (see Figure 3.6B) and snakes occupying the low end of the range, and ravens (see Figure 3.6C) and parrots the other extreme (see Chapter 4). Pigeon and songbird brains have been studied in great detail (Pearson, 1972), primarily because these birds exhibit such fascinating and well-studied behaviors as sun-compass navigation and song learning. Chick brains, on the other hand, have been the object of many developmental studies because they are so readily available and accessible to experimental manipulation. Compared to the great volume of work on avian brains, relatively little is known about reptilian brains. Turtle brains have been studied in some detail, however, because they are often thought to be "primitive" and, less often, because they are extremely resistant to oxygen deprivation (which means that their brains can be studied in vitro).

This concludes our whirlwind tour of the major vertebrate groups, their brains, and their phylogenetic interrelationships. As I stated previously, knowledge of how the various vertebrates are related to one another is fundamental to reconstructing the course of vertebrate brain evolution. This means that, to a large extent, our knowledge of vertebrate brain evolution is only as good as our knowledge of vertebrate phylogeny. Fortunately, phylogenetic systematics has advanced tremendously in the last few years, particularly because of the rise of cladistics (see Chapter 2) and the influx of new molecular data. Therefore, the phylogeny shown in Figure 3.1 is based on solid data and sound logical principles. Nonetheless, some aspects of this phylogeny may turn out to be incorrect, and if that happens, some of the evolutionary scenarios based on that phylogeny may have to be revised. This does not mean that evolutionary neurobiology is "mere speculation"; rather, it means that the field is like most others in science in that its hypotheses are not without assumptions and may have to be revised if the assumptions fail.

Comparing Adult Brains

The first step in comparing the brains of all these different vertebrates is to look for universals. In this book, I will refer to a collection of all universally present features as an archetype, following the usage established by Owen (1846), who considered an archetype to be a transcendental schema, or Platonic ideal, according to which all animals of a given type were created. One danger of using the term archetype, however, is that in common parlance the

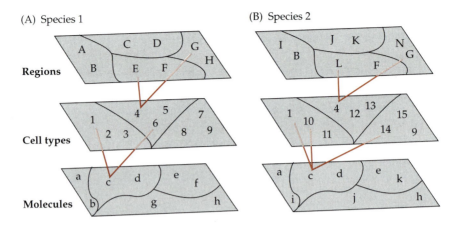

(A) Species 1

(B) Species 2

Regions

Cell types

Molecules

word *archetype* is often defined as "an original pattern or model of which all things of the same type are representations or copies" (Mish, 1993). This was not Owen's definition; to him, the prefix "arch-" in archetype meant "principal" rather than "primitive" (both are legitimate interpretations). A related complication is that, although Darwin's "common ancestor" can explain the existence of Owen's archetype, the two concepts are not equivalent. Any ancestral features that were lost during the course of evolution are not universal and, hence, they are not part of the archetype. For example, the last common ancestor of all vertebrates almost certainly had electroreceptors, but these were lost in most extant lineages (Northcutt, 1985b). Therefore, electroreceptors are not part of the vertebrate archetype as it is defined in the present book.

The next question facing any seeker of archetypes is: what to compare? Because brains are so complex, they can be studied at several different levels of analysis, and these levels form three logically distinct (but interacting) hierarchies of brain structure: regions, cell types, and molecules (Figure 3.7). The first level analyzes brains in terms of their major divisions (such as the telencephalon and diencephalon), which are themselves divided into subdivisions (the dorsal thalamus, ventral thalamus); the latter are in turn composed of cell groups (cortical laminae, brain nuclei, or diffuse formations), and many of these have several subnuclei or sublaminae. This hierarchy of brain regions is the most frequent subject of comparative analyses that span all vertebrates. The second level of analysis focuses on brains as being composed of individual cells, which form a hierarchy of different types (with neurons and glial cells at the highest level). A vast array of different cell types has been described in the neurobiological literature, but there have been relatively few attempts to compare these cell types across species. Finally, the third level

(C) Archetype

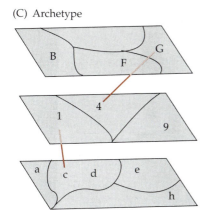

Figure 3.7 Levels of Analysis of the Brain Brains can be analyzed and compared across species at three major levels of analysis: regions, cell types, and molecules. Each of these major levels is itself hierarchically organized (as indicated by letters or numbers within bounded regions). Comparing across (A) species 1 and (B) species 2, we find that characters at one level may be homologous to one another even though some of their lower-level constituent characters are not homologous. This kind of phylogenetic discordance between levels occurs, for example, when a conserved molecule (indicated by a lowercase "c") becomes expressed in a novel set of cell types or if a conserved cell type (indicated by a number "4") forms in a novel set of brain regions (these relationships are indicated by red lines). (C) Archetypes can be constructed for brain regions, cells, or molecules. Because they contain only universally present characters and relationships, they tend to lack detail.

of analysis views brains as complex compositions of various molecules, which can be grouped into families and superfamilies. Although comparative genomics is prominent in modern biology, phylogenetic analyses of neurobiologically important molecules are still relatively rare.

Because these three hierarchies—of regions, cells, and molecules—are ontologically distinct from one another, I use separate sections to review what is conserved in each of them. My review is brief, but a more comprehensive treatment would burst the limits of this book and would not serve its central aim, which is to isolate *principles* of brain evolution rather than to review facts. Accordingly, I stress not the details of what has been conserved, but the *patterns* that are formed by conserved features at the different levels of analysis.

Adult brain regions

All adult vertebrate brains are divisible into telencephalon, diencephalon, mesencephalon, and rhombencephalon. Telencephalon and diencephalon are collectively referred to as "forebrain," the mesencephalon is commonly known as "midbrain," and the rhombencephalon corresponds to what many authors call "hindbrain." Each of these regions is, in turn, composed of two or more highly conserved major divisions. In the following paragraphs, I will briefly describe the major brain divisions in the sequence in which you encounter them as you work your way rostrocaudally through an adult brain (i.e., from nose to tail for all animals except humans), starting with the olfactory bulb and ending with the medulla (Figure 3.8). Within each division, I proceed from dorsal to ventral (as those terms are generally defined in adult animals—that is, from back to belly). Note that this sequence does not follow the "true" longitudinal and transverse axes of the brain as recognized by

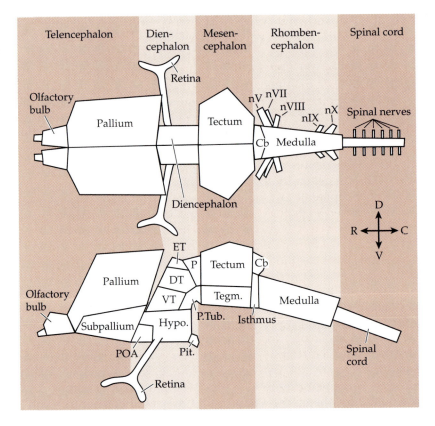

Figure 3.8 The Vertebrate Brain Archetype Schematic drawings of the vertebrate brain region archetype from dorsal (top) and lateral (bottom) perspectives. Some elements have been omitted for the sake of clarity. Abbreviations: C = caudal; Cb = cerebellum; D = dorsal; Dienc. – diencephalon; DT = dorsal thalamus; ET = epithalamus; Hypo. = hypothalamus; nV = trigeminal nerve; nVII = facial nerve; nVIII = octaval nerve; nIX = glossopharyngeal nerve; nX = vagal nerve; P = pretectum; Pit, = posterior pituitary; POA = preoptic area; P. Tub, = posterior tuberculum; R = rostral; Tegm. = tegmentum; V = ventral; VT = ventral thalamus.

embryologists, but it is of historical interest (since that is how most neuroscientists were trained to "march" through brains) and makes some sense as long as we are dealing solely with adults.

The most rostral component of the telencephalon is the olfactory bulb, which receives inputs from the olfactory epithelium and projects primarily to the lateral portions of the more caudal telencephalon. Although an olfactory bulb is lacking in all marine mammals, I include it in the vertebrate brain archetype because the olfactory bulb is such an integral part of the telencephalon in all major vertebrate groups. The remaining telencephalon contains a dorsally located pallium (which means "cloak" in Latin) and a more ventral subpallium. The pallium is relatively small in lampreys, lungfishes, and urodeles, but large and well developed in all of the large-brained vertebrates (see Chapter 4). In mammals, the pallium consists of the hippocampal formation, neocortex, olfactory cortex, claustrum, and parts of the amygdala. Likely homologues of these regions have been identified in all jawed vertebrates, but several of these homologies remain controversial and different patterns of pallial organization may exist in hagfishes and lampreys. Therefore, I have not included any pallial subdivisions in the adult vertebrate brain archetype. The subpallium generally receives dopaminergic projections from the caudoventral diencephalon and/or the mesencephalic tegmentum and is heavily interconnected with both thalamus and the pallium. It is relatively small in teleost fishes but well-developed in most mammals, where it includes the septum, globus pallidus, and striatum. Again, homologues of these subdivisions have been identified in other vertebrates but remain controversial (though less so than the pallial homologies).

The diencephalon is composed of numerous divisions (see Figure 3.8). Its most dorsal division is the epithalamus, which consists mainly of the habenula and the pineal gland (only hagfishes lack a pineal). Ventral to the epithalamus lies the thalamus, which is composed of dorsal and ventral divisions in all vertebrates (except, apparently, hagfishes [Wicht and Nieuwenhuys, 1998]). The dorsal thalamus is large in amniotes but small in most other vertebrates; it receives inputs from many different brain regions and projects mainly to the telencephalon. The ventral thalamus, in contrast, is relatively small in all vertebrates and defies functional generalizations (in mammals, it contains the thalamic reticular nucleus, the subthalamic nucleus, and the ventral lateral geniculate nucleus). Ventral to the thalamus lies the hypothalamus, which is large and highly differentiated in most vertebrates, especially teleosts and cartilaginous fishes. It functions in many homeostatic behaviors (such as feeding, thermoregulation, and reproduction) and controls, among other things, hormone secretion via the posterior pituitary. The most extraordinary part of the diencephalon is the retina, which develops as a laterally directed evagination of the ventralmost hypothalamus. Just rostral to the optic chiasm (where the optic tracts enter the brain) lies the preoptic area, which is generally considered to be part of the diencephalon, though some authors consider it part of the telencephalon. Finally, there are two lesser-known, caudal regions of the diencephalon: the pretectum and the posterior tuberculum. Both are relatively large in nonmammals (the posterior tuberculum is huge in teleosts), but they are small in mammals.

The mesencephalon is divisible into a dorsally located tectum (from the Latin term for "roof") and a subjacent tegmentum in all vertebrates. The tectum can be further divided into an optic tectum (or superior colliculus) and a torus semicircularis (or inferior colliculus). The optic tectum always receives at least some visual information from the retina and integrates it with information from other sensory modalities; it also controls a number of motor behaviors, particularly those that have to do with orientation in space. The torus semicircularis, on the other hand, is concerned primarily with conveying auditory and mechanosensory information to the diencephalon; it lies either ventral or caudal to the optic tectum, depending on the animal group. The mesencephalic tegmentum, finally, contains several important and highly conserved cell groups, including the midbrain reticular formation (which plays an important role in motor control) and the oculomotor nucleus (which innervates most of the external eye muscles in all vertebrates except hagfishes).

The rhombencephalon consists of the isthmus, cerebellum, and medulla (also known as the medulla oblongata). The isthmic region lies just caudal to the mesencephalon and contains, among other things, the interpeduncular nucleus (which receives inputs from the habenula) and the trochlear nucleus (which innervates some of the eye muscles). The cerebellum develops in the dorsal portion of the rhombencephalon just caudal to the isthmic region (from the "rhombic lip"). The cerebellum seems to be lacking in hagfishes but is such a characteristic feature of most other vertebrate brains that I assign it honorary membership in the adult brain archetype. Although the cerebellum is generally regarded as a motor-control structure, it also functions in "purely cognitive" behaviors in at least some vertebrates (Roland, 1993; Parkins, 1997). Thus, despite a mountain of data on cerebellar anatomy and physiology, the behavioral functions of the cerebellum defy simple classification. The medulla, finally, contains numerous vital cell groups, including (most rostrally) the locus coeruleus, which plays a role in regulating arousal. Developmentally, the medulla is divisible into a dorsal alar plate and a more ventral basal plate (Nieuwenhuys, 1998). Functionally, neurons in the alar plate tend to process sensory information coming from either the cranial nerves or the spinal cord, while neurons in the basal plate generally contain either motor or premotor neurons. The size of individual cell groups within these medullary sensory or motor columns tends to correlate positively with an animal's sensory or motor abilities (see Chapter 5).

Thus far, we have talked mainly about what I call major brain divisions, but broadly conserved elements can also be found at lower levels of the brain region hierarchy, among what we generally call cell groups or nuclei; for example, all vertebrates possess a suprachiasmatic nucleus that lies in the hypothalamus, receives retinal inputs, and helps to control circadian rhythms. Some of these highly conserved cell groups are mentioned later in the text. However, I must stress that, in contrast to the brain's major divisions, many smaller cell groups or nuclei cannot be identified in all vertebrates. It is quite unclear, for example, whether birds have a homologue of the mammalian central amygdaloid nucleus (Wild et al., 1990) or the auditory cortex (see Chapter 8). Therefore, the adult vertebrate brain archetype consists mainly of major divisions and subdivisions, not of individual cell groups and their components. This lack of detail in the adult vertebrate brain archetype is due to the fact that the major brain divisions and subdivisions often develop differently in at least some vertebrate groups. That is, they differentiate into different, nonhomologous cell groups in different taxonomic groups, and non-homologous elements cannot, by definition, be part of an archetype.

To obtain archetypes with more detail, we can limit our analysis to smaller taxonomic groups. The archetype for mammalian brains, for example, would be more detailed than the vertebrate archetype described above, and a primate archetype would be more detailed still. In other words, the archetypes are themselves hierarchically organized and become more detailed at lower taxonomic levels, which is why experienced neuroanatomists can quickly identify which class of vertebrates a stained brain section is taken from, but need more time, and more detailed analyses, to determine which order or genus they are examining. At the level of all vertebrates, however, the adult brain archetype is painted with a broad brush.

Before going on to other kinds of structural conservation, let us revisit briefly the question of how brain regions and cell groups are homologized to one another. As mentioned in Chapter 1, any features that are most parsimoniously interpreted as having evolved just once, rather than multiple times independently, are homologous to one another (at least within the cladistic framework that currently reigns in comparative neurobiology). Therefore, any features that are conserved across all vertebrates are automatically homologous, because it would be unparsimonious to think that they evolved independently in the various vertebrate groups. More complicated is the more fundamental question of how two or more features in different species are identified as being comparable in the first place, as being "the same character" (Striedter, 1999). This question is often overlooked because it seems so easy to answer—how hard can it be to identify a fish's cerebellum as a cerebellum? But many characters are not so easy to identify, and different sets of criteria may be used to identify different kinds of characters.

Major divisions are generally identified in relation to major landmarks or to one another (e.g., the ventral thalamus lies ventral to the dorsal thalamus), but for cell groups, additional criteria, such as cytoarchitecture, histochemistry, and neuronal connections, come into play. Comparative neuroanatomists sometimes argue about which criteria are most useful for comparing neural characters across species (Nieuwenhuys and Bodenheimer, 1966; Campbell and Hodos, 1970), but I think that this debate is irresolvable, because different criteria tend to be better for identifying different kinds of characters. Histochemical criteria, for instance, are only rarely useful for identifying major brain divisions (e.g., Swanson and Petrovich, 1998), but they are often used to help identify cell types (e.g., Wicht et al., 1994). Once we realize that different criteria can be used to identify different kinds of characters, but that parsimony is the ultimate arbiter of true homology, we can avoid at least some of the confusion that has previously plagued the concept of homology. We are also ready to consider homologies and archetypes of cells and molecules.

Adult cell types

All vertebrates possess both neurons and glia. In this discussion, I will review glia first, as their classification is fairly straightforward compared to neurons. Although neurobiologists tend to be "neuronocentric," glia are clearly vital to brain function (Vernadakis and Roots, 1995). Three major classes of glia are commonly recognized: microglia, oligodendroglia, and astroglia. Microglia, which are macrophage-like cells that remove damaged cells, have been described in various amniotes and ray-finned fishes (Velasco et al., 1999). They appear to be absent in hagfishes (Wicht and Nieuwenhuys, 1998), but whether they exist in lampreys, cartilaginous fishes, or amphibians remains unclear. Myelin-producing oligodendroglia are found in all vertebrates except hagfishes and lampreys. Consequently, they probably evolved with the origin of the jawed vertebrates and have been retained since then (although the protein composition of myelin has changed significantly; see Jeserich et al., 1997). Astroglia, finally, are divisible into star-shaped astrocytes and radial glia, which generally extend radially from the ventricle to the brain surface (Figure 3.9A). Radial glia tend to disappear during development in both birds and mammals, but persist to adulthood in at least some parts of all vertebrate brains; therefore, they can be considered part of the adult brain archetype. For astrocytes, the story is more complex. Ramón y Cajal (1909) had proposed that astrocytes are unique to birds and mammals, but astrocytes have now been described also in reptiles and various anamniotes, including lampreys and hagfishes (Wasowicz et al., 1994; Wicht et al., 1994). Therefore, astrocytes appear to be a universal feature of adult vertebrate brains. Interestingly, astrocytes tend to be rare in species with thin cerebral walls (e.g., lampreys and salamanders), while radial glia tend to be rare in species with thick cerebral walls. One potential explanation for this distribution is that radial glia cannot function effectively when the cerebral wall gets too thick and may, consequently, transform into astrocytes (both ontogenetically and phylogenetically) (see Wasowicz et al., 1999).

Compared to glial cell classification, neuron classification is a bit of a nightmare. At the simplest level of analysis, neurons can be categorized as unipolar, bipolar, or multipolar, depending on how many primary dendrites emanate from their cell body, or soma. Theoretically, these basic neuron types could then be subdivided according to the number of secondary and tertiary dendrites (i.e., the finer branches of the dendritic tree). However, given the complexity of most real neurons, any rigorous classification system that is based mainly on dendritic branch topology quickly becomes unwieldy. Moreover, neurons differ not only in dendritic branching pattern but also in size, transmitter content, membrane properties, response to stimulation, and axon

(A) Radial glia in lamprey

(B) Purkinje cell in a mammal

(C) Purkinje cell in a perch

(D) Purkinje cell in a mormyrid

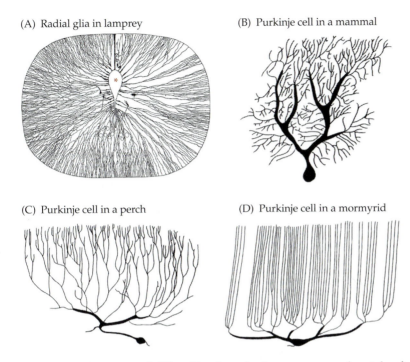

Figure 3.9 Neurons and Glia Vertebrate brains are composed mainly of glia and neurons. (A) Radial glial cells extend radially away from the brain's inner, ventricular surface toward the brain's outer surface (the ventricle is marked by an asterisk). During early development, radial glia form a scaffold for neuronal migration; later on, this scaffold often becomes distorted and/or disappears (see Nieuwenhuys et al., 1998). (B–D) An intriguing, highly conserved type of neuron is the Purkinje cell, which has a large dendritic tree that is primarily confined to a single plane (the plane of these drawings). Note that the dendritic structure of Purkinje cells does vary somewhat between species. (A from Retzius, 1893; B–D from Meek and Nieuwenhuys, 1991.)

morphology. In fact, there is an almost unlimited number of descriptors that can be used to characterize neurons. So, how can we turn these multifarious descriptors into a logically sound taxonomy of neurons (Tyner, 1975)? In my opinion, the most rigorous way to classify neurons is to examine as many of their features as possible and then use statistical analyses to determine which neurons form discrete clusters in the multidimensional parameter space of all possible neuronal attributes (Harping et al., 1985; Bazzaz, 1993). The advantage of such quantitative analyses is that they eliminate observer bias and avoid dichotomizing what is essentially continuous variation. Their major disadvantage, of course, is that they are labor intensive. Consequently, multidimensional analyses of neuronal variation are rare and have never, as far as I know, been used to compare neurons across species.

Of course, less formal classifications may sometimes suffice. For example, one does not need to perform a multidimensional analysis to realize that cerebellar Purkinje cells are found in all jawed vertebrates (see Figure 3.9B–D) and that all major types of retinal neurons (photoreceptors and bipolar, horizontal, amacrine, and retinal ganglion cells) are common to all vertebrates. Beyond these unambiguous cases, however, comparisons become more difficult. For example, an intriguing type of tectal neuron with distinctive "bottlebrush" endings can be homologized across reptiles, birds, and mammals (Luksch et al., 1998), but whether anamniotes also have such neurons remains anybody's guess. Because there are so few truly comparative studies of neuron types, we simply cannot know how many neuron types are conserved across all vertebrates. Given the available data, I suspect that vertebrate neurons comprise a hierarchy of types (much like brain regions form a hierarchy) and that, once we get beyond the first few levels of that hierarchy, most cell types have a rather limited phylogenetic distribution. In other words, I suspect that most neuronal cell types are not conserved across all vertebrates but are restricted to various subsets of vertebrates. Note, for instance, that pyramidal neurons can be identified in mammals and reptiles, but not in birds or most anamniotes (Nieuwenhuys, 1994; see Chapter 8). Similarly, the distinctive octopus cells of the mammalian cochlear nucleus most likely cannot be identified—and hence lack homologues—outside of mammals. Far more comprehensive studies will be needed, however, to determine the degree to which these and other types of neurons are conserved or variable across the vertebrates.

An important question to be answered by those further studies is whether some neuronal attributes prove more useful than others in our search for cellular homologies; at this point, the issue is far from resolved. The shape of a neuron's dendritic tree, for instance, is widely used to classify neurons within a species, but it is developmentally plastic (McAllister, 2000) and might, therefore, be rather malleable across species as well. Moreover, the fact that dendritic geometry is a major determinant of how neurons process their inputs (Mel, 1994) suggests that dendrite geometry might be highly subject to convergent evolution, with nonhomologous neurons tending to evolve similar dendritic trees. That would decrease the utility of similarity in dendrite shape as a criterion of homology. What about neuronal connections? They are often well conserved in evolution (see Chapter 7) but not immutable. Olfactory bulb mitral cells, for instance, have more extensive projections in hagfishes and lampreys than in other vertebrates, suggesting that their projections were "pruned" during the course of evolution (Wicht and Northcutt, 1993). Therefore, it would be wrong to use similarity of neuronal connections as an *essential* criterion of neuron homology or, conversely, to homologize neurons *solely* because they are connected to homologous brain regions (Webster, 1979). What about a neuron's position within the brain? It is clearly a fairly good criterion because experience tells us that homologous neurons are generally found in homologous regions. However, there are exceptions to this rule. The motor neurons of the trochlear nerve, for instance, occupy very different positions in the brain stem of reptiles and birds (Medina et al., 1993; Medina and Reiner, 1994). Therefore, it is probably unwise to demand that homologous neuron types must *always* be located in homologous adult brain regions. As I said, further studies are needed to determine the degree to which the various attributes of neuronal cell types are variable across species. Until then, we should probably refrain from ranking neuron attributes in terms of their utility for finding cellular homologies.

Neuron-typical molecules

Neurons harbor many different molecules, all of which can be compared across species, but neurobiologists tend to be most interested in those molecules that allow neurons to perform their typical functions. Therefore, the following discussion centers on molecules that mediate synaptic transmission and electrical excitability.

All of the principal neurotransmitters, including glutamate, gamma-aminobutyric acid (GABA), acetylcholine, dopamine, noradrenaline, and serotonin, are conserved across all vertebrates. In fact, these same molecules act as transmitters also in invertebrate nervous systems (Messenger, 1996), and most of them probably functioned as signaling molecules even before central nervous systems evolved (e.g., Turlejski, 1996). Although this high degree of conservation is remarkable, it becomes less surprising when we consider that these molecules are all derived (by only one or a few simple steps) from single amino acids. Given this simple structure, it seems likely that they were relatively "easy" to evolve but, thereafter, difficult to modify without impairing their signaling function. Evolution loves to tinker (Jacob, 1977), but the molecular structure of the principal transmitters cannot, apparently, be tinkered with very much. On the other hand, evolution has often modified where a particular transmitter is released, what other substances are co-released, and how the postsynaptic cells respond to the released mix of neuroactive substances (Harris-Warrick, 2000). These issues are addressed in turn in the following sections.

Some of the principal neurotransmitters, particularly dopamine, noradrenaline, and serotonin, are used by neurons in highly circumscribed brain regions that can be homologized fairly easily across vertebrates. A noradrenergic locus coeruleus, for example, has been identified in most major vertebrate groups (Ma, 1994; Adrio et al., 2002). In such cases, we can speak of broadly conserved transmitter systems. Glutamate and GABA, however, are used by so many different neurons, in so many different locations, that it makes little sense to homologize them across the major vertebrate groups, or to speak of them as conserved transmitter systems. As a matter of fact, even the broadly conserved transmitter systems exhibit variation. Among hindbrain noradrenergic and dopaminergic neurons, for example, only those belonging to locus coeruleus are easily homologized between the major vertebrate groups (Ma, 1997), and not all of the serotonergic cell groups in teleost fishes have ready homologues in mammals (Kaslin and Panula, 2001). How to interpret all this variation remains unclear. One possibility is that neurons may switch from one transmitter to another during the course of evolution, causing homologous neurons to release different transmitters. Such transmitter switching may explain why motor neurons use acetylcholine in vertebrates but mainly (though not exclusively) glutamate in invertebrates (Keating and Lloyd, 1999). However, further work will be needed to determine whether transmitter switching has also occurred within vertebrates.

In addition to the principal transmitters, neurons often use neuropeptides that are co-released with the principal transmitters and modulate their effects. These neuropeptides come in many varieties but can be grouped into a handful of major neuropeptide families (Figure 3.10). The tachykinin family, for example, includes the neuropeptides substance P and neurokinin A. Generally speaking, each neuropeptide family can be traced back to a single gene that first evolved among invertebrates and then duplicated at least once during vertebrate phylogeny (Holmgren and Jensen, 2001; Conlon, 2002). The gonadotropin-releasing hormone (GnRH) family, for example, can be traced back, at least tentatively, to an ancestral gene that duplicated just prior to the origin of vertebrates (King and Millar, 1995; Lin et al., 1998).

Of these two original GnRH genes, one encodes the peptide called chicken GnRH-II (named after the species in which it was first described), which has barely changed at all during the subsequent course of evolution. The other GnRH gene, in contrast, accumulated a variety of point mutations that led to the production of modified GnRH molecules (it may also have duplicated again within the ray-finned fishes). Because these modified GnRHs accumulated different mutations in different vertebrate lineages, we now observe a variety of so-called species-specific (really lineage-specific) GnRH molecules in addition to the highly conserved chicken GnRH-II. A similar pattern obtains in the oxytocin/vasopressin neuropeptide family (see Figure 3.10). Again, one copy of a duplicated gene remains highly conserved (at least within the peptide-encoding region), while the other one diverges relatively rapidly. This pattern suggests that the conserved molecule performs some essential function, while the modified copy is free to adopt some novel function or functions (Ohno, 1970).

When we think about how neuropeptides evolved, we must first consider that evolutionary changes in a neuropeptide's amino acid sequence may lead to changes in how that peptide interacts with its receptor. This may, in turn, lead to evolutionary changes in the receptors—to maintain a high affinity for the peptide, for instance. Indeed, a considerable body of evidence now indicates that neuropeptides have co-evolved with their receptors, which themselves form large receptor families (Hoyle, 1999). In addition, we must con-

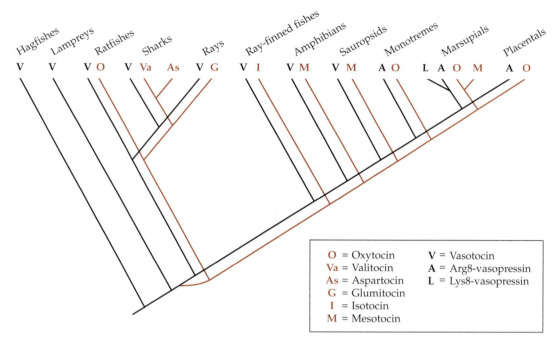

Figure 3.10 A Neuropeptide Family This cladogram depicts a putative phylogeny of peptides within the oxytocin/vasopressin gene family. The last common ancestor of all vertebrates probably possessed only vasotocin, and extant vertebrates retain this peptide with only minor modifications. Just prior to the origin of jawed vertebrates, the ancestral vasotocin gene probably duplicated and gave rise to a new lineage (shown in red) that includes a variety of oxytocin-related peptides. Within this more variable oxytocin-related lineage, we see both convergent evolution (the evolution of oxytocin in both ratfishes and mammals) and a phylogenetic reversal (the re-evolution of mesotocin in marsupials). (After Hoyle, 1999.)

sider that evolution may change a neuropeptide's function by changing when and where it is expressed. Duplicated genes, in particular, are likely to have diverged in their expression patterns because their upstream gene sequences (e.g., signal peptides and promoters) are also likely to have diverged in structure. We still know relatively little about how evolution has tinkered with peptide expression patterns, but we can note that chicken GnRH-II and the species-specific GnRHs are expressed in different brain regions, and that the species-specific GnRHs are more variable in terms of where they are expressed (Muske, 1993; Dubois et al., 2002). Even more striking divergences are seen within the neuropeptide Y superfamily, for neuropeptide Y is expressed in the brain, pancreatic polypeptide (PP) in the pancreas, and peptide YY in the gut (Hoyle, 1999). These changes are remarkable, but even relatively minor evolutionary changes in a peptide's expression pattern can be functionally significant. Arginine vasopressin, for example, is expressed differently in different species of voles, and these differences probably account for major differences in the voles' social behavior (Insel et al., 1998; Young et al., 1998).

More diverse than the principal neurotransmitters and neuropeptides are the receptors that bind them. Again, these receptors can be grouped into families and superfamilies. The superfamily of ligand-gated ion channel, or ionotropic, receptors consists of two major lineages. The first includes the GABA, glycine, and ionotropic glutamate receptors; the second includes mainly the nicotinic acetylcholine receptors. All of these receptor families predate the origin of vertebrates. Even the NMDA-type glutamate receptors, which were once thought to be unique to vertebrates, exist in some invertebrates (Chiang et al., 2002). The other major receptor superfamily, the G-protein–coupled, or metabotropic, receptors, also evolved long before vertebrates arrived on the scene. It includes the catecholamine and muscarinic acetylcholine receptors, which diverged from one another not only in the transmitters they bind but also in the second messenger systems they activate (Fryxell, 1995). One obvious advantage of having so many different receptors is that it increases the number of ways that neurons can communicate with one another. Another advantage is that different receptors can be expressed in different brain regions and/or at different times. Expression of the various NMDA receptors, for example, differs from region to region and during development (Watanabe et al., 1992). We know very little about how conserved these expression patterns are during phylogeny, but some species differences certainly exist. Humans and marmoset monkeys, for example, exhibit different patterns of serotonergic and muscarinic acetylcholine receptor binding within their hippocampi (Kraemer et al., 1995). Unfortunately, the functional significance of most species differences in receptor distribution remains unknown.

The voltage-gated and cyclic nucleotide–gated ion channels, finally, form a giant superfamily that originated among prokaryotes. First to evolve were the potassium (K) and cyclic nucleotide–gated channels; next came the calcium (Ca) channels. Because Ca channels all consist of four separate molecular domains, each of which is similar to a single K channel, it is likely that Ca channels evolved from primitive K channels by two rounds of tandem gene duplication (Anderson and Greenberg, 2001). Sodium (Na) channels probably evolved from primitive Ca channels around the time that the first nervous systems appeared (Hille, 1984). If we now survey all extant species, we find an enormous variety of different Na, Ca, and K channels but, again, we must realize that much of this diversification occurred independently in different lineages. The roundworm *C. elegans*, for example, evolved an unusually large number of different K channels but apparently lost its Na channels (see Anderson and Greenberg, 2001). Among vertebrates, mammals have more Na channel types than teleost fishes do (10 versus 6), but some of the teleostean Na channels are not homologous to any of the mammalian genes because they resulted from independent gene duplication events (Lopreato et al., 2001). Unfortunately, we still know little about the functional significance of all this channel diversity. Perhaps it is advantageous for different neuron types to express different combinations of ion channels, for then they can respond differently to the same stimuli (Salkoff et al., 2001). Or perhaps the diversity of different channel types allows individual neurons to express different combinations of ion channels in different parts of their dendritic tree and, thus, to generate useful physiological properties (Stuart and Spruston, 1998; Cook and Johnston, 1999).

To summarize, we see that neuron-typical molecules, neuronal cell types, and brain regions are all hierarchically organized and all exhibit a great deal of conservation at the highest levels of their individual hierarchies (see Figure 3.7). As we consider successively lower levels of each hierarchy—specific ion channels, neuron types, or neuronal cell groups—the degree of conservation wanes. Therefore, all adult brain archetypes are perforce painted with a broad brush. If we want archetypes with more detail, then we must restrict our attention to more taxonomically limited groups of species such as tetrapods or mammals. An alternative approach to obtaining archetypes with more details might be to compare embryonic rather than adult brains, for the embryos of different species tend to resemble each other more than the adults do. As discussed in the next section, this comparative embryological approach is powerful but must be applied judiciously.

Comparing Embryonic Brains

Karl Ernst von Baer (1828) first noted that the morphological similarity between different vertebrates is highest during early embryogenesis and then diminishes progressively as development proceeds. Subsequent work has supported von Baer's main conclusion but has also shown that the very earliest stages of vertebrate development differ considerably between the major vertebrate groups (e.g., in terms of egg size, cleavage patterns, and mechanisms of gastrulation). From these diverse beginnings, vertebrate embryos seem to converge toward a highly conserved phylotypic stage, or phylotype, after which they diverge again to generate their various adult forms. This convergent–divergent aspect of vertebrate development is nicely captured by the image of a developmental hourglass (Figure 3.11). Most investigators think of the phylotypic stage at the center of this developmental hourglass as the pharyngula stage, the stage when embryos have a tail, limb buds, and pharyngeal pouches (often misidentified as gill slits). Recent studies have raised important questions about whether a phylotypic stage for vertebrate embryos really exists (Richardson et al., 1997), but the idea that vertebrate embryos pass through a highly conserved period of development is firmly entrenched in the biological literature, and it is regularly reinforced by reports

Figure 3.11 The Developmental Hourglass The metaphor of the developmental "hourglass" highlights that vertebrates resemble one another most during the middle stages of development. Shown at the top are early gastrula stages of a zebra fish, a frog, a mouse, and a chicken. From these diverse beginnings, development converges onto a phylotypic stage or period, represented here by one of Ernst Haeckel's famous but oversimplified drawings. After this highly conserved period, development again diverges between the major vertebrate groups. (After Richardson et al., 1997.)

that many developmentally important genes exhibit highly conserved expression patterns in pharyngula-stage embryos (Slack et al., 1993).

Given this context, we might expect that vertebrate brains should likewise pass through a highly conserved stage or period of development. In the first half of the twentieth century, this fundamental notion motivated Harry Bergquist, Bengt Källén, and several other Swedish neurobiologists to embark on an ambitious set of comparative neuroembryological investigations (Bergquist and Källén, 1953a,b) (see Chapter 2). Their principal conclusion was that all vertebrate brains at the pharyngula stage (or thereabouts) have a similarly organized elongated brain that is bent at its anterior end and divisible into a series of transverse, ringlike domains (see Figure 2.15). Bergquist and Källén referred to these transverse domains as "neuromeres" to reflect their apparently segmental arrangement and defined them as distinct zones of increased proliferative activity, separated by regions where cells divide more slowly. These neuromeres were easiest to see in the hindbrain, or rhombencephalon, where they formed distinct bulges, or "rhombomeres" (Figure 3.12A), but Bergquist and Källén also identified a number of more ros-

(A) Gene expression boundaries

(B) Lineage restriction boundaries

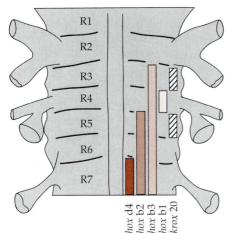
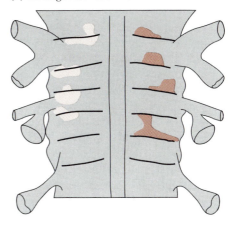

Figure 3.12 The Hindbrain Neuromeres Hindbrain neuromeres, or "rhombomeres" (R1–R7), are evident as ringlike bulges in early embryos. They are shown here from a dorsal perspective and are partially obscured by the cranial nerve trunks. Evidence supporting their existence comes from molecular and cellular studies. (A) Molecular studies have revealed that many gene expression boundaries coincide with the postulated rhombomere boundaries (thick black lines). Illustrated here are the expression domains of several *hox* genes and the transcription factor *krox* 20 (for clarity, the domains are shown as narrow strips, but they actually span the hindbrain's width, at least at early stages of development). (B) Cellular fate mapping studies in chicks revealed that the boundaries between adjacent rhombomeres are lineage restriction boundaries. In these studies, embryonic precursor cells were injected with a fluorescent tracer substance so that the cell's progeny could be identified at later stages of development. Shown here are the results of four injections made before the rhombomeres had formed (four light red patches) and four injections made after rhombomere formation (four dark red patches). In the former cases, labeled cells were spread across multiple rhombomeres; in the latter cases they tended to respect rhombomere boundaries. (A after Wilkinson et al., 1989; B after Fraser et al., 1990.)

tral neuromeres, which they called "mesomeres" and "prosomeres." Bergquist and Källén believed that all neuromeres could be further subdivided into smaller (transverse and/or longitudinal) zones and that this complex patchwork of embryonic divisions and subdivisions was broadly conserved across all vertebrates (Bergquist and Källén, 1954).

Although Bergquist and Källén's observations were never seriously challenged, their significance was called into question. C. J. Herrick, in particular, argued that neuromeres are rather ephemeral structures that have little bearing upon adult structure and function (Herrick, 1933). Consequently, interest in neuromeres waned and only hard-core neuroanatomists continued to be aware of what Bergquist and Källén had described (Braford and Northcutt,

1983; Puelles et al., 1987a). In the early 1990s, however, neuromeres were rediscovered (see Figure 3.12). Experimental studies on cell proliferation and migration showed that hindbrain rhombomeres are, indeed, distinct proliferative zones and give rise to cells that tend not to intermingle with cells from other rhombomeres (Fraser et al., 1990; Lumsden, 1990). Around the same time, numerous transcription factors, including many of the *hox* genes that are also involved in body segmentation, were found to have expression domains that are at least partially coincident with the postulated rhombomere boundaries (Wilkinson and Krumlauf, 1990). Moreover, these expression domains seemed to be highly conserved across species. In both mice and zebra fish, for example, the zinc finger gene *Krox-20* is expressed specifically in rhombomeres 3 and 5 (Nieto et al., 1991; Kimmel, 1993). These data reaffirmed Bergquist and Källén's notion that the vertebrate hindbrain is divided into a series of highly conserved rhombomeres (Gilland and Baker, 1993), but what about the more rostral neuromeres?

Luis Puelles and his collaborators had shown in 1987 that neuronal differentiation in the midbrain and forebrain of early chick embryos proceeds in a segmentlike manner (Puelles et al., 1987a), but the *hox* genes that so nicely revealed rhombomere boundaries did not seem to be expressed in the more anterior brain regions. This changed in the early 1990s, when information became available on the expression patterns of other genes that, like the *hox* genes, bind DNA and control transcription (Rubenstein and Puelles, 1994). Many of these transcription factors were expressed in the developing forebrain, and many of their expression boundaries seemed to correspond to the prosomeric and mesomeric boundaries of Bergquist and Källén (Figure 3.13).

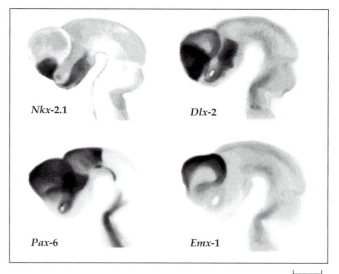

Figure 3.13 Transcription Factors in the Developing Forebrain
Lateral views of four embryonic mouse brains that were stained to reveal the expression domains of *Nkx-2.1, Dlx-2, Pax-6,* or *Emx-1.* Each of these transcription factors is expressed in different regions of the developing telencephalon and diencephalon, but some of their expression domains overlap. All brains are shown in the same orientation, with the telencephalon on the left and the spinal cord in the bottom right corner. (From Puelles et al., 2000.)

500 μm

This correspondence was summarized and codified by Puelles and Rubenstein, who referred to their schema as the "prosomeric model" of vertebrate forebrain development (Puelles and Rubenstein, 1993; Rubenstein et al., 1994). In general, Puelles and Rubenstein argued that not just the hindbrain, but the entire brain of vertebrate embryos, is divisible into a series of neuromeres, and that these neuromeres are highly conserved across vertebrates. In other words, they revived (and revised) the neuromeric model of Bergquist and Källén and argued that it constitutes an archetype for embryonic vertebrate brains. Because this neuromeric model has been influential, I review it here in some detail and then subject it to some critical analysis.

The neuromeric model

At the heart of the neuromeric model lies the idea that the brain's long axis is curved and ends rostrally near the optic chiasm (Figure 3.14A), rather than at the olfactory bulbs, as an inspection of only adult brains might suggest (Puelles et al., 1987b; Puelles and Rubenstein, 1993). An important consequence of this idea is that the terms "dorsal," "ventral," "rostral," and "caudal" can be defined relative to that long axis. These directional definitions can be confusing because they differ from those that are traditionally used to describe some adult forebrain structures (e.g., one now has to say that the dorsal thalamus develops caudal to the ventral thalamus), but they certainly simplify any description of the neuromeric model and will therefore be used in the remainder of this section.

The next most important idea of the neuromeric model is that the entire embryonic brain is divisible into a rostrocaudal series of neuromeres, each of which forms a complete ring around the brain's long axis (see Figure 3.14B). The seven most caudal neuromeres are called rhombomeres. They are bordered rostrally by an isthmic rhombomere, which lies just caudal to a single midbrain mesomere. In the forebrain, Puelles and Rubenstein recognized six prosomeres. The three caudal prosomeres (P1–P3) develop mainly into pretectum, dorsal thalamus, and ventral thalamus, respectively. The three more rostral prosomeres (P4–P6) do not correspond so well to major adult divisions, for each of them gives rise to part of the hypothalamus ventrally and to part of the telencephalon dorsally (Puelles and Rubenstein, 1993).

The model's third major component is that each neuromere comprises two or more longitudinal domains. At the most basic level, each neuromere is divided into a basal region that lies ventral to the brain's long axis and a more dorsal alar region. Such alar and basal regions, or plates, had long been recognized in the hindbrain and spinal cord, but the neuromeric model of Puelles and Rubenstein (1993) extended this concept into the forebrain (without implying that forebrain basal regions are more motor in function than alar regions). These alar and basal regions were subdivided further, particularly in the 3 most rostral prosomeres, where, according to Puelles and Rubenstein, at least 8 distinct longitudinal domains can be recognized. Within the telen-

Figure 3.14 The Neuromeric Model Continues to Evolve
Although most neuroembryologists agree that neuromeres exist, their ideas about them continue to change, particularly when it comes to the forebrain neuromeres, or pro-someres. (A) At the heart of the neuromeric model lies the idea that the brain's longitudinal axis is curved. This idea is not controversial, but there are disagreements about where that axis ends rostrally. (B) Relative to the brain's longitudinal axis, individual neuromeres are arranged as transverse rings. According to the original neuromeric model of Puelles and Rubenstein (1993), the dorsal parts of P4–P6 each develop into specific parts of the telencephalon, and the pallial–subpallial boundary (dashed red line) is orthogonal to the interneuromere boundaries. (C) According to more recent versions of the neuromeric model, only P5 contributes to the evaginated portions of the telencephalon, and the various intratelencephalic zones are neither orthogonal nor parallel to the brain's long axis. Abbreviations: DP = dorsal pallium; Ist. = isthmus; LP = lateral pallium; Mes. = mesomere; MP = medial pallium; OB = olfactory bulb; Pa. = pallidum; Str. = striatum; VP = ventral pallium.

(A) Embryonic brain with curved longitudinal axis

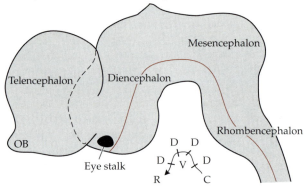

(B) The original neuromeric model

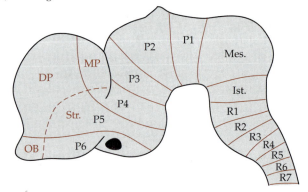

(C) The revised neuromeric model

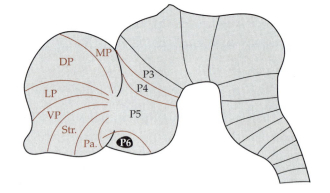

cephalon, for example, the pallium and subpallium were thought to be 2 longitudinal domains that were, in turn, divided by several transverse boundaries (see Figure 3.14B). All in all, Puelles and Rubenstein (1993) proposed that embryonic vertebrate brains are divisible into about 60 embryonic zones that form a topologically deformed, but nonetheless clearly identifiable, checkerboard pattern.

When Puelles and Rubenstein first presented their model, they admitted that "the position and number of subdivisions is tentative, particularly within the secondary prosencephalon [i.e., P1–3]" (Rubenstein et al., 1994, p. 580). Indeed, their model has changed as new data have come in (Puelles and Rubenstein, 2003), and the most significant changes all involve the three most rostral prosomeres (see Figure 3.14C). The mammillary hypothalamic region, for example, was originally thought to develop from the ventral portion of P4 but has now been placed within P5 (Puelles and Medina, 2002). Similarly, the hippocampal formation was originally thought to develop from the dorsal portion of P4 but is now considered part of P5. Thus, P4 was displaced caudally as the neuromeric model was revised (L. Puelles, personal communication). In addition, P6 no longer includes the presumptive olfactory bulb, so that virtually the entire telencephalon (i.e., the entire pallium and the lateral subpallium) is now thought to develop from P5. Another significant modification concerns the boundary between the telencephalic pallium and the subpallium, which runs parallel to the brain's long axis in the original model, but assumes a curved trajectory in the revised version (see Figure 3.14C). Indeed, Puelles and his collaborators now divide the telencephalon into a series of curved zones that are most likely patterned by a set of mechanisms unique to the telencephalon (Cobos et al., 2001). Thus, in the revised prosomeric model, the classic neuromeric checkerboard pattern does not extend into the telencephalon (see Puelles and Rubenstein, 2003).

In sum, the evidence supporting the neuromeric model, in its various manifestations, is extensive and diverse. The original model was based mainly on classic histological observations (by Bergquist, Källén, and others) and the expression patterns of about 45 different genes (Bulfone et al., 1993; Puelles and Rubenstein, 1993). Since then, many more genes have been examined and many of these also were found to have expression boundaries that coincide, at least partially, with some of the proposed inter- and intraneuromeric boundaries. Additional support for the neuromeric model has come from experimental studies on the development and ultimate fate of embryonic brain regions (Marín and Puelles, 1995; Cobos et al., 2001). Most of these data were gathered in chickens and mice, but recent studies indicate that many aspects of the neuromeric model are applicable also to anamniotic vertebrates. Because of all this supporting evidence, the neuromeric model of Puelles and Rubenstein is becoming widely accepted, at least within the community of developmental neurobiologists, as representing the phylotypic stage of brain development for vertebrates. However, as I now review, this interpretation must be scrutinized with diligence.

Criticisms of the neuromeric model

One criticism that has been leveled against the neuromeric model of Puelles and Rubenstein (1993) is that the prosomeres may not be true segments (Larsen et al., 2001). According to traditional ideas of body segmentation, which are derived from studies on invertebrates, true segments should form (or be derived from) a linear series of repeating elements that do not commingle developmentally. Rhombomeres tend to fulfill these criteria because they contain serially repeating cell types in at least some taxonomic groups (Metcalfe et al., 1986) and are separated by fairly tight lineage restriction boundaries (see Figure 3.12B). The more rostral neuromeres, however, contain no serially repeating cell types and are not always separated by lineage restriction boundaries. Cells from the isthmic neuromere, for example, frequently migrate into the rostrally adjacent mesomere (Jungbluth et al., 2001), and some cells from prosomere 4 (and/or prosomere 3) reportedly migrate into the telencephalon (P5)(Larsen et al., 2001). Therefore, if we use lineage restriction as a defining criterion for true segments, then only the rhombomeres qualify.

But Puelles and Rubenstein did not define the zones in their neuromeric model in terms of either repeated elements or lineage restriction. Instead, they defined them as embryonic "fields" that adopt a specific adult phenotype because of the regulatory genes that are expressed within them (Puelles and Medina, 2002). Such fields need not be separated by lineage restriction boundaries as long as any cells that migrate across field boundaries reliably adopt the fate of their new neighbors. The midbrain–isthmus boundary, for example, separates two embryonic fields with very different gene expression patterns, but it is not a lineage restriction boundary; presumably, any cells that migrate across this boundary quickly integrate into their new environment (Jungbluth et al., 2001). Therefore, evidence of tangential migration prior to final regional specification is not evidence against the neuromeric model. Nor is the neuromeric model falsified by tangential migration after regional specification, since the model does not require its embryonic fields to remain coherent throughout development. Indeed, Rubenstein and his collaborators have argued that some of the complexity in adult brains is due to the relatively late commingling of cells that were born in different embryonic fields (Marin and Rubenstein, 2001). Thus, the most crucial element of the neuromeric model is not that its zones are immiscible but that they are molecularly distinct at the time of regional specification. Unfortunately, we still know little about which genes specify regional identity in the brain, or about when regional specification occurs, but further research into these mechanisms could lead to a falsification of the neuromeric model.

Another potential criticism of the neuromeric model is that many developmental gene expression patterns are more dynamic than one might think from the examination of single images. Especially in the forebrain, several genes are first expressed in relatively restricted locales and then expand their expression domains as development proceeds; others restrict their expression domains during development, and some seem to "flicker" on and off in different regions at different times of development (Alvarez-Bolado et al., 1995; Bell et al., 2001). But if gene expression patterns are dynamic, the critics might charge, how can stable neuromeric zones exist? This is indeed an interesting question, but the neuromeric model does not require its embryonic zones to be stable throughout embryogenesis. Indeed, the model clearly implies that these zones emerge from previously undifferentiated tissue at some point during development (around the pharyngula stage of embryogenesis) and, if we review the theoretical and experimental literature on how such zones might be formed, we see that they almost certainly result from spatiotemporally dynamic patterns of gene expression (Meinhardt and Gierer, 2000; Wurst and Bally-Cuif, 2001). Therefore, the discovery of dynamic gene expression patterns is not necessarily in conflict with the existence of discrete embryonic zones. What matters is how the dynamic gene expression patterns are related to the mechanisms that give cells their regional identity and, thus, specify their fate. As I mentioned earlier, we still know little about these relationships and mechanisms, but research in this area is proceeding rapidly and may soon clarify which gene expression patterns, if any, are inconsistent with the neuromeric model.

My own most serious concern with the neuromeric model is the difficulty of obtaining an objective assessment of how well the data match the scheme. By now, hundreds of gene expression patterns have been examined, and many of these exhibit complex boundaries that correspond for only part of their length to boundaries in the neuromeric model. This raises the possibility, at least in theory, that investigators might be focusing their attention selectively on those expression boundaries that are consistent with the model, neglecting those that are not. If that were the case, then the model would become unfalsifiable. To state the criticism more generally, we must remember that even careful observers may glimpse relatively simple patterns where, in reality, the situation is far more complex.

Ideally, one might answer this criticism by performing a statistical analysis of all available data and showing that a majority of the gene expression boundaries indeed lines up with the model's proposed boundaries. Such an analysis would be tedious and expensive, because data from many different embryos would have to be mapped onto a "standard embryo" before they could be compared quantitatively. Moreover, one might argue that the proponents of the neuromeric model have already performed this kind of comprehensive analysis "in their head" and have adjusted the model's boundaries to fit the data. Indeed, I believe that Puelles and Rubenstein have done this, and that most aspects of their revised neuromeric model (Puelles and Rubenstein, 2003) will stand the test of time. At this point, however, it is difficult for other observers to ascertain which of the model's boundaries are strongly supported by the data and which are tenuous. One way to tackle this problem is for independent investigators to focus on specific boundaries, study the expression of several relevant genes simultaneously, and then to ask how well the various markers line up with one another and with the putative boundaries. In one nice example of this approach, Fernandez et al. (1998) discovered that two gene expression boundaries were, contrary to expectations, separated by a band of tissue that expresses neither gene. This finding ultimately led to a significant change in the neuromeric model (i.e., the recognition of a ventral pallial zone; Puelles et al., 2000), and further studies of this type might well force similar modifications.

Finally, any critical analysis of the neuromeric model must ask whether the model really fits all vertebrates. As mentioned earlier, the model is based mainly on data from chickens and mice, but recent studies of frog embryos have revealed generally similar gene expression patterns (Bachy et al., 2002; Brox et al., 2003). Outside of tetrapods, more serious species differences may exist, especially within the telencephalon (Figure 3.15). In a study of zebra fish embryos, for example, the three rostralmost prosomeres could not be identi-

(A) Zebra fish

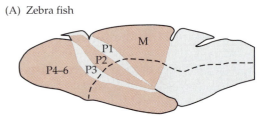

Figure 3.15 Neuromeres in Anamniotes
Attempts to extend the neuromeric model to (A) zebra fish and (B) lampreys have, by and large, been successful. However, one should note that there are major differences in the size and shape of individual neuromeres between species. Moreover, it remains unclear whether or how P4–P6 are subdivided in zebra fish and whether the medial pallium (asterisk) in lampreys really develops from P4 (since it seems to develop from P5 in birds and mammals; see Figure 3.14C). Abbreviation: M = mesomere. (A after Wullimann and Puelles, 1999; B after Pombal and Puelles, 1999.)

(B) Lamprey

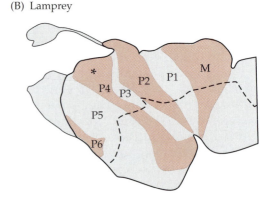

fied with certainty (Wullimann and Puelles, 1999; see also Hauptmann and Gerster, 2000), and the lamprey telencephalon may differ significantly from that of other vertebrates (Pombal and Puelles, 1999; Myojin et al., 2001). Thus, considerably more work will have to be done before we can be certain that the neuromeric model of Puelles and Rubenstein really represents the embryonic brain archetype for all vertebrates. For the moment, we may want simply to assume, as our null hypothesis, that the brains of all vertebrate embryos are indeed quite similar to one another.

This assumption should be tested, however, because recent studies have shown that the embryos of different vertebrate groups may not be as similar as we tend to believe. Indeed, the similarities between different vertebrate embryos have often been exaggerated, and Haeckel's famous drawings of early vertebrate embryos are obviously inaccurate (Richardson et al., 1997). As one comparative embryologist noted long ago when he compared chicks and sharks: "There is no stage of development in which the unaided eye would fail to distinguish between them" (Sedgwick, 1894, p. 36). Most of the variation between different vertebrate embryos is due to differences in the size of individual body parts, their time of appearance, and their rate of development (Richardson et al., 1997). However, it is also possible for genuinely novel structures to appear early in development. For example, early in the development of turtles, we find the carapacial ridge, which is found only in turtles and is required for the development of their carapace (Burke et al., 1989). Given findings such as these, we should cease to assume that evolution tinkers only with relatively late stages of development (Richardson, 1999), and students of brain development and evolution should be on the lookout for species differences in early embryos. Clearly (though this has not been studied in detail) most neuromeric zones differ in shape and size across the vertebrates (see Figure 3.15), but more striking, qualitative differences may yet be in the offing.

Even if embryonic brains turn out to be more variable across vertebrates than the neuromeric model suggests, comparative neuroembryology will remain exciting because it allows us to construct testable hypotheses about how species differences in adult brain structure have come about. For example, if we notice that the embryos of two species differ in the spatial expression of a particular gene, then we might be able to duplicate that embryonic difference experimentally and ask whether the manipulation duplicates important aspects of the adult species differences. Such experiments are difficult and fairly futuristic, but some investigators have already developed fairly specific hypotheses about how evolutionary changes in brain development might explain species differences in adult brain structure (e.g., Letinic and Rakic, 2001). More generally, research along these lines will deepen our understanding of how similarities and differences in embryonic organization map onto similarities and differences in adult structure.

Mapping Embryos onto Adults

Back in the seventeenth century, the Dutch microscopist Jan Swammerdam discovered that the wings, legs, and antennae of an adult moth are contained within the cuticle of its caterpillar. This finding is sometimes cited as the beginning of preformationism, according to which all adult body parts pre-exist in the zygote, but Swammerdam was careful to point out that primordia of a moth's adult structures could be found only in relatively late-stage cater-pillars, and that they were fragile and incomplete (Cobb, 2000). Von Baer (1828) and others later confirmed that in vertebrates as well, most major organs and body parts can be traced back to specific structures or fields in embryos that had reached the phylotypic period of embryogenesis (discussed earlier in this chapter). As development proceeds, these embryonic fields appear to subdivide and gradually give rise to the smaller, more specialized body parts. Hence, we now believe that embryogenesis generally proceeds in a treelike manner, starting with a few major limbs that divide repeatedly to form ever finer branches (see Striedter, 1998b, for a discussion of other metaphors for organismal development, including Waddington's notion of "epigenetic landscapes").

The basic notion of embryogenesis as a branching hierarchy also applies to many aspects of brain development. The neural tube divides into the spinal cord and three cerebral vesicles (forebrain, midbrain, and hindbrain), the vesi-cles divide into the neuromeres, and the neuromeres then divide into smaller embryonic zones, which in turn divide again and again until the adult nuclei and cells have formed (Puelles and Medina, 2002). Direct evidence for this progressive compartmentalization model of brain development is difficult to obtain because individual embryonic zones have to be labeled experimentally so that their adult fate can be mapped. However, some fate-mapping experi-ments have been done, and they generally support the idea that each neu-romere becomes progressively compartmentalized (Inoue et al., 2000; Cobos et al., 2001). These studies have revealed that some cells migrate tangentially and thus "emigrate" to other zones (Marin and Rubenstein, 2001), but most cells seem to migrate radially (following the radial glia; see Figure 3.9A). This means that any embryonic zone, once formed, tends to remain distinct from other zones. It also means that the relative position of individual neuromeric zones should be identical to the relative position, or topology, of their adult derivatives. Further studies will be needed to confirm that tangential migra-tion is indeed relatively rare and that brain development conserves topology (see Burke et al., 1989), but for now the progressive compartmentalization model of brain morphogenesis seems to be well supported by the evidence (see Striedter, 2003).

One appealing aspect of the progressive compartmentalization model of brain development is that it facilitates comparisons across species. As every comparative biologist knows, some adult structures are difficult to homolo-gize across species, but what if we can trace these troublesome adult struc-

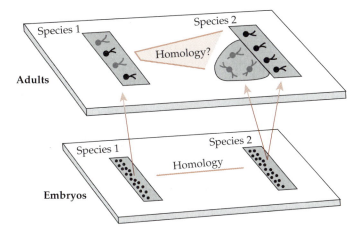

Figure 3.16 Homologies in Embryos and Adults Homologies can be established for both adult and embryonic characters (here depicted on separate planes), but mapping embryonic homologies onto adult ones (red arrows) can be difficult. In particular, when homologous embryonic characters develop into a single adult region in one species (left) and into two or more regions in another (right), then the adult homologies become dubious. If one of the two regions in the second species is very similar to that in the first species, then we can identify those regions as being strictly homologous between the two species; the other region would then be without a homologue. Often, however, the homologies are far from clear (see Chapter 6).

tures back to homologous embryonic zones (Figure 3.16)? According to Smith (1967), adult structures that develop from homologous embryonic structures are field homologues of one another; that is, they are homologous to one another as derivatives of homologous embryonic fields. This notion of field homology remains contentious, mainly because its logical basis and explanatory value are debatable (Northcutt, 1999; Puelles and Medina, 2002). However, few would dispute that the homological relationships of many adult structures have at least been illuminated by comparative analyses of their ontogenies. For example, comparative embryological data helped to clarify that a large ridgelike structure in the avian telencephalon is not homologous to the mammalian striatum (as comparisons of the adult structures had suggested), but to part of the pallium (Källén, 1953) (see Chapter 8). Similarly, comparative developmental data helped to demonstrate that birds have a homologue of the mammalian subthalamic nucleus (Jiao et al., 2000). Successes such as these suggest that homologies that are based on where an adult structure develops from are generally superior to those that are based exclusively on adult features. As E. S. Russel (1916) put it, "homologies observed in embryonic life are to be upheld even if the relations in adult life seem to indicate different interpretations" (p. 168).

But let us be careful here. If we determine adult homologies *solely* on the basis of developmental origin, then we are saying, in essence, that homologous structures cannot develop from nonhomologous precursors. Yet, we already have evidence that homologous structures sometimes do develop from nonhomologous precursors. The vertebrate gut, for example, is clearly homologous across vertebrates, yet it develops from very different, nonhomologous, precursor tissues in different vertebrates (see Striedter and Northcutt, 1991). Apparently, a gut can form from several different kinds of tissues, and evolution has taken advantage of that flexibility. Whether brain development is similarly flexible remains to be seen, but we should be open to that possibility. For example, dopaminergic neurons with ascending projections to the subpallium develop in the caudoventral forebrain of all vertebrates, but in some mammals, very similar cells develop also in the midbrain tegmentum (Puelles and Medina, 1994; Verney et al., 2001; Rink and Wullimann, 2002). If we argue that homologous adult structures must originate from homologous embryonic zones, then we must argue that the midbrain dopaminergic cells in adult mammals (generally referred to as the substantia nigra) cannot possibly be homologous to the more rostral cells in nonmammals. But isn't it possible that the same (homologous) constellation of genetic mechanisms that induce substantia-nigra-like cells in the caudoventral forebrain of one species (see Wurst and Bally-Cuif, 2001) also induces homologous cells in the ventral midbrain of another? Currently we have little evidence for or against this hypothesis. However, there is no a priori reason to think that evolution cannot tinker with a neural structure's embryonic origin, particularly since we already know that many embryonic cells are pluripotent (i.e., capable of adopting several alternative fates).

Given these considerations, I believe it would be rash to assume that homologous brain regions must *always* develop from homologous precursors. Embryonic origins may be conserved as a rule but, like most rules in biology (see Chapter 1), the rule of conserved embryonic origins is likely to have exceptions. Conversely, we should probably not define the adult derivatives of homologous brain regions as being *necessarily* homologous to one another. As I discuss in Chapter 8, developmental similarities can be enormously helpful when it comes to testing homology hypotheses. However, they are neither necessary nor sufficient as criteria of homology.

Conclusions

The question of what constitutes the vertebrate brain archetype is more difficult to answer than one might have thought. First, we must clarify our level of analysis. Brain regions are obviously composed of cells, and these are obviously composed of molecules, but brain regions, cells, and molecules may all evolve independently of one another. For example, evolution may change where a particular transmitter is expressed, which means that homologous transmitters may come to be used by nonhomologous cell types or, conversely, that homologous cell types may use nonhomologous transmitters (Harris-Warrick, 2000). Similarly, homologous cell types may (at least theoretically) be located in nonhomologous brain regions, and homologous brain regions may contain nonhomologous cell types (Reiner, 1993). Therefore, we must recognize at least three separate adult archetypes: one for brain regions, one for cell types, and one for molecules. Second, we must realize that an archetype for embryonic brains may differ from that for adult brains, since embryonic homologies need not always correspond neatly to adult homologies. All of these difficulties, if one chooses to view them as such, arise because evolution has repeatedly found different ways to build the same (homologous) structures and regularly invents new ways to combine old materials. It is a prolific tinkerer.

Because of evolution's propensity for change, all vertebrate brain archetypes are vague. They are but shadows on the wall of Plato's cave. In order to obtain more detailed archetypes, we must limit our analysis to smaller taxonomic groups, but even those more limited archetypes remain abstractions, not real brains. This is well known, but neurobiologists routinely write about "the rodent brain," "the mammalian brain," or even "the vertebrate brain" as if they were talking about a specific brain rather than a highly generalized abstraction. Such linguistic sleight of hand cannot hide species differences for long. Provided that neuroscientists continue to study more than a single "standard" animal (Bolker and Raff, 1997), they will encounter differences between species. Evolutionary theory helps to explain those species differences, but even evolutionary neuroscientists tend to emphasize species similarities rather than differences. This preferential emphasis on similarities is understandable because similarities can always be interpreted as "conserved elements" that bolster Darwin's theory of common descent. It is also true, however, that species differences tend to receive less attention because they are more difficult to explain. As Pere Alberch (1984) put it, "evolutionary morphology has been unable to deal with problems concerning the transformations of complex systems" (see Chapter 7). To overcome that limitation, we must recognize species differences as well as similarities. Therefore, I devote the following 4 chapters to the search for rules and principles that may help us to explain how brains became diverse.

4 Evolutionary Changes in Overall Brain Size

All things being equal, the small animals have proportionately larger brains.
—Georges Cuvier, 1805–1845

Overall brain size is relatively easy to measure and has, consequently, been measured many times, in many species. Since humans are widely regarded as "the paragon of animals" (see Chapter 9), Aristotle and his followers expected humans to exceed all other species in terms of absolute brain size. However, we now know that the brain of an adult blue whale, for example, is roughly five times the size of an adult human brain (7 kg versus 1.3 kg). What does that say about our "place in nature"? Are blue whales five times smarter than we are? Probably not, for whales generally have simpler neocortices than primates (Glezer et al., 1988) and larger bodies to control with those brains (see Chapter 10). In fact, in blue whales the brain comprises only 0.01% of the body's weight, whereas in humans the brain:body ratio approximates 2%. So, we might ask, do humans have a higher brain:body ratio than other animals? Again we are disappointed, for many small animals have brain:body ratios that exceed 2%. In pocket mice and harvest mice, for example, the brain comprises roughly 10% of the body's mass (Mace et al., 1981). These observations present a conundrum. Most of us feel that humans are the most intelligent animals, and we strongly suspect that brain size has something to do with intelligence. But if humans have neither the largest brains nor the highest brain:body ratio, how can the human brain be any kind of paragon?

In order to answer this question, comparative neurobiologists have analyzed brain and body size data from many species and have discovered, first of all, that "small animals have proportionately larger brains" (Cuvier, 1805–1845; see also von Haller, 1762). Moreover, Cuvier and others realized that proportional brain size increases quite predictably as body size decreases. This can be seen most readily when brain size is plotted against body size in logarithmic coordinates, for in such a log–log plot, the data points all tend to

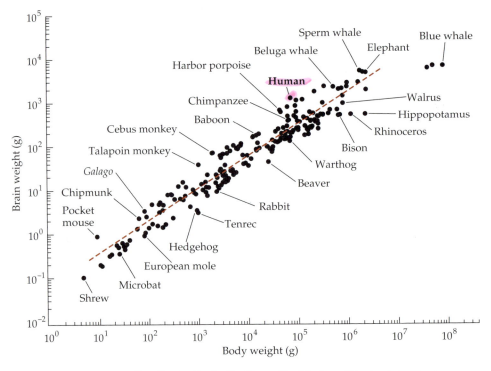

Figure 4.1 Mammalian Brain–Body Scaling Double-logarithmic plot of brain weight versus body weight for 213 species of placental mammals. The dashed line indicates the best-fit allometric line for mammals, as calculated by van Dongen (1998). (Data from Mangold-Wirz,1966; pocket mouse data from Mace et al., 1981.)

fall along a straight line with a slope of less than one (Figure 4.1). In technical terms, this means that brains scale allometrically, rather than isometrically, with body size. This is not particularly surprising, as most body parts and physiological processes scale allometrically with body size (Schmidt-Nielsen, 1984). However, comparative neurobiologists were struck by the fact that in a log–log plot of mammalian brain and body sizes, the data point for humans lies farther above the best-fit line than any other data point. If we take the best-fit line to indicate the brain size of "average" mammals at various body sizes, then we can say that human brains are four to five times larger than one would expect for an average mammal of the same body size. No other mammals exceed their expected brain size by as much. In fact, humans exceed all other mammals—and indeed all animals—in terms of what is generally called relative brain size.

This fundamental insight prompted comparative neurobiologists to collect and analyze a vast array of brain and body sizes (e.g., Crile and Quiring, 1940). The most important finding to emerge from this work is that the log-transformed brain–body data tend to form straight lines in all major taxonomic groups. Since straight lines in log–log plots correspond to power functions in linear coordinate systems, this statement is equivalent to saying that in all major taxonomic groups, brain–body scaling tends to obey a simple power law of the form

$$M_{brain} = a * M_{body}{}^{b}$$

(Note that the power function $y = ax^b$ is mathematically equivalent to $\log(y) = \log(a) + b \log(x)$, which describes a straight line with slope b.)

Many authors have tried to discern the biological significance of the scaling exponents in these power functions (i.e., the slope of the best-fit lines in log–log plots). Harry Jerison (1973), for example, noted that for many data sets the scaling exponent (b) is remarkably close to 2/3. Since the surface area of any body also tends to scale with the two-thirds power of body weight (at least if we consider only the smooth objects of Euclidean geometry), Jerison (1973) and others suggested that brain size might be causally related to body surface area (Snell, 1891). This idea has captured many imaginations but does not explain, for instance, why brain regions that are not related to somatic information processing also scale with body size (Fox and Wilczynski, 1986). In a related effort, Jerison and others calculated "encephalization quotients" (defined as observed brain weight divided by expected brain weight) for many different species and then sought correlations between these encephalization quotients and various behavioral variables such as "intelligence" (which is discussed in more detail later in this chapter). These calculations provided an important mathematical framework within which one could think about how brain size has changed during the course of evolution and how it might be linked to behavior.

Despite these contributions, brain allometry has long suffered from rather intractable difficulties (Gould, 1975; Deacon, 1990a). First among these is that best-fit lines often vary greatly depending on which data are included in the analysis. The allometric line for primates, for example, has a much higher slope when humans are included in the analysis than when they are not. Similarly, the slope of the mammalian line decreases steadily as the number of data points for baleen whales and small rodents is increased. Although these sampling problems can be minimized (Harvey and Pagel, 1991; Smith, 1994), they plague essentially all allometric studies. Another serious problem is that different taxonomic groups often exhibit different scaling exponents (Pagel and Harvey, 1989). This diversity of exponents calls into question any attempts to find truly general explanations for the exponent values. For example, how can brain weight be linked to body surface area if many animal groups exhibit scaling exponents other than 2/3? A related difficulty is that, because different taxonomic groups have different scaling exponents, an animal's encephalization quotient may vary depending on which taxonomic group is used to calculate its expected brain weight. For example, because the allometric line for rodents has a lower slope than the allometric line for all mammals, small rodents have higher encephalization quotients than large rodents when the allometric line for all mammals is used to calculate encephalization quotients, but small rodents appear just as encephalized as large rodents when the rodent allometric line is used as a reference. Whenever methodological choices can sway results so much, most scientists become concerned. Finally, some authors have pointed out that most log-transformed brain–body data are fitted better by downwardly concave curves (i.e., quadratic functions) than by straight lines (Count, 1947). This problem is generally ignored because data can always be fitted better with more complex functions, but the concern remains.

Because of these difficulties, it is tempting to dismiss overall brain size as an old-fashioned and relatively uninformative variable. Instead, one might prefer to focus on the size of individual brain areas and neuronal circuits, particularly since these are more readily linked to specific behaviors and physiological functions (Harvey and Krebs, 1990). This argument has considerable appeal, but comparative analyses of individual brain areas and circuits must also take into consideration overall brain size. For example, if you want to determine whether humans or echolocating bats have more highly developed auditory systems, it is not particularly useful to know that the human subcortical auditory system is 63 times larger than that of a little brown bat (187 mm^3 versus 3 mm^3)(Glendenning and Masterton, 1998). Given that echolocating bats depend on their sense of hearing more than humans do, it makes more sense to consider that in bats the subcortical auditory system occupies about 1.6% of total brain volume, while in humans this percentage is only 0.015. Even this figure is misleading, however, because small mammals generally have small cerebral cortices (see Chapter 5) and, hence, large subcortical fractions. Therefore, it is most informative to plot auditory system size versus overall brain size for many different mammals, fit an allometric line to these data, and then ask how much the various data points deviate from that line (Figure 4.2). Such an analysis reveals that the subcortical auditory system is larger than expected in bats (and cats, for example) but smaller than expected in humans and other primates. Therefore, measuring overall brain size and then analyzing the size of individual brain regions or systems allometrically can be quite informative. Several such analyses are detailed in Chapters 5, 6, and 9.

In the present chapter, I review how overall brain size varies in and among the major vertebrate groups. The chapter begins with a broad overview of how brain size has changed as vertebrates diversified. My focus in this first section is on relative brain size (rather than proportional or absolute brain

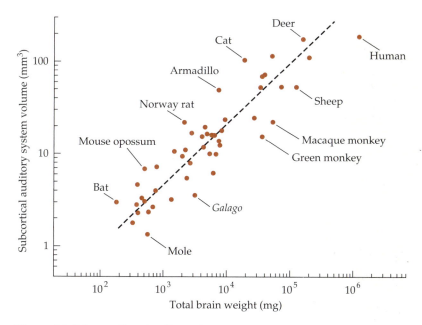

Figure 4.2 Mammalian Auditory System Scaling   The volume of the subcortical auditory system in a variety of mammals, scaled against total brain weight. Based on an allometric analysis, the subcortical auditory system in humans and in other primates (e.g., the bush baby, or *Galago*) is smaller than one would expect, whereas in cats and bats, it is larger than expected. (After Glendenning and Masterton, 1998.)

size), because each major taxonomic group contains species of various body sizes and because we often do not know which body size is primitive (ancestral) for the group. Given that focus, the data clearly show that relative brain size has increased and decreased repeatedly during the course of vertebrate phylogeny. Next, I discuss some of the mechanisms that control overall brain size and its relationship to body weight. Since we know little about these mechanisms, this section remains rather sketchy, but the next few years are likely to see some progress on this front. After that, I review some of the correlations that exist between relative brain size and behavioral variables such as "foraging complexity" and "intelligence." These correlations are controversial and complex, but they are interesting and can enhance our understanding of why vertebrate brains vary in relative size. With all this emphasis on relative size, however, it is easy to overlook that brains vary enormously in absolute size and that this may be of consequence. To explore that possibility, I devote this chapter's final section to the correlates of absolute brain size, a topic that reverberates throughout this book.

encephalized = $\frac{brain\ mass}{body\ mass}$

Changes in Relative Brain Size

The major groups of vertebrates differ significantly from one another in relative brain size. One convenient way to appreciate this variation is to plot all available brain and body size data in a log–log plot and then draw minimum convex polygons around the data from each major vertebrate group (Figure 4.3A; see also Figure 4.1). This kind of polygon plot reveals, for example, that in terms of relative brain size, birds exceed reptiles but are similar to small- and medium-sized mammals. Another way of examining the variation in brain–body scaling among vertebrates is to plot the best-fit lines for all major taxonomic groups on a single graph (Figure 4.3B). In such a plot it is relatively easy to see, for example, that avian brains are, on average, 4–6 times larger than reptile brains at the same body size. It is important to note, however, that such quantitative comparisons become more complex when the taxonomic groups that are being compared have best-fit lines with different slopes. This is particularly clear when one compares birds and mammals. Because the best-fit lines for these two groups intersect, small birds end up being more encephalized (having higher encephalization quotients) than small mammals, while large birds are less encephalized than large mammals. Whether such comparisons are biologically meaningful is debatable. A more interesting lesson to be learned from this example is that species may differ not only in their degree of encephalization (i.e., the elevation of their best-fit lines), but also in the very nature of their brain–body scaling (i.e., their allometric slopes).

Next, we might want to know at what points in evolution, or in which lineages, relative brain size increased or decreased. This sounds like a simple question, but it is full of complexities. First, we must realize that, because of the above-mentioned problem with best-fit lines of different slopes, not all changes will be simple factorial increases or decreases in relative brain size (i.e., changes in y-intercepts). Some authors (e.g., Jerison, 1989) have tried to avoid this problem by forcing all best-fit lines to identical slopes, but this is bound to generate at least some artificial results. A second difficulty is that

Figure 4.3 Vertebrate Brain–Body Scaling   Variation in relative brain size can be illustrated by two different kinds of graphs. (A) Drawing the smallest (minimum) convex polygon that encompasses all of the brain–body data points for a taxonomic group is one convenient way to compare brain–body scaling between the major vertebrate groups. The polygon for amphibians is not shown here, but it would be largely coincident with the left portion of the polygon for ray-finned fishes. (B) Alternatively, one may fit allometric lines to the data points. This method hides much of the variation within groups, but it is useful for comparing average scaling relationships between groups. In this diagram, allometric lines with slopes of 0.5–0.6 are shown in red; lines with other slopes are shown in black. The dashed line indicates an isometric line—that is, a line with a slope of 1. Data points above this line are "inadmissible" since brain weight cannot exceed body weight. (A after Jerison, 1973; allometric functions in B calculated by van Dongen, 1998.)

(A) Minimum convex polygons

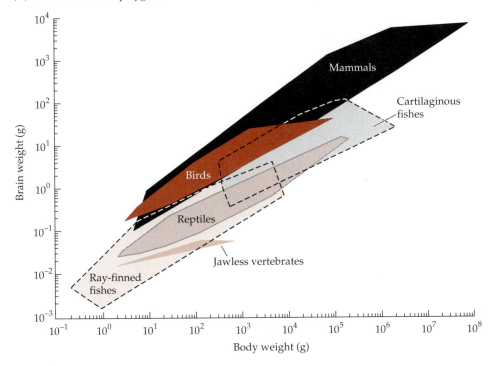

(B) Allometric lines

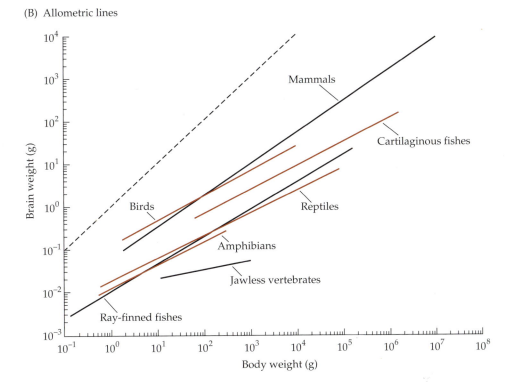

relative brain size varies significantly within all of the major vertebrate groups (e.g., among mammals). This means that, in order to identify phylogenetic changes between major groups (e.g., birds and mammals), one must first determine which patterns of brain–body scaling are primitive within each group. This may sound easy, but many of the smaller taxonomic groups (e.g., marsupials) also exhibit significant variation in relative brain size and, thus, present similar difficulties. Moreover, as one considers smaller and smaller taxonomic groups, sample sizes shrink and it becomes more difficult to establish best-fit lines with confidence. One can try to avoid the latter problem by comparing only animals of similar body size (Smith, 1980), but such "narrow" allometric comparisons do not allow us to identify phylogenetic changes in allometric slope. Because of all these difficulties, I here discuss only some particularly obvious changes in relative brain size. However, even this cursory analysis leads to some general insights about how brains have changed in size.

Let us begin with jawless vertebrates (Figure 4.4). Contemporary lampreys have brains that are smaller, relative to body size, than those of any other vertebrates (Ebinger et al., 1983; Platel and Vesselkin, 1989). The brain of a 55-g lamprey, for instance, weighs only about 30 mg. Hagfish brains are larger, but not by much. Because they have unusually well-developed telencephalons and trigeminal systems, it is likely that hagfishes underwent an evolutionary increase in relative brain size. However, it is also possible that lampreys experienced a decrease in relative brain size, for lampreys are ectoparasites, and parasitic species are likely to exhibit reduced brain sizes (since most parasites presumably need little "intelligence" to obtain their food; see the discussion of intelligence later in this chapter). Neither hypothesis can be considered well established and both may contain elements of truth. In any case, we can safely say that the brains of both hagfishes and lampreys are larger, in terms of relative size, than those of their nearest invertebrate relatives (e.g., amphioxus; see Figure 3.2A). Therefore, we can conclude that ancestral vertebrates probably underwent a significant increase in relative brain size.

In order to determine more precisely how large the brains of the earliest jawed vertebrates might have been, we need to know what relative brain sizes prevailed in early bony and cartilaginous fishes. The most primitive bony fishes—those lineages that branched off early during the evolution of bony fishes—are coelacanths, lungfishes, sturgeons, gars, the bowfin (*Amia*), and the bichir (*Polypterus*). In a log–log plot of relative brain size, the best-fit line for these primitive bony fishes lies well above that for jawless vertebrates (see Figure 4.4). The brain of a bichir, for example, is about four times as large as the brain of an equally heavy hagfish (Platel et al., 1977). Among cartilaginous fishes, the ratfishes (also known as chimeras) branched off first, and their brains are roughly three times larger than the brains of primitive bony fishes at the same body size. Since most other cartilaginous fishes have even larger relative brain sizes, we can conclude that the earliest cartilaginous fishes probably had brains that were larger than those of primitive bony

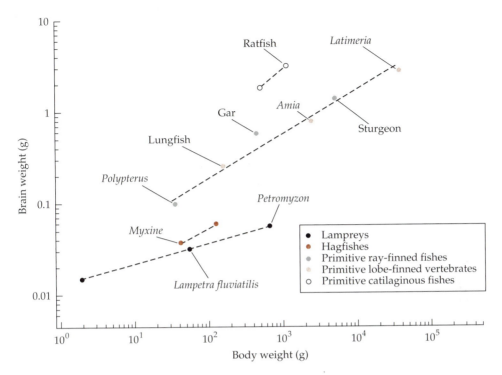

Figure 4.4 Brain–Body Scaling in Various Primitive Vertebrates   Plot of brain weight versus body weight for selected primitive vertebrates, defined here as taxa that evolved early during vertebrate phylogeny (see Figure 3.1). Note that jawless vertebrates (hagfishes and lampreys) have relatively small brains compared to other vertebrates; primitive cartilaginous fishes have brains that are 3–4 times larger than the brains of primitive ray-finned fishes and primitive lobe-finned vertebrates of similar body size. (Data from Platel et al., 1977; Ebinger et al., 1983; Platel and Vesselkin, 1989; Bauchot et al., 1995.)

fishes at the same body size. Collectively, these data suggest either that relative brain size increased once with the origin of jawed vertebrates and then again in cartilaginous fishes, or that it increased (to levels seen in primitive cartilaginous fishes) with the origin of jawed vertebrates and then decreased in primitive bony fishes. I favor the former hypothesis because it involves smaller changes in relative brain size, but it is hard to be sure. In any case, there must have been at least one significant increase in relative brain size near the origin of jawed vertebrates.

Within cartilaginous fishes, relative brain size varies enormously (Figure 4.5). Many sharks, skates, and rays have brains that are similar to those of ratfishes in terms of relative size, but some sharks have much larger brains (Bauchot et al., 1995; Demski and Northcutt, 1996). Some carcharhiniform sharks

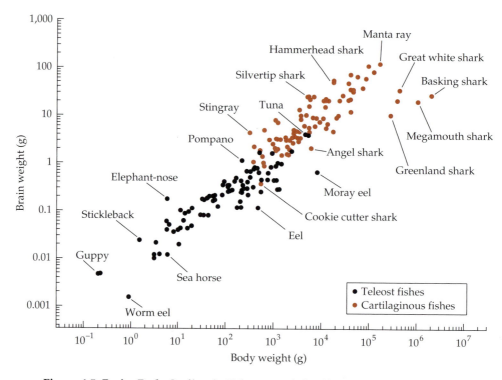

Figure 4.5 Brain–Body Scaling in Teleosts and Cartilaginous Fishes Plot of brain weight versus body weight for selected teleosts (black data points) and cartilaginous fishes (red data points). Note that cartilaginous fishes generally exceed teleosts in terms of relative brain size. Also interesting is that the manta ray (or devil ray) has the largest brain of any cartilaginous fish, even though its body is considerably smaller than that of the largest sharks (e.g., the megamouth; see Figure 4.6). (Data from Bauchot et al., 1976; Bauchot et al., 1989; Demski and Northcutt, 1996; Nilsson, 1996; Ito et al., 1999; supplemented with data by Roland Bauchot published at http://www.fishbase.org.)

(e.g., hammerhead and requiem sharks) have brains that are about 5–10 times larger than the brains of other sharks at the same body size. Similarly, large brains are found in mylobatiform rays (e.g., stingrays and devil, or manta, rays; Figure 4.6A). Since mylobatiform rays and carcharhiniform sharks are only distantly related to one another (Compagno, 1999), relative brain size probably increased independently in these two groups. In contrast, basking sharks and other large filter-feeding sharks have surprisingly small brains. The brain of a 1,000-kg megamouth shark, for example, weighs only 20 g (Figure 4.6B) (Ito et al., 1999). These small relative brain sizes are most likely due to several independent phylogenetic decreases in relative brain size, but,

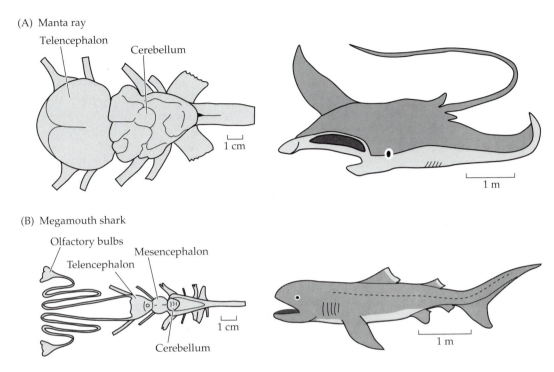

Figure 4.6 Variation in Brain Size within the Cartilaginous Fishes   The manta ray (A) is smaller than the megamouth shark (B) in terms of body size, but its brain is much larger (see also Figure 4.5). In manta rays, the telencephalon and cerebellum are particularly large and cover the other brain regions (the olfactory bulbs are not shown for the manta ray). Remarkably, both species are planktivorous. The manta ray does, however, appear to have more complex social behaviors. Scale bars for the body drawings are approximate. (A after Ito et al., 1999; B after Ebbesson, 1980.)

given the present uncertain state of cartilaginous fish taxonomy, this remains unclear. Additional, unrelated decreases in relative brain size probably occurred in cookie cutter sharks, named for their habit of gouging flesh out of larger animals, and in angel sharks, which ambush unsuspecting prey; their brains weigh less than half of what one would expect in sharks that size. Overall, we can conclude that relative brain size increased and decreased several times independently within cartilaginous fishes.

Among the ray-finned fishes, we can distinguish between primitive ray-finned fishes (see Figure 4.4) and the enormously diverse teleosts (see Figure 4.5). Teleosts are similar to primitive ray-finned fishes in terms of relative brain size (Bauchot et al., 1989), but they tend to be smaller in absolute body size (the world's smallest vertebrates, the 10–15-mm-long dwarf pygmy gobies are teleosts). Since small animals tend to have proportionately large brains (recall Cuvier's observation), it is not surprising that some of the highest brain:body ratios in nature are found in teleosts. Guppies, for example, have brain:body ratios of 2.5%, considerably greater than the typical human ratio of 2%. Yet guppy brains are roughly the size one would expect in a teleost that small. Of course, there are some teleosts that deviate markedly from allometric expectations. Some of the lowest relative brain sizes among teleosts (indeed, among all vertebrates) are found in eels; the brain of an 8-kg giant moray eel, for example, weighs only 650 mg. Relatively small brains are also seen in some benthic teleosts that ambush unsuspecting prey (e.g., stargazers and frogfishes). Pompanos and other fast-moving predators, on the other hand, have brains that are approximately twice the size one would expect in teleosts of their body size. Relatively large brains are also seen in triggerfishes and other active inhabitants of coral reefs. The most highly encephalized teleosts are the weakly electric mormyrid teleosts (see Figures 3.4 and 4.5). The brain of a 6-g elephant-nose, for example, weighs 180 mg and, thus, occupies 3% of the body's mass (Nilsson, 1996). Clearly, relative brain size both increased and decreased several times within the teleosts.

The most primitive lobe-finned vertebrates, the lungfishes and coelacanths, do not differ significantly from primitive ray-finned fishes in terms of relative brain size (see Figure 4.4). Therefore, brain size probably did not change significantly when lobe-finned vertebrates first evolved. It did change, however, when tetrapods came on the scene. Among tetrapods, let us first consider amphibians.

Most frogs and toads resemble ray-finned fishes and primitive lobe-finned vertebrates in terms of relative brain size (Bauchot et al., 1983), but arboreal frogs tend to have larger brains than terrestrial species of the same body weight (Figure 4.7) (Taylor et al., 1995). Since ancestral amphibians did not live in trees, relative brain size probably increased in those lineages that became arboreal. Salamanders and newts, in contrast, underwent at least one phylogenetic decrease in relative brain size, for their brains are significantly smaller than those of frogs and primitive lobe-finned fishes at the same body size (see Figure 4.7) (Thireau, 1975). The brain of a 100-g salamander, for example, weighs only about 100 mg, less than half the weight of a similarly sized lungfish's brain. Some very small salamanders have even smaller brains (ranging down to 0.5 mg), but whether these miniaturized brains are smaller than one would expect in salamanders of that body size is difficult to determine from the published literature (Roth et al., 1995). Still, it is clear that during the evolution of salamanders relative brain size decreased at least once.

The brains of reptiles are generally similar to those of terrestrial frogs, toads, and primitive lobe-finned vertebrates in terms of relative size. This suggests that no major change in relative brain size occurred with the origin of reptiles. Within reptiles, however, snakes and slow worms (legless lizards) have unexpectedly small brains for their body size (Figure 4.8A) (Platel, 1979).

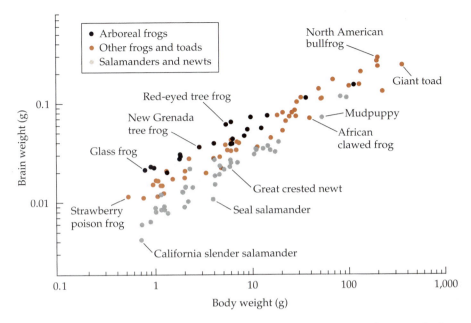

Figure 4.7 Brain–Body Scaling in Amphibians   Plot of brain weight versus body weight for selected amphibians. (Data from Thireau, 1975; Bauchot et al., 1983; Taylor et al., 1995.)

The brain of a 6 kg python, for example, weighs only about 1 g—less than half the weight of a similarly sized turtle's brain. Because snakes and slow worms are only distantly related to one another, their small relative brain sizes are likely to be the result of independent evolution. Indeed, it seems that elongate, wormlike vertebrates generally have relatively small brains (recall the moray eel's tiny brain). On the other end of the spectrum, iguanas, varanids (e.g., grey monitors), and dwarf tegu lizards all have relatively large brains. Although the phylogenetic relationships of lizards remain controversial, large-brained lizards are found in several seemingly unrelated lineages (Northcutt, 1978). Therefore, relative brain size probably increased several times independently among reptiles. Since crocodilians are closely related to birds, which have relatively large brains (see the discussion later in this chapter), it is interesting to note that the brains of alligators and crocodiles are not larger than one would expect in reptiles of their body size. Similarly, the brains of most extinct dinosaurs were, as far as we can tell from fossil finds, about as large as one would expect them to be (Jerison, 1973; Hopson, 1979). Nonetheless, reptilian phylogeny was clearly characterized by at least some significant increases and decreases in relative brain size.

(A) Reptiles

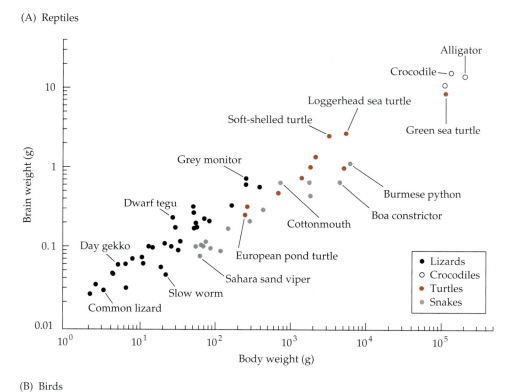

(B) Birds

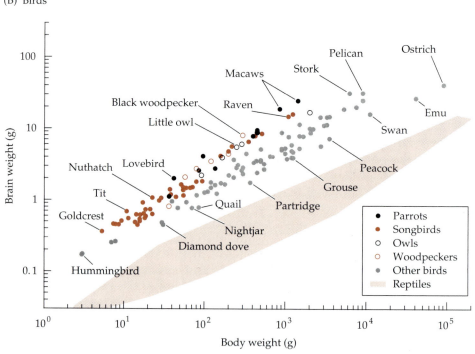

◀ **Figure 4.8 Brain–Body Scaling in Reptiles and Birds** Plot of brain weight versus body weight in selected (A) reptiles and (B) birds. Note that snakes and legless lizards (i.e., slow worms) tend to have relatively small brains and that the brains of all birds are significantly larger than the brains of similarly sized reptiles. Among birds, the parrots, songbirds, owls, and woodpeckers have unusually large brains for their body size. (A from Platel, 1979; B from Portmann, 1947.)

Birds, the "flying dinosaurs" (see Chapter 3), have brains that are significantly larger than the brains of terrestrial reptiles at the same body size (see Figure 4.8B) (Portmann, 1947). The brain of an ostrich, for example, is roughly four times as large as the brain of an equally heavy crocodile (41 g versus 11 g). Because ostriches belong to the oldest group of birds, the ratites, it is reasonable to assume that the earliest birds already had relative brain sizes considerably above those of extant reptiles. This hypothesis is supported by the available fossil record (Jerison, 1973) and by the fact that another primitive group of birds, the Galliformes (chickenlike birds), is similar to ratites in terms of relative brain size. Within the remaining orders of birds, relative brain size increased again several times. The largest avian brains are found in parrots. The brain of a 1-kg macaw, for example, weighs about 20 g—more than the brain of a 200-kg alligator and about as much as the brain of a 1,000-kg megamouth shark. Owls, woodpeckers, and songbirds (passerines) also have relatively large brains. Among songbirds, the corvids (crows, jays, and their relatives) have the largest brains. The brain of a 1-kg raven, for example, weighs approximately 14 g. Because owls, woodpeckers, songbirds, and parrots are only distantly related to one another (Sibley et al., 1988), relative brain size probably increased independently in each of these four lineages. With the exception of some domesticated birds (e.g., turkeys; see Ebinger and Röhrs, 1995) there are no clear examples of birds with phylogenetically reduced relative brain sizes.

Mammal brains are also larger than reptilian brains at the same body size. Within mammals, marsupials tend to have smaller relative brain sizes than placental mammals (Figure 4.9). The brain of a 30-kg kangaroo, for example, weighs about 60 g (Möller, 1973), roughly two-thirds the weight of an equally heavy placental mammal's brain (e.g., a sheep). In contrast, the two surviving groups of monotremes (the platypus and the echidna) both have relative brain sizes that are similar to those of "average" placental mammals (van Dongen, 1998). Since monotremes are the sister group of marsupials and placental mammals, it is not unreasonable to suppose that relative brain size increased with the origin of all mammals and then decreased in marsupials. The fossil record strongly suggests, however, that 50–150 million years ago mammalian brains were all similar in relative size to the brains of today's marsupials (Jerison, 1973). If these fossils represent the full range of early mammals and were interpreted accurately, then today's monotremes are not

(A) Marsupials and monotremes

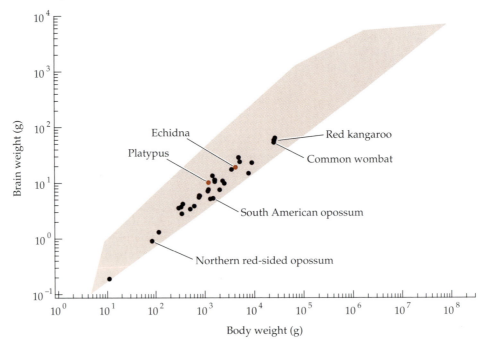

(B) Placental mammals

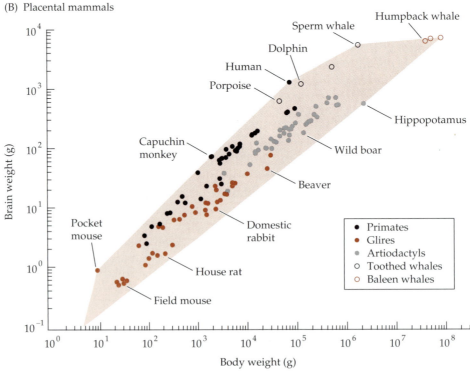

◀ **Figure 4.9 Brain–Body Scaling in Mammals**   Plots of brain weight versus body weight in (A) monotremes and marsupials and (B) placental mammals. In both plots, the data are placed within the minimum convex polygon for all mammals (red area). Part B emphasizes data on primates, whales, and some of their closest relatives (glires and artiodactyls, respectively; see Figure 2.11). Baleen whales feed on plankton or small schooling fish, whereas toothed whales (e.g., dolphins and killer whales) eat mainly fish. (Data from Mangold-Wirz, 1966; Möller, 1973; McNab, 1989.)

representative of early mammals, at least in terms of relative brain size, and monotreme phylogeny must have involved at least one increase in relative brain size. Given that only a few rather odd monotremes survive today, I favor this interpretation and conclude that relative brain size probably increased with the origin of all mammals and then increased again in both placental mammals and some monotremes (see also Chapter 8).

Within placental mammals, the largest relative brain sizes are found in primates, elephants, and toothed whales (e.g., dolphins and sperm whales)(see Figures 4.1 and 4.9). Since these three groups are only distantly related to one another, they most likely experienced independent increases in relative brain size. Additional increases in relative brain size occurred among primates, as we shall see in Chapter 9. More difficult to determine is whether relative brain size ever decreased among placental mammals. Baleen whales (e.g., blue whales and humpback whales) have brains that are much smaller, relative to body size, than the brains of their closest relatives, the toothed whales. This suggests that baleen whales might have undergone a phylogenetic decrease in relative brain size. However, the closest living relatives of all whales are probably the hippopotami (Gatesy and O'Leary, 2001), and hippopotamus brains are no larger, in terms of relative size, than the brains of baleen whales (see Figure 4.1). This suggests that baleen whales did not decrease their relative brain size, but this hypothesis also remains uncertain because cetacean phylogeny remains a subject of considerable debate. The next most likely candidates for mammals that underwent a reduction in relative brain size are the insectivorous bats, shrews, and tenrecs (Eisenberg, 1981; Mace et al., 1981). Since these relatively small-brained mammals are dispersed throughout the mammalian phylogenetic tree, it is unlikely that they *all* retained their small relative brain size as a primitive feature. More likely is that one or two of these lineages underwent a decrease in relative brain size, but more work is needed to flesh out this hypothesis (see Chapter 8).

Considering all these increases and decreases in relative brain size (Figure 4.10), we may ask whether any general pattern of change can be discerned. Clearly, relative brain size did not increase linearly "from fish to man," as Huxley and others had supposed (see Chapter 2). Rather, it increased independently in several different lineages (Northcutt, 1981) and it decreased in others. The pattern is complex. However, if we step back and look for an overall trend, we can see that, *on average*, relative brain size has increased

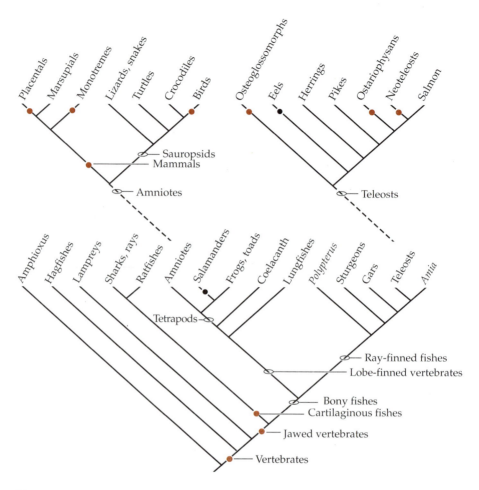

Figure 4.10 Evolutionary Increases and Decreases in Relative Brain Size   Cladogram of the vertebrates, depicting nodal points at which relative brain size probably
increased (red dots) or decreased (black dots). Apparently, increases in relative brain
size occurred independently in several different lineages and outnumbered the decreases in relative brain size. Since the cladogram shows only a subset of all vertebrate
lineages, not all changes in relative brain size are shown.

over the course of vertebrate evolution. A closer look suggests that this trend
exists not because the increases are, on average, larger than the decreases but
because the increases outnumber the decreases. In addition, we may note
that several lineages (e.g., birds and mammals) experienced major radiations
soon after they underwent a major increase in relative brain size, while none
of the taxonomic groups that underwent significant decreases in relative
brain size are very species rich today. Even the fairly successful but small-

brained urodeles (salamanders and newts) comprise only a few hundred species and are much less diverse than their more encephalized anuran relatives. Thus, we might speculate that larger-brained species were better able to invade new niches or survive changing environments than their less encephalized competitors, but this hypothesis is probably too general and vague to be testable (see Sol et al., 2002 for more concrete, related ideas). Instead, we might ask more modestly whether relative brain size correlates with any interesting behaviors. Before we discuss such functional questions, however, let us examine some of the mechanisms that underlie changes in relative brain size.

Mechanisms of brain–body scaling

Evolutionary changes in relative brain size may be due to changes in absolute brain size, but they also may be due to changes in absolute body size. It is important, therefore, to know how many of the inferred increases in relative brain size were caused by dwarfism (phylogenetic decreases in body size that were not accompanied by decreases in absolute brain size) and, conversely, how many decreases in relative brain size were caused by gigantism (increases in body size unaccompanied by changes in brain size). Occasionally, these questions are relatively easy to answer. Dwarf tegu lizards and pocket mice, for instance, are much smaller than their less encephalized relatives, strongly suggesting that they are phylogenetic dwarves. Similarly, the Malayan sun bear is both highly encephalized and extremely small (for a bear; see Kamiya and Pirlot, 1988), indicating that it probably increased its relative brain size mainly by reducing body size. At the other end of the spectrum, eels, snakes, megamouths, and basking sharks probably ended up with relatively small brains because their bodies increased dramatically in size. In all these cases, the changes in body size are obvious because the species in question are much larger or smaller than their closest relatives. In many other cases, however, it is difficult to prove either dwarfism or gigantism because the most closely related species exhibit so much variation in body size that it becomes difficult to establish an ancestral body size. Even fossils don't help much because we never know which fossils are bona fide ancestors and which are offshoots or dead-ends. For instance, the fossil record strongly suggests that birds evolved from a group of dinosaurs (theropods) that includes *Tyrannosaurus rex*, but no one would suggest that *T. rex* is ancestral to birds (Dingus and Rowe, 1997). Most likely, birds evolved from some cat-sized theropod, but which fossil (if any) represents "the" ancestor of birds remains unknown. Therefore, we often cannot tell how closely phylogenetic changes in relative brain size are linked to changes in body size.

Nevertheless, it seems to me that most major increases in relative brain size are due to factors other than dwarfism. More generally, we can safely say that the overall trend of increasing relative brain size among vertebrates is not due to a parallel trend of decreasing body size. In fact, the opposite seems to be true, for absolute body size has, on average, increased in all of the lineages that underwent significant increases in relative brain size (especially birds and mammals, Stanley, 1973; Alroy, 1998). The one major vertebrate lineage that reduced its adult body size when it first evolved, the ray-finned fishes, did not simultaneously increase in relative brain size. Therefore, we can say with some assurance that most increases in relative brain size were associated with either increases or stasis (no change) in absolute body size. On the other hand, most decreases in relative brain size (e.g., those seen in eels, snakes, and planktivorous sharks) were probably linked to significant increases in body size. Although this rule is fairly general, it has exceptions. Frogfishes and cookie cutter sharks, for example, are poorly encephalized but not particularly large, implying that they probably decreased their relative brain size by some mechanism other than gigantism. In conclusion, we can say that, despite some exceptions, most increases and at least some decreases in relative brain size are not simply due to changes in absolute body size. Instead, absolute brain size often changed independently of body size.

Now, to understand how brain size can change independently of body size, we must first understand how brain–body scaling during embryonic development relates to brain–body scaling in adults (Figure 4.11A). This issue, in turn, requires a basic understanding of brain and body growth.

Figure 4.11 Ontogenetic Brain–Body Scaling   By understanding how brains grow relative to the rest of the body, we can better understand what causes species differences in adult relative brain size. (A) Brains generally grow isometrically with body size up to a point, after which the body continues to increase in size while brain size plateaus. Shown here are idealized curves (based on a compound Gompertz model of brain and body growth) for two hypothetical species. (B) Replotting these data as brain weight versus body weight (thereby taking time out of the picture), we see that the curves for both species exhibit an initial steep phase, where brain and body grow proportionately, and a later flat phase, where only the body continues to grow. When you connect the end points of these ontogenetic allometry curves for individuals from different species, you obtain a classic adult interspecies allometry line (dashed line). (C) In order to determine how some species became more encephalized, we need comparative developmental data on brain and body growth. The data illustrated here, first culled from the literature by Count (1947), suggest that proportional brain size is probably greater in primates, from the earliest stages of brain development, than in several domesticated animals. In addition, the data show that the flat phase of brain–body growth is significantly shorter in humans than it is in chimpanzees. This probably means either that our brains grow at their fast embryonic rate for a longer period of time than chimpanzee brains do, or that our bodies stop growing before they are as large as they would be if we were scaled-up chimpanzees. Most authors support the former alternative, but the issue has not been settled yet. (D) A theoretical analysis reveals that increases in adult relative brain size may be obtained by diverse changes in growth function parameters. In addition to increasing the brain's proportional size at the onset of growth, relative brain size can be increased by increasing the rate at which the brain grows relative to the body. Thus far, there are no empirical data supporting the latter possibility.

Briefly, we need to know that, around the time of gastrulation, molecular signals cause part of the embryo to become neuronal precursor tissue. These precursor cells then go through several rounds of cell division, exponentially expanding the neuronal precursor pool. As development proceeds, more and more of the newly generated cells stop dividing and begin to differentiate into neurons and glial cells. Soon, the precursor pool is exhausted and neurogenesis stops (neurogenesis persists in some adult brain regions, but it con-

(A) Brain and body growth curves

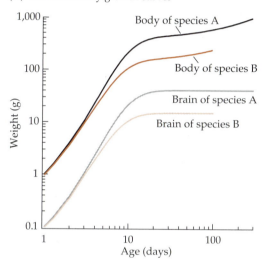

(B) From developmental to adult allometry

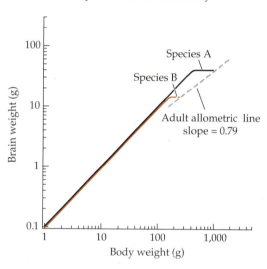

(C) Real data on brain and body growth in mammals

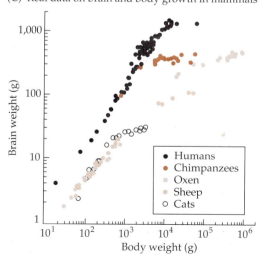

(D) Ways to increase relative brain size

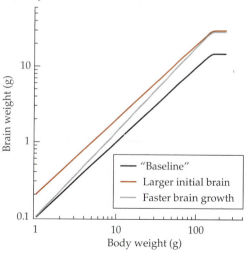

tributes little to overall brain growth in most amniotes). After that, brains grow mainly because: (1) individual neurons grow in size and complexity; (2) glial cells continue to be born; and (3) axons become myelinated (by oligodendroglia). In most species, brain growth is largely completed by the time of birth, but the body continues to grow. Therefore, the fraction of the body that is occupied by brain decreases steadily during the late stages of development. This relationship becomes apparent in log–log plots of brain size versus body size at different stages of development (see Figure 4.11B). In such ontogenetic allometry plots (Count, 1947), one can always identify an early "steep" phase, where brain and body both grow at similar rates, and a later (generally postnatal) "flat" phase where brain weight remains roughly constant while body weight continues to increase.

Comparing the ontogenetic allometry curves for cats, sheep, and oxen (ignoring primates for the moment), we see that they diverge primarily at the larger brain and body sizes (see Figure 4.11C). This means that during early development the relative rates of brain and body growth are similar in all three species. Eventually, however, the curves diverge because in the larger animals both brain and body grow for longer periods of time. This extension of the steep phase of brain and body growth in the larger species accounts for most of the increase in their absolute brain and body sizes. This is not the only difference, however, for the larger animals also have longer flat phases. That is, the larger animals undergo more postnatal body growth that is not accompanied by a commensurate amount of brain growth. This difference is important because it may explain why larger adult animals generally have proportionately smaller brains than smaller animals do (Cuvier, 1805–1845). To grasp this connection better, consider that a line drawn through the rightmost data points in each ontogenetic allometry curve corresponds to the adult brain–body line (see Figure 4.11B). If the slope and y-intercept of the steep phases are conserved between small and large animals, and if the flat phases are longer in the larger animals, then the slope of this adult allometry line must be lower than the slope of the steep phase in the ontogenetic curve. Since the slope of the steep phases is generally close to one, the slopes of the adult allometry line must be less than one. Thus, the difference in proportional brain size between cats, sheep, and oxen is due primarily to the fact that the larger species experience more postnatal body growth (relative to weight at birth). More generally, we can infer that brain size always scales against body size with a slope of less than one because the young in larger species always tend to grow more after birth.

Next, we can contemplate how development might be modified to generate adults with increased relative brain sizes (Count, 1947; Deacon, 1990b). Theoretically, there are many different changes that could heighten the trajectory of an ontogenetic allometry curve to yield an adult with a larger relative brain size (see Figure 4.11D), but which of these possible mechanisms actually occurred? Reasonably good data are available for brain and body growth in humans and chimpanzees (see Figure 4.11C). Remarkably, the ontogenetic allometry curves for these two species lie well above the nonprimate curves from the very beginning; that is, from the earliest developmental stages that were examined. This suggests that adult primates have larger relative brain sizes than nonprimates, mainly because in primate embryos, a greater proportion of the body is specified (by molecular signals) to become neuronal precursor tissue. A very different kind of change accounts for the increased encephalization of humans relative to chimpanzees. Inspection of the ontogenetic allometry curves for these two species shows that in humans the steep phase is prolonged beyond birth, while the flat phase is abbreviated relative to its length in chimpanzees. That is, in humans, the brain keeps growing at its fast embryonic rate past the point at which it normally stops in chimpanzees, but human body growth does not proceed much further than it does in chimpanzees. Of course, more data will be needed to confirm and elaborate these ideas, but the available data already illustrate how ontogenetic allometry curves can be used to construct fairly specific hypotheses about the cellular and molecular mechanisms that underlie evolutionary changes in brain–body scaling.

To test these hypotheses, one must compare and/or manipulate the cellular and molecular mechanisms that govern brain and body growth. One useful strategy in this enterprise is to generate transgenic animals that differ from normal animals in how they express molecules that function in cell division, cellular differentiation, and/or cell death. Pasko Rakic and his collaborators, for example, created transgenic mice that lack one kind of caspase molecule (Kuida et al., 1998). Because this caspase mediates normally occurring cell death in the developing brain, the transgenic mice grew up with many more cells in their brain and, hence, an increased relative brain size. Other investigators produced more encephalized mice by mutating other genes, but thus far all large-brained mutants have either died before birth or have lived with serious behavioral abnormalities (as some commentators were surprised to discover, these mutants were hardly "smarter" than their normal relatives!). Therefore, these simple mutation-induced increases in relative brain size are not equivalent to the kinds of changes that occurred during vertebrate phylogeny. Perhaps several different mutations must be combined to increase relative brain size without diminishing an animal's viability; or perhaps different molecules must be targeted. In any case, what becomes clear from these considerations is that we still know virtually nothing about the molecular mechanisms that underlie those changes in relative brain size that actually occurred during the course of evolution. Robert Williams (2000) and his collaborators are attempting to fill this gap by mapping genetic mutations that correlate with relative brain size in various strains of mice but, thus far, the gap remains.

Also unknown are the mechanisms that coordinate brain and body growth. That such coordinating mechanisms must exist is evident from the fact that brains scale so predictably (and allometrically) with body size. Terrence Deacon has proposed that the allometric scaling of brain and body size is due to predictable changes in growth hormone secretions from the hypothalamus (Deacon, 1990b). This hypothesis is plausible because the hypothalamus does secrete growth hormones, but there is no direct evidence that changes in hypothalamic hormone secretion actually account for a significant proportion of the phylogenetic changes in body size. Moreover, Deacon's hypothesis is not consistent with data from Evan Balaban and his collaborators, who transplanted embryonic brains from quails into chickens (Balaban et al., 1988). Since chickens are much larger than quail, the hypothalamic hormone hypothesis predicts that the chickens with transplanted quail brains should grow up to be abnormally small, but the so-called "chimeric" chickens turned out to be of normal size (Balaban, personal communication). That is, the brain transplants did not alter adult body size. Therefore, the adult sizes of both body and brain are probably specified early in embryogenesis. Perhaps precursor cells at some point become programmed to undergo a fixed number of cell divisions before they differentiate (Williams and Herrup, 1988), but whether such a mechanism can explain brain–body scaling remains unknown. In conclusion, we must admit that the developmental mechanisms behind adult brain–body scaling remain amazingly mysterious. Even in insects, where so many developmental phenomena are well understood, and where we might look for inspiration, the molecular mechanisms of allometric growth are quite unclear (Stern and Emlen, 1999).

Functional Correlates of Relative Brain Size

Biologists often answer questions about *why* structures changed in evolution by identifying the selective "pressures" that may have acted upon those structures. As I discussed in Chapter 1, however, we have virtually no data on the effects of natural selection on brain structure. Moreover, something as crude as overall brain size is unlikely to be explicable in terms of a single selective pressure. Nor is it likely that brain size evolution is subject to any single, readily specifiable constraint. Therefore, I will postpone questions about natural selection and constraints to the end of this chapter and begin by focusing on the simpler (but still nontrivial) task of identifying some behavioral attributes that correlate with relative brain size.

Searching for correlations between brain size and behavior is not for the faint of heart. First, there is the difficulty of gathering comparable data from a broad range of species. This is labor intensive, especially for behavioral data, and one may argue about which measures are truly "comparable" across species (this is discussed in more detail later in this chapter). Second, one must worry about getting a random species sample of the available variation, since biased sampling can obviously yield biased results (Ackerly, 2000). Third, and this has received a great deal of attention in the comparative biological literature, one must realize that different species cannot be treated as independent data points, since some of them will be closely related to one another and, therefore, similar simply because of shared ancestry. For example, if we find that 64 primate species are large brained and prone to play, while 23 rodent species are relatively small brained and less playful, then we might get an impressive-looking correlation between relative brain size and play behavior, but a statistician might well argue that we really have only two phylogenetically independent data points, one for primates and one for rodents; and with only two data points, it is impossible to obtain a statistically significant correlation. Several methods have been proposed to deal with this phylogenetic nonindependence, and application of one such method shows that play behavior does correlate significantly with relative brain size, at least when we compare vertebrate *orders* (Iwaniuk et al., 2001). For details on these methods, however, I must refer interested readers to the rather thorny technical literature (Felsenstein, 1985; Garland et al., 1992; Lorch and Eadie, 1999; Martins, 2000; Martins et al., 2002). Finally, we must keep in mind that correlations between brain and behavior may be limited to only a subset of taxonomic groups. For example, Adolf Portman noted long ago that altricial birds (birds with hatchlings that require extensive parental care) tend to be more encephalized as adults than birds with precocial (less helpless) young (Portmann, 1962; see also Iwaniuk and Nelson, 2003), but this correlation does not seem to hold in mammals (Bennet and Harvey, 1985a).

Despite these difficulties, the literature abounds with reported correlations between brain size and behavior. Most of these are, perhaps not surprisingly, quite controversial. For example, much has been made of the observation that in mammals absolute brain size and resting metabolic rate scale similarly with body size (Figure 4.12A) (Martin, 1981; Mink et al., 1981). Later authors pointed out, however, that the amount by which species differ from their expected brain weight (i.e., their *relative* brain size) is poorly correlated with the deviations from their expected metabolic rate (see Figure 4.12B) (McNab, 1989). This lack of correlation does not prove that metabolic rate is unrelated to brain size, but it means that the relationship cannot be simple. Similarly controversial is the observation that brain size correlates with life span in some mammals, notably primates (Sacher, 1973; Allman et al., 1993). This correlation is interesting because it suggests that large-brained species may be better equipped to survive crisis situations and, unlike the correlation with metabolic rate, it persists if we look only at the deviations from what is "expected" on the basis of body weight. However, species-typical body weights are notoriously difficult to estimate (partly because brains are often collected from dead zoo animals, which tend to be either overweight or emaciated). This is important because, if brain weight and life span correlate better with species-typical body weight than measured body weight does, then the observed correlation between relative brain size and life span might be artifactual (see Economos, 1980; Barton and Dunbar, 1997). In addition, it is still unclear whether brain size correlates with life span in taxonomic groups

(A) Brain weight, body weight, and metabolic rate

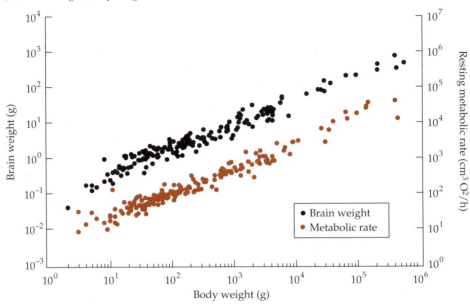

(B) Brain weight and metabolic rate residuals

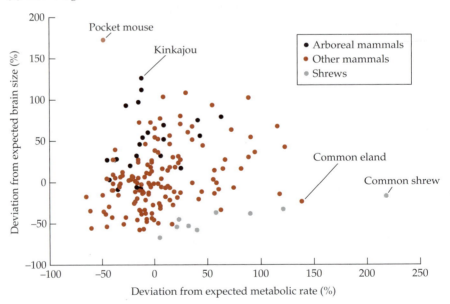

◀ **Figure 4.12 Brain Size and Metabolism in Mammals**   The relationship between brain size and metabolic rate is complex. (A) When you plot brain weight (left axis) and basal (resting) metabolic rate (right axis) versus body weight (abscissa), you obtain very similar allometric slopes, suggesting that there is a causal link between absolute brain size and metabolic rate. (B) However, when you plot the residuals for those two parameters (i.e., the deviations of each data point from the best-fit allometric line), you see no obvious correlation between brain size and metabolic rate. Shrews, for instance, have unusually high metabolic rates but not unusually large brains (their extra energy is probably devoted mainly to thermoregulation). Arboreal mammals, on the other hand, have relatively large brains, but their metabolic rates are about what you would expect in mammals of their body size (most likely, arboreal creatures generally tend to reduce body size without making commensurate reductions in brain size). These data indicate that relative brain size and metabolic rate are not linked in a tight, obligatory manner. (After McNab, 1989.)

other than primates. Given all these difficulties, I here focus only on the taxonomically most general and, to me, most interesting correlates of relative brain size, namely foraging behavior and intelligence.

Numerous studies have reported correlations between diet and relative brain size. Among bats, the insect-eating species tend to be less encephalized than those that eat meat, fish, blood, fruit, or flowers (Figure 4.13) (Eisenberg and Wilson, 1978; Hutcheon et al., 2002), and among primates leaf-eaters tend to be less brainy than insectivores or frugivores (Clutton-Brock and Harvey, 1980). In addition, filter-feeding sharks and whales are less encephalized than those that pursue individual prey, and large herbivores tend not to be brainy. At first glance, these data suggest that relative brain size might generally correlate with dietary quality, but this hypothesis is not consistent with the bat data, for insects are relatively rich in nutrients, while fruits and flowers tend to be protein-poor. A somewhat better rule is this: species that dine on hard-to-find, large foods tend to be more encephalized than those that eat abundant, small morsels (e.g., insects and plankton). Even this rule must be broadened, however, since it fails to account for the small brains of parasites (e.g., lampreys) and ambush predators (e.g., frogfishes). Therefore, I suggest a very general principle: highly encephalized species tend to forage (or hunt) strategically, taking into account the habits of their food (or prey), while less encephalized species tend to graze (or hunt) opportunistically. This "clever foraging" hypothesis (see also Parker and Gibson, 1977) is consistent with the bat and primate data, since frugivores must learn where to find fruit trees and when their fruit is likely to be ripe (Fleming et al., 1977; Clutton-Brock and Harvey, 1980). It also fits much of the avian data (Bennet and Harvey, 1985b), since the most encephalized birds tend to use rather complex, individually learned strategies to obtain food (Volman et al., 1997; Clayton et al., 2001; Hunt et al., 2001), while the least encephalized birds are relatively simple-

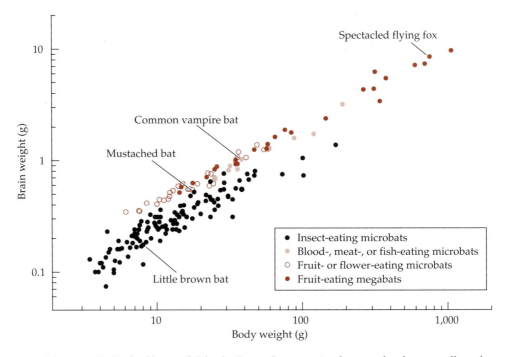

Figure 4.13 Brain Size and Diet in Bats Insect-eating bats tend to have smaller relative brain sizes than bats that eat fruit, flowers, fish, meat, or blood. The fruit-eating bats can be divided into the large, highly visual megabats (Megachiroptera) and their smaller echolocating cousins, the microbats (Microchiroptera). Apparently, relative brain size in bats correlates more strongly with diet than with taxonomic group. Most likely, this correlation involves differences in the amount of foraging intelligence required to obtain different kinds of foods. (After Stephan et al., 1981b.)

minded about finding food (see Lefebvre et al., 1997). However, even the clever foraging hypothesis has exceptions. Manta rays, for instance, are large-brained but planktivorous, and elephants are far more encephalized than your typical herbivore. Most likely, these exceptional species use their large brains for feats of "intelligence" that have little to do with food. Which brings us to the next topic, namely the correlation between brain size and intelligence.

By far the most famous (and infamous) correlate of relative brain size is what we generally call intelligence, or intellect. As Charles Darwin wrote in his *Descent of Man*:

> *No one, I presume, doubts that the large proportion which the size of man's brain bears to his body, compared to the same proportion in the*

gorilla or orang, is closely connected with his higher mental powers. . .
On the other hand, no one supposes that the intellect of any two animals
or of any two men can be accurately gauged by the cubic contents of their
skulls. (Darwin, 1871, p. 37)

Since then, many have argued fiercely about how to define Darwin's "higher mental powers" and about whether these are correlated with relative brain size. Most famous is Harry Jerison's proposition that relative brain size correlates with "biological intelligence," which he defined as "the capacity to construct perceptual worlds in which sensory information from various modalities is integrated as information about objects in space and time" (Jerison, 1976, p. 101). Nowadays the word "intelligence" is usually replaced by terms like "cognitive ability," but the underlying concept has barely changed. Thus, Hubert Markl's (1985) definition of cognition as "the ability to relate different unconnected pieces of information in new ways and to apply the results in an adaptive manner" (see Cheney and Seyfarth, 1990, p. 9) is quite similar to Jerison's definition of biological intelligence. Whichever definition one prefers, it certainly seems that humans are more intelligent (at least on average) than other animals. Since we also exceed other animals in terms of relative brain size, it is reasonable to suggest that intelligence might be positively correlated with relative brain size. The challenge, of course, is to support this hypothesis empirically.

Anecdotes regarding animal intelligence abound, but often they are colored by what we expect to find and biased in favor of species that are emotionally or phylogenetically close to us (such as horses, dogs, and chimps)(see Morgan, 1894). Therefore, rigorous comparisons of animal intelligence must be conducted in the laboratory, where different species can be tested in a variety of learning and problem-solving tasks. This approach has its own complications, however. As Niko Tinbergen, the great ethologist, remarked, "one should not use identical experimental techniques to compare two species because they would almost certainly not be the same to them" (Tinbergen, 1951, p. 12). Most reptiles, for example, perform much better in learning tasks if they are given warmth, rather than food, as a reward; this is not surprising since reptiles eat less frequently than mammals or birds and are cold-blooded. Similarly, chimpanzees are better at learning human sign language than spoken language partly because their vocal tract does not allow them to produce well-formed speech sounds (and partly because chimpanzees naturally use their hands to gesture). Because it is so difficult to give the same intelligence test to different species, most students of animal behavior nowadays focus on so-called domain-specific intelligences (e.g., the learned navigational skills of homing pigeons: see Hauser, 2000). Others have proposed that all vertebrates (except humans) possess the same level of "general intelligence" (Macphail, 1982). The latter conclusion is clearly at odds with most people's intuitive notions about how "higher mental powers" vary across vertebrates, but it is difficult to refute because the above-mentioned definition of general intelligence (or intellect) has fallen out of favor (see also Chapter 9).

Some investigators, however, have managed to think differently about the question of animal intelligence. Nicholas Humphrey, in particular, asked himself why great apes exhibit more intelligence in laboratory tests than they seem to need in order to obtain food in the wild (Humphrey, 1976). Nature surely would not waste excess intelligence on them. This paradox led Humphrey to consider that the most intelligent animals (e.g., the great apes) are highly social and that "the life of social animals is highly problematical [because] there are benefits to be gained for each individual member both from preserving the overall structure of the group and at the same time from exploiting and out-manoeuvring others within it" (Humphrey, 1976, p. 309). Social animals, Humphrey argued, frequently learn from and about each other, form alliances, deceive competitors, and otherwise manipulate conspecifics. To the extent that these behaviors promote an individual's survival and reproductive fitness, they would be selected for. This, in turn, could set up an evolutionary "ratchet" that acts like "a self-winding watch to increase the general intellectual standing of the species" (Humphrey, 1976, p. 311). Thus, the social intelligence hypothesis was born (it is also known as the Machiavellian intelligence hypothesis)(Byrne and Whiten, 1988; Whiten and Byrne, 1997). Since then, many studies have examined various aspects of social intelligence, including social learning, tactical deception, and the ability of one animal to understand another's motivations, beliefs, and states of mind. Predictably, most of this work has focused on primates, but some data are available for birds as well. Collectively, these data confirm Humphrey's suspicion that highly social animals tend to excel at the various forms of "social intelligence."

Because of Tinbergen's warning that different species may experience identical experimental tasks differently, and because "natural" social intelligence (e.g., deception of conspecifics) has been well described for only a few species, we cannot correlate relative brain size directly with measures of social intelligence. However, if we accept that social intelligence generally correlates with social complexity, then we may be able to use social group size as a proxy for social intelligence. Simple measures of group size are problematic because even asocial species may live in huge aggregates (think of herring or anchovies), but *social* group size, defined as the mean number of individuals that interact socially, may indeed be a plausible stand-in for social complexity and, hence, social intelligence. Unfortunately for this hypothesis, Robin Dunbar showed that, in a sample of 24 primate species, social group size correlates only weakly with relative brain size (Dunbar, 1998). This finding soured some investigators on the idea that relative brain size might be linked to social intelligence (Allman, 1999), but perhaps even social group size is not a very good indicator of social intelligence. This idea is supported by the finding that "grooming clique size," which presumably captures more of a group's social complexity than social group size, correlates positively with relative brain size (Figure 4.14A) (Kudo and Dunbar, 2001). If we then look beyond the primates, we find additional support for the social intelligence hypothesis. Particularly striking is that the toothed whales, elephants, parrots, and corvids all lead complex social lives and all are well-endowed in terms of relative brain size. Even the highly encephalized manta rays and hammerhead sharks form large social groups (though admittedly we know very little about what goes on in those gatherings). Therefore, I conclude that social intelligence probably is a fairly general correlate of relative brain size.

If that is true, how do we reconcile the social intelligence and clever foraging hypotheses? Clearly, the two are not incompatible. Indeed, we might combine them both into a more sweeping principle, stating that an animal's relative brain size correlates with its ability to manipulate other individuals, be they of another species or its own. This idea appeals in its simplicity, but risks being overly general and vague. Alternatively, we might hypothesize that social intelligence and foraging intelligence constitute fundamentally different skill sets that may vary independently of one another. This hypothesis is consonant with the views of many evolutionary psychologists, who tend to view the mind as being modular (see Chapter 5). In order to decide between these two hypotheses, we can ask whether social intelligence and foraging intelligence are mediated by different brain regions, and whether highly social animals have enlarged the region mediating the former while clever foragers have enlarged the region mediating latter. If both answers were affirmative, then the "modularity hypothesis" would be supported. Unfortunately, however, we have little data relevant to these questions and reasons to believe that things may not be that simple. Even if we accept that social intelligence and foraging intelligence are *behaviorally* distinct modules, it cannot be assumed that they map onto distinct regions (or circuits) in the brain. Although different behaviors are generally controlled by different *sets* of brain regions, those sets typically include shared elements. Thus, it is conceivable that social intelligence and foraging intelligence might be distinct at the behavioral level of analysis but involve overlapping brain regions. Indeed, I think that in primates, at least, they do. As I elaborate in Chapter 9, the disproportionate enlargement of the lateral prefrontal cortex in large-brained primates probably enhanced their ability to manipulate food (or inedible objects) as well as peers. The idea is speculative and I will not deal further with it here. My main point is simply that social intelligence and foraging intelligence might well share neurobiological substrates. More generally, I want to emphasize that the social intelligence and clever foraging hypotheses are not mutually exclusive. Further studies will be needed to sort out their interrelationships.

Now, there is one more important twist to the correlation between brain size and intelligence, for thus far, with all the emphasis on allometric scaling and relative brain size, we have tended to ignore absolute brain size. This

(A) Relative brain size versuus grooming clique size

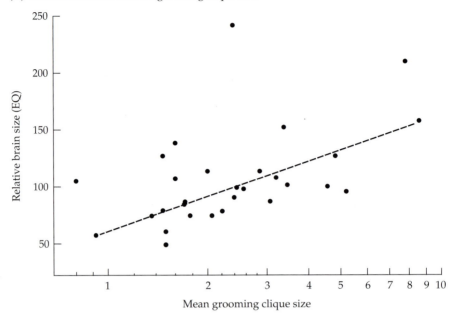

(B) Absolute brain size, neocortex ratio, and clique size

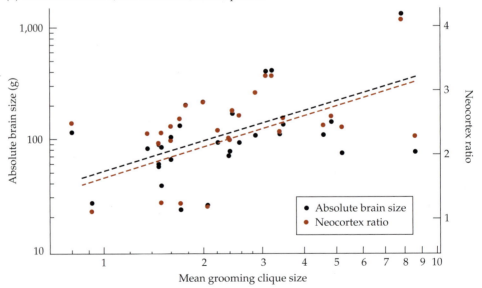

◀ **Figure 4.14 Brain Size and Clique Size in Primates**   The social intelligence hypothesis predicts that brains should be larger in highly social species. This prediction has been tested by analyzing the relationship between brain size and grooming clique size in primates. (A) A plot of relative brain size (here depicted as "encephalization quotient" = "EQ") versus grooming clique size for 29 primate species reveals a weak but positive correlation between the two parameters. (B) Plotting the same grooming clique size data against absolute brain size (left axis) and neocortex ratio (the ratio of neocortex to the rest of the brain; right axis) yields stronger correlations. The correlation with neocortex ratio probably arises because neocortex ratio is itself tightly correlated with absolute brain size (see Chapter 5). (Data from Stephan et al., 1981a; Harvey et al., 1987; Kudo and Dunbar, 2001.)

may have been misguided (Smith, 1980). As Bernhard Rensch noted long ago, the largest species of ants, bees, and wasps tend to have the most complex social instincts; rats tend to perform better than mice in learning tasks; ravens learn better than jackdaws (small corvids); and large parrots seem smarter than parakeets (Rensch, 1960). These remarks were widely distrusted and ignored, but they mesh rather nicely with my own knowledge of animal behavior. More important, students of social intelligence have recently begun to ask, once again, whether "perhaps, after all, larger animals are simply more intelligent" (Byrne, 1997). These authors pointed out that, within a given taxonomic group, larger animals almost always have larger brains and that these additional neurons may well provide more "computing power." Even if one accepts the argument that larger bodies require more neurons to control them (Snell, 1891), the number of neurons *not* dedicated to muscular control or sensory processing is almost certainly larger in larger brains and, so the argument goes, these extra neurons may permit more complex computations. Indeed, Kudo and Dunbar's data show that grooming clique size correlates more strongly with absolute brain size than it does with relative brain size (Figure 4.14B). Similar correlations seem to hold in carnivores and bats (Barton and Dunbar, 1997). Therefore, we can say that, at least within taxonomic groups, absolute brain size is generally better than relative brain size at predicting a species' degree of social intelligence (the ratio of neocortex:rest of brain may be an even better predictor of social intelligence, but this ratio also correlates with absolute brain size, as we will see in Chapter 5). Given this conclusion, it behooves us to look in more detail at how absolute brain size has changed during the course of evolution, and what the functional consequences of these changes might have been.

Changes in Absolute Brain Size

Just as relative brain size has increased and decreased several times independently during the course of vertebrate phylogeny, so has absolute brain size (Figure 4.15). In fact, the evolutionary changes in absolute brain size were even more dramatic than those in relative brain size, because most of the taxonomic groups that underwent significant increases in relative brain size (e.g., mammals, primates) also underwent significant increases in body size (discussed earlier) which, as we have seen, always involve some allometric increases in absolute brain size. But what are the functional consequences of increasing absolute brain size?

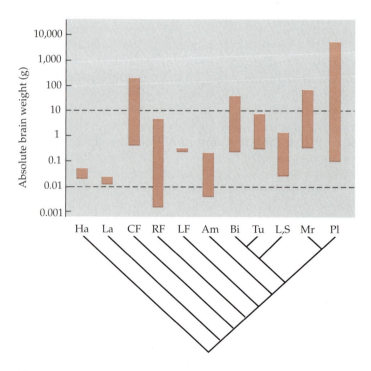

Figure 4.15 Variation in Absolute Brain Size   Absolute brain size varies significantly across the major vertebrate groups. Shown here is the range of absolute brain weights for the various data sets described in Figures 4.1–4.9. A cladistic analysis of this variation suggests that absolute brain size increased independently in cartilaginous fishes, birds, and mammals, and decreased in both amphibians and ray-finned fishes. Abbreviations: Ha = hagfishes; La = lampreys; CF = cartilaginous fishes; RF = ray-finned fishes; LF = lungfish; Am = amphibians; Bi = birds; Tu = turtles; L = lizards; S = snakes; Mr = marsupials; Pl = placental mammals.

To answer that question, let us begin by thinking metaphorically. Are brains like computers that improve in performance as you add memory? Or are they more like companies, which tend to become more powerful as they expand in size? Both metaphors are limited, but I prefer the second one because brains, like companies, must reorganize as they increase in size. As employees are added to a company, tasks that were initially handled by a single person become distributed across several different people, who can then become highly specialized and thus enhance the company's performance. However, and this is the most interesting aspect of the analogy, increasing the size of a company also makes it more difficult for the various employees to communicate with one another, and that can create conflicts and impede innovation (which is why large companies typically create human resources and/or communication departments). For brains the problem is similar: as they add neurons, it becomes more and more difficult for these neurons to talk to one another. To see more clearly why this is true, let us conduct a thought experiment.

Imagine a small artificial brain in which every neuron is connected to every other neuron via an axon. If such a brain contained 4 neurons, then it would have to have 12 axons (Figure 4.16A). If there were 8 neurons, there would have to be 56 axons, and 16 neurons would require an impressive 240 axons. Expressed mathematically, we can say that the number of axons in such a fully interconnected brain would scale with the square of the number of neurons. Because of that scaling law, any fully interconnected network quickly becomes axon-dominated and unwieldy (Deacon, 1990c; Ringo, 1991). As Mark Nelson and Jim Bower put it, "if the [human] brain's estimated 10^{11} neurons were placed on the surface of a sphere and fully interconnected by individual axons 0.1 μm in radius, the sphere would have to have a diameter of more than 20 km to accommodate the connections" (Nelson and Bower, 1990, p. 408; see also Murre and Sturdy, 1995). Now imagine the other extreme: a network in which every neuron contacts only three other neurons, regardless of the total number of neurons (see Figure 4.16B). In that case, the number of axons is directly proportional to the number of neurons and, consequently, axon number does not "explode" as neuron number is increased. More generally, any networks that maintain absolute connectivity as neuron number is increased can be scaled up far more easily, in terms of both physical space and energy requirements, than networks that maintain proportional connectivity.

However, scaling up a network that maintains absolute connectivity is not trivial either. Specifically, if you keep the connections randomly distributed, then it becomes more difficult for neighboring neurons to communicate with one another. On the other hand, if you minimize connection lengths and interconnect only neighboring cells (Figure 4.16B), then the network tends to fractionate into multiple clusters or modules that are only sparsely interconnected with one another. That, in turn, makes it difficult to exchange infor-

Figure 4.16 Neural Network Allometry   As a neural network increases in size, how do its connections scale? If the network maintains the same degree of *proportional* connectivity (e.g., 100% interconnectedness), then the number of axons increases exponentially with the number of neurons (top). On the other hand, if the network maintains the same degree of *absolute* connectivity (e.g., each neuron connects to 3 other neurons), then the number of axons increases linearly with the number of neurons (bottom). Obviously, the latter solution is more economical in terms of wiring costs. See Chapter 7 for further details. (After Deacon, 1990c; Ringo, 1991.)

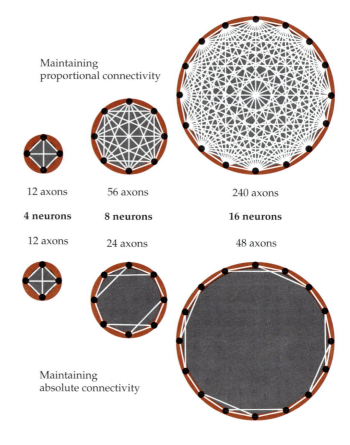

Maintaining proportional connectivity

12 axons	56 axons	240 axons
4 neurons	**8 neurons**	**16 neurons**
12 axons	24 axons	48 axons

Maintaining absolute connectivity

mation between distant cells. Overall, this means that a network's "degrees of separation" (i.e., the average number of steps required to get from any one point to any other point) must increase as neuron number increases (even if the network has a "small world" architecture; see Chapter 7). Thus, our thought experiment reveals that, as brains increase in size, their connectivity and functional organization must change as well. With that in mind, let us leave the thought experiment behind and examine how neuronal connectivity scales with neuron number in real brains.

Counting neurons is tedious and difficult, but we can estimate neuron number from brain volume and neuron density. If we examine the neocortex of various mammals that differ in absolute brain size, we see that the size of individual cell bodies changes relatively little as brain size goes up, but that cell density declines (Figure 4.17A). The extra space between the cell bodies in larger brains is filled mainly by dendrites, which tend to be larger in larger

(A) Neuron density scaling

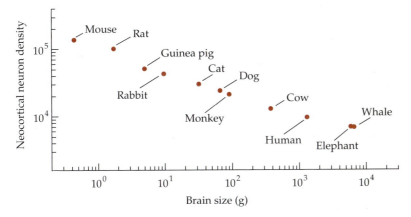

(B) Gray matter versus white matter scaling

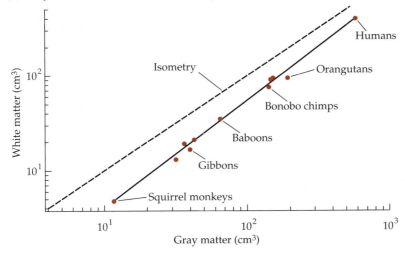

Figure 4.17 Neocortical Neuron Density, Number, and Connectivity The available data suggest that mammalian brains maintain absolute connectivity rather than proportional connectivity (see Figure 4.16) as they increase in size. (A) Neocortical neuron density decreases by a factor of roughly 10 as overall brain size increases by a factor of more than 10,000, implying that increases in overall brain size mainly result from increases in neuron number. (B) The ratio of neocortical white matter:gray matter also increases as overall brain size goes up. However, the slope of this allometric line is only slightly steeper than the isometric line. This suggests that the neocortex is closer to maintaining absolute connectivity than proportional connectivity as it increases in size, because maintaining absolute connectivity would yield a slope of roughly two. (Data in A from Tower and Elliott, 1952 and Tower, 1954; B after Rilling and Insel, 1999b.)

brains (Bok, 1959). We will come back to this extra space, or "neuropil," but for now let us focus specifically on neuron density, which has been measured by several investigators, all of whom were careful to count only neurons, not glial cells (the ratio of glia to neurons may increase in larger brains, but the data on this are equivocal; see Jerison, 1973 and Haug, 1987). If we plot neuron density against overall brain volume in log–log space, the data form a virtually straight line with a slope of approximately −1/3 (there is some controversy about the value of this slope, but see Prothero, 1997). Considering only the extremes, we see that neuron density varies from about 142,000 cells per mm^3 in a mouse to about 6,000 cells per mm^3 in elephants (Tower, 1954). This 24-fold difference is considerable, but it is dwarfed by the 16,000-fold difference in brain volume between mice and elephants (0.4 g versus 6–7 kg). Therefore, we can conclude that elephant brains are larger than mouse brains *mainly* because they contain many more neurons (about 800 times as many in the neocortex alone). If we assume that these considerations apply not only to neocortex but also to other brain regions and to nonmammalian brains, then we can use basic mathematics to show that neuron number generally scales against brain volume with a slope of +2/3. That is, whenever evolution increased absolute brain size, it apparently did so *mainly* by increasing neuron number, not neuron size. This makes sense, because dramatic increases in neuron size would quickly lead to both computational and metabolic difficulties (Bekkers and Stevens, 1970). In that context, we may note that the largest known neurons actually have doughnut-shaped cell bodies, presumably to overcome metabolic constraints (see Rensch, 1960). But let us return to the question of connectivity.

A brain's average degree of interconnectedness is even harder to measure than neuron number but, again, we can derive some estimates. A good starting point is the remarkable finding that neocortical synapse size and density barely differ between hedgehogs and macaques (Schüz and Deminanenko, 1995), despite a 120-fold difference in neocortical volume (Stephan et al., 1981a). If we combine this finding with the above-mentioned finding that neocortical neuron density scales with a slope of −1/3, then we can conclude that the number of synapses per neuron scales with a slope of +1/3. This conclusion is built on a number of assumptions, and other investigators have made different assumptions (Stevens, 1989), but what is clear from all these considerations is that the number of connections per neuron does not increase with the square of neuron number, as we would expect if brains maintained proportional connectivity as they increased in size. Instead, the number of connections per neuron increases roughly as if absolute connectivity were maintained. A second line of reasoning leads to essentially the same conclusion. If we look at how neocortical white matter (i.e., the myelinated axons) scales against neocortical gray matter (the cell bodies and neuropil), we find a scaling exponent of 1.1–1.2 (Rilling and Insel, 1999b; Zhang and Sejnowski, 2000). That is, as mammalian brains increase in size, the ratio of white matter:gray matter increases (partly because axons must, on average, be longer in larger brains) (Zhang and Sejnowski, 2000), but this increase is not nearly as great as one would expect if these brains maintained proportional connectivity (i.e., the scaling exponent is not near 2). Finally, Malcolm Young and his collaborators (1995) have found that the visual cortical areas are less densely interconnected with one another in macaques than in domestic cats (15% versus 27% of all possible connections, with macaque brains being roughly three times as large as those of cats).

Collectively, these findings and analyses suggest that whenever brains increase in absolute size, their proportional connectivity decreases. This, in turn, implies that increases in absolute brain size are always associated with at least some changes in brain connectivity and function. I will come back to this point in later chapters. For now, I want to mention only that these size-related changes in connectivity *might* impose some limits on how large brains can get without becoming functionally incoherent; at least, they might drive evolutionary innovations that overcome or extend those limits. This idea is speculative but there are also several other, more concrete limitations on brain size, to which we shall now turn.

Constraints and Compromises

The most straightforward limit on brain size in vertebrates is that the brain must fit inside the head (in some very small invertebrates, the cerebral ganglia may spill into the thorax [Beutel and Haas, 1998], but for vertebrates this seems not to be an option). This constraint sounds silly but is easily forgotten. For instance, when we plot brain versus body weight in log–log coordinates and marvel at the strength of the observed correlation (usually in excess of 0.9), we easily forget that large portions of such a graph are in fact "inadmissible" (see Figure 4.03), for brain size certainly cannot be larger than body size! Because vertebrate brains scale with negative allometry (i.e., slopes <1), large vertebrates do not usually have to worry about fitting their brains inside the head (humans may be an exception; see Chapter 9), but for small vertebrates, with their large proportional brain sizes, this constraint can become serious. It is difficult to imagine, for example, how an echolocating bat, a hummingbird, or a small teleost could survive if its brain were much larger than it is (Northcutt and Striedter, 2002). Therefore, from a purely physical perspective, it should be "easier" (i.e., less problematic) for large vertebrates to increase their relative brain size than it is for small-bodied vertebrates. This differential may partly explain why the overall trend of increasing relative brain size is accompanied by a roughly parallel trend of increasing body size. However, physical space is not the only constraint on absolute brain size.

A second, more widely discussed constraint on overall brain size is the amount of energy an animal has available to feed its brain. This constraint comes into play because the brain is a metabolically "expensive" organ (Aiello and Wheeler, 1995). An average human brain draws about 15 watts of energy (yes, its energy consumption is equivalent to that of a dim light bulb!), which is about 16% of the entire body's energy consumption; only the gut consumes more energy (about 70% of the total energy supply). Because the brain requires so much energy it is reasonable to expect that, if brains become too large, then there might not be enough metabolic energy available to run them effectively. Therefore, we would expect absolute brain size to scale with an animal's metabolic rate, and to a first approximation it does (see Figure 4.12A; we're really interested in *average* metabolic rate, but scientists usually measure *resting* metabolic rate; let us assume the two scale similarly). That is, small vertebrates not only have proportionately larger brains than large vertebrates, they also have higher metabolic rates (per unit body mass). All this makes sense but, as I pointed out above, the deviations from an animal's expected brain size do not correlate well with the *residual* variation in metabolic rate (see Figure 4.12B). So, are we to conclude that metabolic rate does not, after all, constrain brain size? Yes we are, but the full answer is more complicated.

The complexity in the nexus between brain size and metabolic rate lies in the fact that vertebrates have evolved a variety of different strategies to accommodate large, energy-hungry brains. One way to increase absolute brain size in spite of limited energy resources is simply to allocate more energy (of the overall budget) toward the brain and less to other organs. A good example of this strategy are the primates, which have reduced the size of their gut (gastrointestinal tract) and therefore have more energy available to support their enlarged brains (Aiello and Wheeler, 1995). This shift toward a smaller gut was only possible, so the argument goes, because primates also began to eat more nutrient-dense foods (e.g., meat), and hence needed to eat less (see Chapter 9 for more details). What I find most intriguing about this argument is that the dietary shift itself was probably made possible by the increase in brain size (recall that brain size correlates with foraging intelligence), which means that we here have a tightly interlocking network of causal relationships (Aiello and Wheeler, 1995); I suspect that evolution often works this way. Also important to note is that whenever different species have allocated different percentages of their total energy budget to the brain, then we would expect exactly what the correlative analyses of brain size, body size, and metabolic rate have shown (see Figure 4.12): the residuals for brain size and metabolic rate should correlate poorly, even though both brain size and metabolic rate are highly correlated with body size. It is unclear how often evolution has changed the percentage of an animal's energy that is allocated to the brain, but we know that some such variations also exist outside of mammals. Some cartilaginous fishes, for example, have brains that require less energy per unit mass than the brains of other vertebrates (Nilsson et al., 2000), and mormyrid teleosts (see Figure 3.4A) devote an astonishing 60% of their metabolic energy to their hypertrophied brain (Nilsson, 1996). Intriguingly, those same mormyrids also (like primates) exhibit smaller than expected guts (Kaufman, 2003).

Thus, it is misleading to argue that metabolic rate tightly constrains overall brain size. At the same time, however, if would be wrong to ignore the fact that brains are metabolically expensive. Clearly, metabolic considerations imply that brain size cannot be increased indefinitely. More important, we have seen that increases in relative brain size often require some kinds of other changes (e.g., in diet, size of other organs, or neural rate of energy consumption) to make those large brains possible. An even better way to realize the costs associated with large brains is to invert the problem and think about those animal groups that underwent significant reductions in relative brain size. Many of these small-brained vertebrates are parasites (e.g., lampreys), solitary planktivores (e.g., megamouths), or fishes that hide until unsuspecting prey saunters by (e.g., angel sharks or stargazers). None of these "lifestyles" requires a great deal of intelligence (be it of the social or foraging type). Therefore, it seems reasonable to speculate that these species reduced their relative brain size simply because it allowed them to conserve metabolic energy. In an evolutionary sense, there is nothing wrong with being "stupid," as long as you can get away with it. In fact, there is a widely recognized principle in evolutionary biology, called symmorphosis, which states that organs and body parts tend generally to be no larger than they need to be (Taylor and Weibel, 1981; Weibel, 2000). Of course, it is not so easy to determine how large any given brain "needs to be," and those needs certainly vary with the stage of an animal's life (e.g., social animals may want particularly large brains when they are trying to attract a mate), but as a first approximation, it seems reasonable to conclude the following: because brains are metabolically expensive, they tend not to be larger than they need to be. Putting it more scientifically, the metabolic cost of having a large brain must be balanced against its advantages, and the details of this equation depend on the overall ecological, physiological, and morphological contexts in which those brains are found. That is why, when we are trying to understand brain evolution in a particular group of animals (see Chapters 8 and 9), we cannot focus *only* on the brain.

Conclusions

So, what have we learned about the evolution of overall brain size? The way I see it, the problem of brain-size evolution can be divided into two questions: what happened, and what does it mean? On the first question, good progress has been made and future progress seems assured. We now know that relative brain size has tended (on average) to increase over the course of vertebrate phylogeny, that similar increases occurred independently in several different lineages, and that relative brain size also decreased in some groups. In most of these cases, the changes in relative brain size are not due to simple changes in body size (dwarfism or gigantism) but to changes in the relationship between early brain and body growth. Much remains to be learned about the molecular and developmental mechanisms that underlie these changes (and brain allometry in general), but we already have most of the conceptual and experimental tools that are needed to advance on this front. The second question is more difficult, and all of the answers that we have discussed thus far seem vaguely incomplete. To put it mildly, we are still far from consensus when it comes to the biological significance of evolutionary changes in overall brain size.

Most frustrating is that there is no general agreement on whether we should focus on relative or absolute brain size (I think we can safely ignore proportional brain size for present purposes). Because humans outrank all other vertebrates in terms of relative brain size, but are outclassed by elephants and whales in absolute brain size, we naturally tend to focus on relative brain size. And indeed, major increases in relative brain size occurred early on in the evolution of several major lineages (e.g., mammals, birds, and probably cartilaginous fishes), suggesting that these increases were biologically significant. It is also true that relative brain size correlates, at least vaguely, with several behaviors that strike us as intelligent, such as hunting for elusive prey and socializing with conspecifics. Yet, it remains unclear *why* increases in relative brain size should lead to increases in intelligence. The notion that increases in relative brain size provide "extra neurons" that can be used for intelligent rather than "somatic" functions has intuitive appeal (Jerison, 1973) but is difficult to translate into concrete neural terms. Are somatic and intelligent functions really separable in brains? And why do the supposedly nonsomatic regions also scale with body size (Fox and Wilczynski, 1986)? Since no answers are forthcoming, I consider this "excess neuron hypothesis" to be most likely false.

Absolute brain size, on the other hand, correlates quite nicely with many aspects of brain structure, such as neuron size, connectivity, and several other features that we will encounter in the following chapters (e.g., neocortex ratio). It also correlates with some measures of social intelligence (e.g., grooming clique size) and, within a given taxonomic group, larger species often seem "smarter" than smaller ones (e.g., parrots versus parakeets). Finally, since most evolutionary increases in relative brain size were not due to dwarfism (decreases in body size without concomitant decreases in brain size), we can conclude that most increases in relative brain size *also* involved significant increases in absolute brain size. For all these reasons, I personally find it simpler (and hence preferable) to think about changes in absolute, rather than relative, brain size.

However, there is an obvious difficulty with absolute brain size, particularly when we think about its relation to behavior. As I asked at the beginning of this chapter: is a whale really more intelligent than a human just because its brain weighs more? And is a cow more intelligent than a macaque (Aboitiz, 1996)? We may reply that it is difficult to compare the behavior (particularly the "intelligence") of humans and whales, or of monkeys and cows, but that is only half the answer. The other (obvious, yet easily forgotten) half is that absolute brain size is not the sole determinant of brain function. Connections, cytoarchitecture, and physiology all matter. As I reviewed above, a brain's degree of interconnectedness indeed correlates with its absolute size, but this correlation involves *average* interconnectedness. When it comes to connectional *specifics*, species differences abound (see Chapter 7). Similarly, *average* neuron size and density may be correlated with absolute brain size, but cytoarchitectural *details* often vary dramatically between different vertebrate lineages (see Chapters 6 and 10). In other words, different lineages may have undergone similar changes in absolute brain size, but these are always associated with at least some divergent changes in connectivity, cytoarchitecture, and physiology or function. Therefore, if we want to compare a cow with a macaque, we must acknowledge that primate brains also differ from artiodactyl brains in many aspects other than size (see Chapter 9) (Aboitiz, 1996). In other words, it is not only a cow's behavior that is difficult to compare to that of a macaque but also its brain. On the other hand, if we are comparing closely related species (e.g., macaques and chimpanzees) then differences in absolute brain size may well account for many differences in behavior because, as we saw in Chapter 2, the details of brain structure tend to be more similar in closely related species than in distant relatives.

Therefore, comparative analyses of relative and absolute brain size must be supplemented by careful study of the individual brain parts and their connectivity. The remaining chapters will elaborate upon this point.

5 Evolutionary Changes in Brain Region Size

The faculties and propensities have their seat, their origin in the brain.... Not only are the faculties distinct from and independent of the propensities, but the faculties and propensities are themselves essentially distinct and independent of one another: therefore, they must have their seat in distinct and independent parts of the brain.... From the differential apportioning of the different organs, and from their differential development, arise different forms of the brain.

—F. J. Gall, 1798

When, during the course of evolution, brains increase or decrease in overall size, what happens to their component parts? Vertebrate brains seem to be composed of structurally and functionally distinct cell groups, but can these cell groups vary in size *independently* of one another, or can they change only in a highly coherent manner? Because most species excel at only some behaviors (e.g., songbirds at singing), and because different brain regions tend to be causally linked to different behaviors, we might suppose that individual brain regions should evolve independently of one another. However, brains are complex, dynamic systems, and changes in the size of one brain region might well force changes in the size of another. Therefore, it remains an open question whether, or to what extent, individual brain regions evolve independently of one another. Similarly unresolved is what evolutionary changes in the size of individual brain regions mean in terms of function and behavior. We tend to think that "bigger is better" even in brains, but what is "better" behavior, and what definition of "bigger" do we want to use? Are we concerned with a region's absolute size, its size relative to body weight, or its size relative to other brain regions?

The first scientists to think seriously about these questions were Franz Joseph Gall and his fellow phrenologists (Figure 5.1). Their well-known practice of deducing someone's personality from the shape of their skull was undeniably ludicrous, but the early phrenologists were prescient in some respects. Gall himself was one of the first truly comparative psychologists, for he noted that different species (as well as different individuals within a species) exhibit different behavioral specializations or "talents." Among humans, he observed that geniuses rarely excel in all domains and that idiots are often skilled at some specific tasks (e.g., idiots savants). From that, he inferred that the mind is composed of several distinct and independent elements. In this, Gall clearly anticipated today's evolutionary psychologists who routinely compare the mind to a highly modular Swiss army knife (Barkow et al., 1992). Since Gall was convinced that the brain controlled the mind, he reasoned that the brain must also be highly modular. Again, this

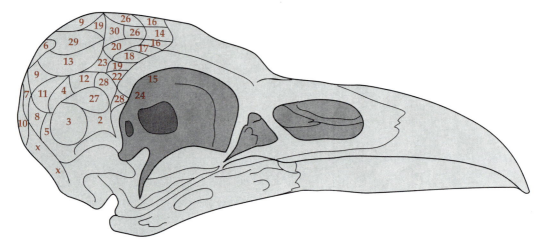

Figure 5.1 A Phrenological "Bust" of a Crow   Most phrenological busts depict human heads, but George Combe (1836) also mapped a crow's behavioral "faculties and propensities" onto its skull (most of them are numbered on the figure and listed in the key). Although most aspects of his drawing are pure fantasy, Combe obviously knew where the brain lies in a crow and, more remarkably, correctly identified many of the mental capacities crows typically possess. Key: 2 = alimentiveness; 3 and 5 = destructiveness; 4 = secretiveness; 6 = inhabitiveness; 7 = concentrativeness; 8 = attachment for life; 9 = adhesiveness; 10 and x = amativeness; 12 = acquisitiveness; 13 = cautiousness; 14 = individuality; 16 = sense of size; 17 = distance; 18 = geometrical sense; 19 = weight; 20 = locality; 22 = order; 23 = time; 26 = eventuality; 28 = musical talent; 29 = imitation. Note that, according to Combe, faculties 16, 18, 23, 28 and 29 are rare among nonhumans. (After Combe, 1836.)

argument reappears, with subtle modifications, in the writings of at least some evolutionary psychologists (Duchaine et al., 2001).

Of course, the early phrenologists knew very little about the brain; histological techniques had yet to be invented (see Chapter 2), and experimental studies of brain function (e.g., those of Bell and Magendie) were in their infancy. However, the early phrenologists did what they could: they compared the gross anatomy of various vertebrate brains and, in the process, discovered that the behavioral specializations of a species are reflected in the size of its cranial nerves. George Combe, one of the most prominent early phrenologists, summarized the findings thus:

> The nerve of smell is small in man and in the monkey tribe; scarcely, if at all, perceptible in the dolphin; large in the dog and the horse; and altogether enormous in the whale [sic: adult whales don't have an olfactory nerve; Combe probably meant the terminal nerve; see Buhl and Oelschlager, 1986.] and the skate, in which it actually exceeds in diameter of the spinal marrow itself. In the mole it is of extraordinary size, while the optic nerve is very small. In the eagle the reverse is observed, the optic nerve being very large and the olfactory small. Most of the quadrupeds excel man in the acuteness of their hearing, and accordingly it is a fact, that the auditory nerve in sheep, the cow, the horse, &c., greatly exceeds the size of the same nerve in man. In some birds of prey, which are known to possess great sensibility of taste, the palate is found to be very copiously supplied with nervous filaments. (Combe, 1836, p. 37)

From these observations the phrenologists inferred (rather audaciously) that each part of the brain "is a separate organ, because its size, *caeteris paribus*, bears a regular proportion to the energy of a particular mental faculty" (Combe, 1836, p. 111). Starting with this fundamental principle, the phrenologists built an inordinately complex system that became famous mainly for its (unfulfilled) promise of applicability: they argued that one could "read" a person's psychological constitution from the size of their various brain regions. Since the brain was not directly accessible, they used the skull as a substitute, arguing that skull shape accurately reflects brain shape. This last step in their argument was fatally flawed, since skull shape does *not* accurately predict the size of individual, functionally distinct brain regions. That is why phrenologists became the laughingstock of subsequent neurobiologists. Nonetheless, it is useful to recall some of the phrenologists' insights as we begin, once again, to contemplate how individual brain regions differ between species.

In the following sections, I first discuss whether individual brain parts evolve "mosaically" or "in concert" with one another. Although this question implies a strict dichotomy, the available evidence suggests that mosaic and concerted evolution both occur, albeit at different frequencies. Next, I review what we know about the relationship between a brain region's size and its functional role. That such a relationship exists is undeniable, but it is more delicate than one might at first surmise; particularly problematic is the relationship between region size and adaptation. In the chapter's final section, I use data on the avian hippocampus to illustrate how a brain region's absolute, proportional, and relative size can be related to a specific behavior, namely the ability to store food for later use. Overall, the chapter's message is that brain region size can change in a variety of ways and that different kinds of changes have different consequences for brain function. As long as those distinctions are drawn, and other possibilities for change are kept in mind (see Chapters 6 and 7), brain region size is an interesting and relatively simple parameter to compare across species.

Concerted versus Mosaic Evolution

Contemporary neurobiologists generally agree that vertebrate brains consist of numerous morphologically and functionally distinct cell groups (Kaas, 1982; Passingham et al., 2002), but do these cell groups vary in size *independently* of one another between individuals or across species? This question,

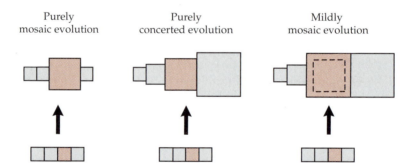

Figure 5.2 Mosaic, Concerted, and Mildly Mosaic Evolution   Three hypotheses of how individual brain regions scale with overall brain size, illustrated schematically. If brains evolved in a purely mosaic fashion, individual brain regions would change size independently of one another (the red square could increase in size without taking all of the other squares "along for the ride"; left diagram). On the other hand, if brains evolved in a purely concerted fashion, individual brain regions would scale according to a single overarching rule (increasing the size of the red square would force concomitant increases in the size of the other squares; middle diagram). Although this is a convenient dichotomy, real brains probably do something in between. They probably evolve in a mildly mosaic fashion, meaning that the size of individual brain regions may be subject to a general rule that is loose, rather than tight (increasing the size of the red square can be increased beyond what one would expect if the rule were tight; dashed line in rightmost diagram).

which Gall answered affirmatively has been debated extensively in recent years (Figure 5.2). Barbara Finlay and her colleagues, in particular, have argued that individual brain regions tend to evolve in concert with one another because they are constrained to do so by the rules of neural development (Finlay and Darlington, 1995; Finlay et al., 2001). In opposition, several authors have protested that brain evolution is far less constrained than Finlay et al. proposed and that, therefore, brain evolution is at least partially mosaic (Barton and Harvey, 2000; de Winter and Oxnard, 2001). This debate is important because, if brain regions do not evolve independently of one another, then natural selection cannot act independently on any one of them. More generally, the concern is that, if features are constrained to evolve in concert with one another, then it becomes difficult or impossible to single out any one feature as being adaptive in the sense of having been specifically "selected for." Thus, the debate about concerted versus mosaic brain evolution becomes part of a larger debate in evolutionary biology about the extent to which natural selection is free to shape organismal form (Gould and Lewontin, 1979; Alberch, 1982; Maynard Smith et al., 1985).

In my opinion, this debate is best resolved not by general arguments about the relative potency of selection or constraints, but by detailed examination of the covariance between different characters in evolution. If we find that some brain regions are tightly correlated in size across species, then some kind of constraints are likely to exist. And if constraints exist, then we can inquire whether they are developmental (due to the rules of development) and/or functional (due to how things have to work if they are to work at all). That is, we can build some sort of model to explain where the supposed constraints come from. We can also ask what mechanisms could, in theory, allow those constraints to be modified or breached. After that, we can gather data to discover whether, where, and how the supposed constraints have, in fact, been breached; generally speaking, such breaches constitute instances of mosaic evolution. Once we are reasonably clear on which constraints are (or were) at work and how those constraints were modified, we can investigate the functional consequences of having those constraints and breaking them. Only then, after the functional analysis has met with some success, can we hope to answer how natural selection has shaped our system of interest, the brain, to achieve functionally significant results. That, at least, is the general strategy I have adopted in the following discussion.

Concerted evolution

Virtually the entire debate about concerted evolution versus mosaic evolution is based on data collected by Heinz Stephan and his collaborators (Stephan et al., 1970; 1981; Baron et al., 1996), who measured brain and body size, as well as the size of more than a dozen different brain regions in numerous primates, insectivores, and bats. Stephan himself thought that the most important con-

Figure 5.3 Neocorticalization in Primates   Transverse sections through the telencephalon of (A) a lemur, (B) a macaque, and (C) a chimpanzee. Comparing these sections, it becomes apparent that the proportional size of the neocortex (its size relative to other brain regions) increases with absolute brain size. (After images on the Comparative Mammalian Brain Collections website [http://brainmuseum.org], from the University of Wisconsin–Madison Brain Collection.)

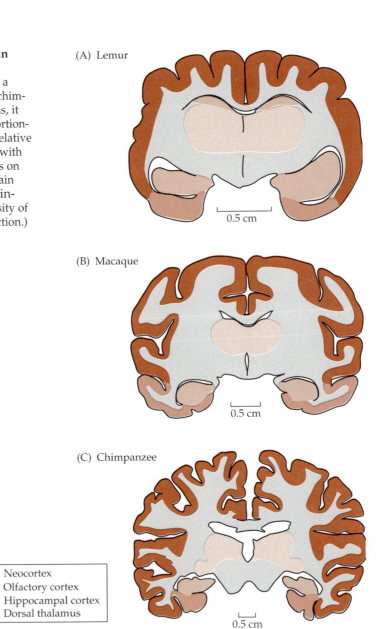

(A) Lemur

0.5 cm

(B) Macaque

0.5 cm

(C) Chimpanzee

- Neocortex
- Olfactory cortex
- Hippocampal cortex
- Dorsal thalamus

0.5 cm

clusion to emerge from his data set was that the proportional size of the neocortex increases along an evolutionary scale from insectivores over prosimians and monkeys to apes and humans (Figure 5.3) (Stephan and Andy, 1970). Such *scala naturae* arguments have ceased to be acceptable (even if they are suffused with caveats) but we can look at Stephan's data without invoking the *scala naturae*. Plotting neocortex size versus overall brain size in double logarithmic coordinates (Figure 5.4A), we see the data fall on a line of slope

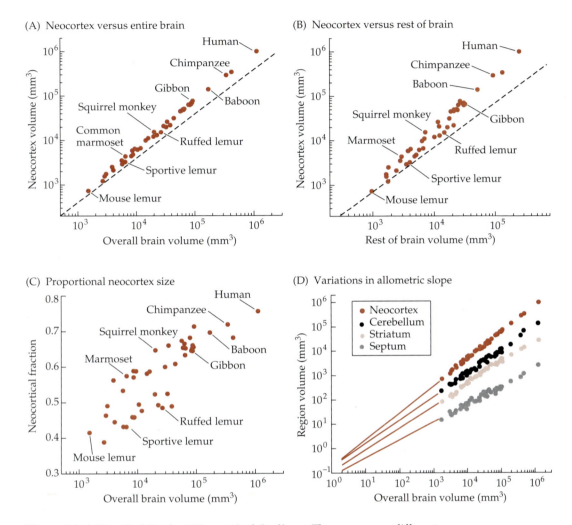

Figure 5.4 A Detailed Look at Neocortical Scaling   There are many different ways to examine neocortical allometry. (A) Plotting neocortex volume against absolute brain size, we see that it scales with a slope greater than one (the dashed line indicates isometry). (B) Essentially the same result obtains when we plot neocortex volume against the size of the remaining brain (though the slope is steeper and the scatter more pronounced). These results imply that the percentage of the brain that is occupied by the neocortex (i.e., the brain's neocortical fraction) increases as overall brain size goes up. This is shown graphically in panel (C). Also apparent from this plot is that lemurs and other prosimians tend to have lower neocortical fractions at a given overall brain size than simians (i.e., monkeys and apes; see Figure 5.7 and Chapter 9). (D) Compared to cerebellum, striatum, septum, and most other brain regions, the neocortex scales against overall brain size with the steepest slope. This is most readily apparent if we extend the best-fit lines toward the origin (red lines). This regional variation in allometric slope is at least partially explained by the rule that "late equals large" (Finlay and Darlington, 1995; see Figure 5.6). (Data from Stephan et al., 1981.)

~1.1 (Passingham, 1975). As we learned in Chapter 4, this means that neocortex size scales allometrically, rather than isometrically, with overall brain size (in this case the allometry is said to be positive because the slope is greater than one). We ought to be suspicious of this correlation because the neocortex is a fairly large structure and therefore contributes significantly to overall brain size (Deacon, 1990), but if we correct for this problem by correlating neocortex size against the size of the rest of the brain (see Figure 5.4B), a remarkably tight correlation remains. Even human neocortex is roughly as large as we would expect it to be, given the size of our remaining, non-neocortical brain. Therefore, Stephan's neocorticalization hypothesis can be rephrased as follows: the proportional size of the neocortex scales allometrically with absolute brain size (see Figure 5.4C) and, to the extent that absolute brain size increased during primate evolution (see Chapters 4 and 9), proportional neocortex size likewise increased.

But if the proportional size of the neocortex increased during primate evolution, what happened to all of the non-neocortical regions? Obviously, as the proportional size of neocortex increases, the proportional size of all non-neocortical regions must, at least on average, decrease; this is not particularly interesting. More interesting is that, when we plot the size of various individual brain regions against absolute brain size in double logarithmic coordinates (see Figure 5.4D), the data tend again, as in the case of neocortex, to form straight lines (technically, the lines cannot all be straight, but that need not concern us here; see Fox and Wilczynski, 1986). Moreover, the slopes of these lines differ from one another, which means that the proportional size of most major brain regions, not just of neocortex, changes predictably with absolute brain size (Finlay and Darlington, 1995). To quantify this degree of predictability, one can correlate the size of each structure against the size of every other structure and perform a "principal components analysis" on the resultant data matrix (Sacher, 1970). The details of this procedure are complex, but the result is simple: If we focus just on primate and insectivore brains, more than 85% of the variation in the size of individual brain regions can be predicted from absolute brain size (Jerison, 1989). If we add bats into the mix, and restrict our analysis to the neocortex, striatum, diencephalon, cerebellum, schizocortex, hippocampus, septum, midbrain, medulla, olfactory cortex, and the olfactory bulb, then the result is even more dramatic: more than 96% of the total variance is predicted by absolute brain size (Finlay and Darlington, 1995). In other words, if I gave you an unknown primate, insectivore, or bat brain, you would be able to predict, with a fair degree of accuracy, how large the major individual components of this brain are likely to be. Such predictions should be impossible if brains evolved in a purely mosaic fashion (see Figure 5.2). Therefore, we can conclude that brains must evolve, at least to some degree, concertedly.

This conclusion is counterintuitive because it suggests that evolutionary changes in the size of individual brain regions are, in some sense, *due* to changes in absolute brain size rather than the other way around, as we tend to think. It seems akin to arguing that giraffes have long necks simply *because* they have large bodies (an unlikely story; see Simmons and Scheepers, 1996). In order to allay this concern, we need to know what kind of mechanism could accomplish the interregional coordination that would have to exist if brain regions really evolve in concert with one another. One possibility is that axonally interconnected brain regions may influence each other such that an increase or decrease in the size of one leads epigenetically (i.e., without requiring additional mutations) to correlated changes in the size of its target and/or input structures. For example, a mutation that reduces eye size might cause retinal target structures to degenerate (decrease in size) because those retinal targets need trophic support from the retina to avert cell death (Katz et al., 1981). Reductions in the size of retinal target structures might, in turn, lead to decreases in the size of other visual system structures, and these might lead to yet further changes, which eventually end up reducing overall brain size. Such "epigenetic cascades" probably do occur (see Chapter 7) but they seem not to cascade very far (Finlay et al., 1987). Moreover, epigenetic cascades must occur relatively late in development, after axonal connections have formed and neurogenesis has largely ceased. Therefore, epigenetic cascades can modulate only cell death, not neurogenesis. Since the allometric differences in region size that we discussed above range over several orders of magnitude, whereas naturally occurring cell death rarely involves more than 50% of a region's cells, the allometric differences cannot be due solely to phylogenetic modulations of cell death. It would be implausible, for example, to argue that mouse lemurs have a 1,300-fold smaller neocortex than humans (see Figure 5.4A) simply because their neocortex undergoes more cell death. It makes far more sense, instead, to say that the neocortex is larger in humans than in mouse lemurs because, in humans, more neocortical cells are generated. Which leaves us with the following question: How can neurogenesis be coordinated between different brain regions such that the observed allometric correlations in final region size (see Figure 5.4D) are obtained?

An ingenious answer to that question was proposed by Finlay and Darlington (1995), who argued that different regions enlarge at different allometric slopes because, when brains increase in overall size, the order in which different regions are "born" is preserved. This explanation seems obscure at first, but is actually fairly straightforward. Consider first the data on neurogenesis. As we saw in Chapter 4, neurons are said to be born when they have completed their last round of cell division and, therefore, cease to be precursor cells. Now, it is well established that, within a given species, neurons in different brain regions tend to have different birth dates (with the most anterior and dorsal regions generally, though not always, being born after those that are located more posteriorly and ventrally) (Finlay et al., 2001). When Finlay and Darlington (1995) compared those spatiotemporal maps, or schedules, of neurogenesis across species, they found that the schedules tend to be

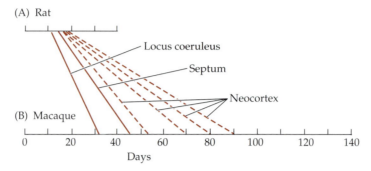

Figure 5.5 Changing Schedules of Neurogenesis According to Barbara Finlay and her collaborators, neurogenetic schedules—the sequence of dates at which different embryonic brain regions cease neuronal cell division—tend to be stretched rather than rearranged as absolute brain and body size increase. Shown here is a comparison between rats and macaques for a few key neurogenetic events. Obviously, macaques have a longer period of embryonic development (~40 days versus ~140 days), but the sequence in which the major brain regions are "born" is similar between the two species. Locus coeruleus is born first, followed by the septum and the various neocortical layers (dashed lines). If more species and structures were included, some sequence rearrangements would become apparent, but the similarity in birth order would remain statistically significant. (After Clancy et al., 2001.)

stretched or squeezed, but not fundamentally rearranged (Figure 5.5). That is, although large mammals tend to have longer gestation periods and longer absolute periods of brain growth than small mammals, the general sequence of regional birth dates is preserved. Thus, it seems that, when evolution changes absolute brain size, it simply expands or contracts a highly conserved schedule of neurogenesis. We cannot yet specify precisely how neurogenetic schedules are tweaked at the molecular level (though we know some candidate molecules; see Bond et al., 2002) but that is not critical to the argument. What is important is that, in the species that Finlay and Darlington examined, the order in which brain regions are born correlates remarkably well with how rapidly they enlarge with increasing overall brain size: the later a region is born, the larger it becomes as overall brain size increases; as they put it, "late equals large." But why *should* stretching or squeezing a neurogenetic schedule cause the proportional size of individual brain regions to change predictably with overall brain size? To answer that, we must delve more deeply into the logic of Finlay and Darlington's argument.

Consider the following idealized scenario (Figure 5.6). Imagine 3 neuronal precursor regions (*a*, *b*, and *c*) that each contain 2 precursor cells at the time that they are first formed (during the phylotypic period, for instance). Imag-

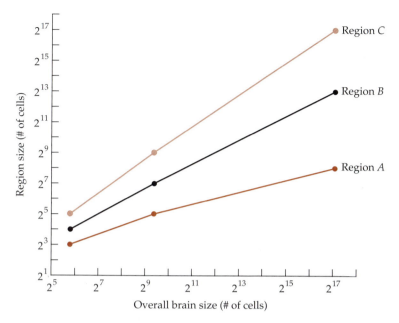

Figure 5.6 Finlay and Darlington's Model of "Late Equals Large" The data points represent three different brain regions that are born at different times (*A* is born first, *C* last) from three different species with small, intermediate, and large brains (assumed to consist only of regions *A*, *B*, and *C*). Given the assumptions of the thought experiment, the allometric lines for the three brain regions end up having different slopes, with the last-born structure having the highest slope. The lines are not all straight because all three allometric lines must sum to a straight line with a slope of one (which would be impossible if all the lines were straight).

ine further that the cells in regions *a*, *b*, and *c* undergo 2, 3, and 4 rounds of symmetrical cell division, respectively, before they become post-mitotic and differentiate into the adult regions *A*, *B*, and *C*. According to this scenario, regions *A*, *B*, and *C* are born sequentially and, in adulthood, contain 2^3, 2^4, and 2^5 cells, respectively. Now, imagine a second scenario—a second species if you will—in which the neurogenetic schedule is doubled in length but the rate of cell division is maintained. In this case, regions *a*, *b*, and *c* would undergo 4, 6, and 8 rounds of cell division, yielding adult regions with 2^5, 2^7, and 2^9 cells, respectively. If we then compute the proportional size of each adult brain region in each of the two scenarios (or species), we see that the later a region was born, the more it increased in proportional size as overall brain size increased. The fractional size of region *C*, for instance, increased from 57% to 76%, while that of region *A* decreased from 14% to barely 5%. And if we extend our thought experiment by doubling the length of the neurogenetic schedule again, we can plot the results in double-logarithmic coordinates (see Figure 5.6) and see that the data for each brain region form a series of (almost) straight lines with different slopes. Most important, the data reveal that the later a region is born, the steeper is its allometric slope. This result follows logically from how the scenario, or model, was set up. Therefore, Finlay and Darlington's hypothesis that structures with late neuronal birth dates must enlarge disproportionately as overall brain size increases, that "late equals large," is at least logically sound. At this point, however, we must ask the crucial empirical question: How closely does the development of real brains obey the assumptions we built into our idealized scenario?

To date, the relevant data are scarce but supportive. Specifically, Clancy et al. (2001) have shown that the sequence in which brain regions are born is indeed maintained fairly well (though not perfectly) across various species that differ dramatically in overall brain size. In contrast to the model, however, the neurogenetic schedules are not stretched or squeezed uniformly in real mammals; instead, the late portions of longer schedules are expanded disproportionately. Nor is the rate of precursor cell division constant across species, as our model assumes; in macaques, for example, neocortical precursor cells divide 2–5 times more slowly than they do in mice (Kornack and Rakic, 1998). These deviations from the model's assumptions are interesting and worthy of further study, but they do not alter the logic of Finlay and Darlington's hypothesis. They affect merely the magnitude of interregional differences in allometric slope. Even if neurogenetic schedules are stretched nonuniformly, and even if rates of cell proliferation are slower in larger species, it still must be true that "structures whose neurons are born late get disproportionately large as absolute brain size increases" (Finlay et al., 2001, p. 272).

So, what kind of developmental data would falsify Finlay and Darlington's "developmental constraint hypothesis"? As far as I can tell, there are three kinds of deviations from our model that would spell trouble for Finlay and Darlington's proposal, namely: (1) major species differences in the proportional size of different precursor regions at the time that they are first set up; (2) major regional differences in the rate and/or onset of precursor cell division; and (3) dramatic species differences in the sequence of regional birth dates. In the data set that Clancy et al. examined, there is little evidence for such fatal deviations, but neurogenetic schedules are known for only a handful of species, and *minor* deviations are certainly evident even in this limited data set (Barton, 2001; Clancy et al., 2001). Therefore, Finlay and Darlington's hypothesis of late equals large is supported by the currently available developmental data, but not immune to challenges. Indeed, as I shall now review, it has been challenged repeatedly.

Mosaic evolution

The argument that brains evolve mosaically generally begins with a critique of Finlay and Darlington's developmental constraint hypothesis. None of the critics advocate what I call purely mosaic evolution (see Figure 5.2), since they all acknowledge that "there may be some constraints on evolutionary change in individual neural systems" (Barton and Harvey, 2000). Since Finlay and Darlington are similarly circumspect and do not advocate purely concerted evolution, the debate really revolves around the *degree* to which evolutionary change is constrained by conserved rules of development. Finlay and Darlington think that developmental constraints are strong and tend to overpower mosaic evolution, whereas Barton and Harvey (2000) argue that "the constraints are evidently insufficiently tight to prevent . . . evolutionary change in individual neural systems" (pp. 1057–1058). How can two groups of authors look at essentially the same data set (Stephan et al., 1981) and arrive at such divergent conclusions? To answer that, let us review some of the principal arguments against concerted evolution and, hence, in favor of mosaic evolution.

Looking closely at how neocortex volume scales against remaining brain volume, Barton and Harvey (2000) observed that the allometric lines for different taxonomic groups have almost identical slopes but different y-intercepts: the line for simians (monkeys and apes) lies slightly above the prosimian line, which lies well above the line for insectivores (Figure 5.7). These differences in y-intercepts, or grade shifts, appear small in double logarithmic plots but are of considerable magnitude. The neocortex of pygmy marmosets, for example, is about 10 times as large as the neocortex of common tenrecs, even though the non-neocortical brain volume in the two species is nearly identical. Similarly, the neocortex of a cotton-top tamarin is about 9 times as large as the neocortex of a solenodon (an odd, shrew-like creature), even though both species have very similar non-neocortical brain volumes. Since these differences are significantly larger than the 2- to 3-fold difference that Finlay and Darlington (1995) allowed for in their developmental constraint hypothesis, we may view them as clear evidence of mosaic evolution. On the other hand, we must admit that the data points for tenrecs and solenodons lie well below the best-fit line for all insectivores and that, in general, 9- to 10-fold differences are rare in Stephan's data set. Moreover, we may note that prosimians and tree shrews are intermediate between simians and insectivores in terms of both phylogenetic position and degree of neocorticalization (see Figure 5.7; for details on primate phylogeny, see Chapter 9). This suggests that the difference in neocorticalization between simians and insectivores is the result of at least two smaller but cumulative grade shifts, each of which remains at least marginally consistent with the developmental constraint hypothesis. So, depending on how we look at it, neocortical evolu-

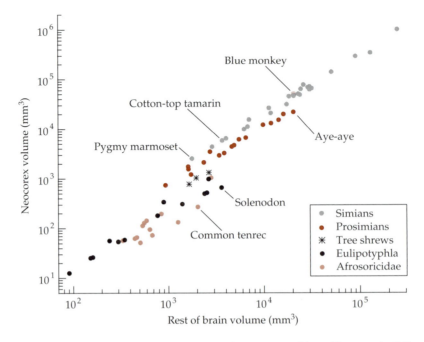

Figure 5.7 Grade Shifts in Proportional Neocortex Size   Taxonomic differences in how the neocortex scales against the remaining brain provide some evidence for mosaic evolution. In simians, the neocortex is roughly twice as large, relative to the remaining brain, as it is in prosimians (e.g., lemurs, galagos, and the aye-aye), and it is larger in prosimians than in insectivores. The latter include the tree shrews (which are closely related to primates), the eulipotyphla (which consist of shrews, hedgehogs, moles, and solenodons) and the afrosoricidae (i.e., tenrecs and golden moles). The eulipotyphla and afrosoricidae are not close relatives and, therefore, probably evolved insectivory independently of one another (Madsen et al., 2001; Waddell et al., 2001). (After Stephan et al., 1981; Barton and Harvey, 2000.)

tion can be consistent with either mosaic or concerted evolution. In other words, the discovery of some isolated instances of mosaic species differences seems insufficient to falsify the developmental constraint hypothesis. I shall return to this idea later in the chapter but, first, let us look at structures other than the neocortex. To what extent do they evolve mosaically?

Consider first the olfactory bulb. In humans the main olfactory bulb occupies only about 0.009% of the total brain volume, whereas in long-eared desert hedgehogs it occupies approximately 10%. To some extent, this difference in proportional olfactory bulb size is not surprising. Since we already saw that humans have proportionately more neocortex than insectivores (such as hedgehogs), we should expect humans to have proportionately smaller olfactory bulbs than insectivores, even if the olfactory bulbs per se had not changed in size. After all, the total of all proportional region sizes must add up to 100%. So, how can we determine whether the olfactory bulb evolved mosaically, independent of the neocortex? One way to solve this puzzle is to plot olfactory bulb size against the size of some reference structure that did not evolve mosaically (in the taxonomic group under consideration). Picking such a reference structure is difficult (see Mangold-Wirz, 1966) but let us assume, for simplicity's sake, that the medulla can serve as a suitable reference in Stephan's data set. Now, if we plot olfactory bulb size against medulla size in double-logarithmic coordinates (Figure 5.8A), we see that, at any given medulla size, simian olfactory bulbs are 3–5 times smaller than the olfactory bulbs of prosimians. Since insectivores have olfactory bulbs that are even larger than those of prosimians, relative to medulla size, we can conclude that simians most likely underwent a major phylogenetic reduction in olfactory bulb size. Even Finlay and Darlington (1995) recognized that this reduction in the size of the olfactory bulb is too large to fit into the simple one-factor model of concerted evolution that I described earlier (the one factor being absolute brain size). There is little doubt, then, that simian olfactory bulbs evolved mosaically relative to the rest of the brain.

Finlay and Darlington (1995) go on to suggest that several other limbic system structures, notably the olfactory cortex (or piriform cortex) and the hippocampus, evolve in concert with the olfactory bulb (Finlay et al., 2001). In support of this hypothesis we may note that the size of the olfactory cortex correlates rather tightly with olfactory bulb size in prosimians and all insectivores (see Figure 5.8B). This correlation breaks down, however, when we look at simians, which have a larger olfactory cortex than one would predict given the size of their olfactory bulb. Apparently, the constraint that controls the size relationship between olfactory cortex and olfactory bulb in most mammals was broken, or at least altered, in simians. Turning to the hippocampus, we see that its size, relative to the rest of the brain, is indeed reduced in simians (see Figure 5.8C), as we would expect if it evolved in concert with the olfactory bulb. However, if we plot hippocampus size against medulla size (see Figure 5.8D), we see that the hippocampus is only slightly smaller, relative to medulla size, in simians than in prosimians or insectivores. If, in addition, we plotted hippocampus size against olfactory bulb size for all primates and insectivores, we would see that the simian hippocampus is considerably larger than we would expect given the size of the simian olfactory bulb. These observations indicate that, in primates, both olfactory cortex and the hippocampus evolved less in concert with the olfactory bulb than with the remaining (non-olfactory) brain (see also Stephan et al., 1988). To put it another way: the primate olfactory bulb evolved more mosaically than two other major components of the limbic system. This, in turn, suggests that some components of the limbic system became developmentally and/or functionally uncoupled from one another as simians evolved. Such phylogenetic

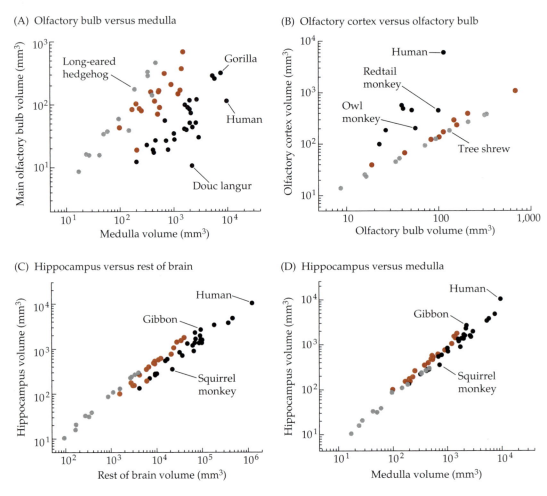

Figure 5.8 Mosaic Evolution in the Limbic System The mammalian limbic system includes some structures that probably evolved mosaically relative to one another and the rest of the brain. (A) The main olfactory bulb is significantly smaller in simians (black data points) than in prosimians (red data points) or eulipotyphlan insectivores (gray data points), relative to what one would expect from the size of their medulla. (B) The size of the olfactory cortex is correlated tightly to the size of the olfactory bulb in insectivores, tree shrews, and prosimians, but in simians, the olfactory cortex is larger than one would expect given the size of simian olfactory bulbs; the functional significance of this fascinating species difference remains unknown. (C) The hippocampus is smaller, relative to the remaining brain, in simians than in prosimians or eulipotyphlan insectivores. However, since the neocortex is part of the remaining brain, this increase in relative hippocampal size could be an artifactual consequence of simians having a proportionately larger neocortex (see Figure 5.7). (D) Indeed, when hippocampal volume is plotted against the volume of the medulla alone, the apparent grade shift in hippocampus size disappears (or is greatly reduced). Therefore, the case for mosaic evolution is easier to make for the olfactory bulb than for the hippocampus or the olfactory cortex. (Data from Stephan et al., 1981.)

decoupling may be associated with changes in neural connectivity (see Chapter 7) but, at this point, no specific hypotheses have been proposed to explain why some parts of the primate limbic system evolve more mosaically than others (see also Barton et al., 2003).

Going beyond Stephan's data set but staying with mammals, we find additional evidence of mosaic evolution. In ground squirrels, for example, the superior colliculus (the major visual region in the midbrain) is approximately 10 times larger than in laboratory rats, even though brain and body sizes are similar in the two species (Figure 5.9) (Kaas and Collins, 2001). And in blind mole rats, the superior colliculus is roughly 38 times smaller than in hamsters, even though both species are equally large and probably have similar overall brain sizes (Cooper et al., 1993b). These differences reflect obvious behavioral specializations, since squirrels are far more visual than rats and mole rats are virtually blind, but let us postpone all questions about functional correlates. For now let us focus on a purely morphological question: Are these 10- to 38-

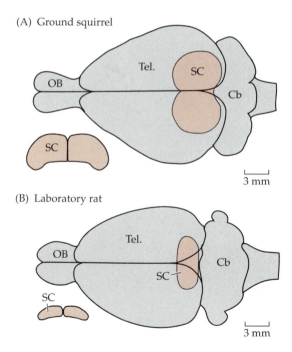

(A) Ground squirrel

(B) Laboratory rat

Figure 5.9 Mosaic Evolution of the Mammalian Midbrain Roof   The midbrain's superior colliculus is approximately 10 times larger in (A) ground squirrels than in (B) laboratory rats, even though overall brain size is similar in these two species. Brains are shown from a dorsal perspective with the superior colliculus (SC) visible through the overlying telencephalon (Tel.). Inserts show the approximate size and shape of the superior colliculus in transverse section. Other abbreviations: OB = olfactory bulb; Cb = cerebellum. (After Kaas and Collins, 2001.)

fold differences so rare that they fail to invalidate Finlay and Darlington's claim that mammalian brains evolve *primarily* by concerted evolution? As far as I can tell, they are indeed. Two- to 3-fold differences are far more common, and most differences are smaller still. For example, the cerebellum is significantly larger in dolphins or nonhuman apes than in monkeys, but only by a factor of 1.5 (Rilling and Insel, 1998; Marino et al., 2000). Differences of this magnitude do not qualify as evidence for mosaic evolution as Finlay and Darlington defined it (recall that they allow for changes by a factor of 2–3 and I accept that criterion here, even though a more precise statistical criterion is clearly desirable). Even the inferior colliculus, which we would expect to be enlarged in echolocating bats because it is heavily involved in processing auditory information, is only about 3 times larger in the echolocating Microchiroptera than in the (generally) nonecholocating Megachiroptera (Baron et al., 1996). If we look at the entire subcortical auditory system (Glendenning and Masterton, 1998), we do find some 5- to 6-fold differences in size (e.g., between mole and mouse opossum; see Figure 4.2), but these differences are between distantly related species. If we restrict our comparisons to within orders, 2- to 3-fold differences predominate. Therefore, I agree with Finlay and Darlington that, in mammals, mosaic evolution is relatively rare.

What about nonmammals? Here we see several spectacular examples of mosaic evolution, notably in the medulla. The neurons that innervate the electric organ in *Torpedo* rays, for example, occupy an astounding 60% of the entire brain, far more than the homologous brain regions in other cartilaginous fishes (Roberts and Ryan, 1975). Among teleost fishes, goldfishes and carps (i.e., cyprinids) have huge vagal lobes that are used to taste and process food inside the mouth (Finger, 1988). Catfishes are closely related to cyprinids but have much smaller vagal lobes (Figure 5.10); instead, they have large facial lobes that process gustatory information from the barbels (fleshy filaments hanging from the mouth) and the trunk. Catfishes also have a large electrosensory lateral line lobe, which processes electrosensory information and has no apparent homologue in cyprinids (Finger, 1986). This electrosensory lateral line lobe is even larger in gymnotoid electric fish, which can not only sense electric fields but also generate them. Compared to catfishes and cyprinids, the vagal and facial lobes of gymnotoids are small and poorly differentiated from one another (see Figure 5.10). Thus, although precise volumetric measurements have not been done, it seems safe to say that these medullary lobes evolved mosaically (see also Leonard and Willis, 1979; Braford, 1986). More rostrally within the brain, we also see some major species differences in region size. In birds, for example, the lateral pallium is much larger than it is in mammals (see Chapter 8), and in teleost fishes, the posterior tubercular region is enormous relative to its size in other vertebrates (see Chapter 6). Perhaps the most impressive example of mosaic evolution in

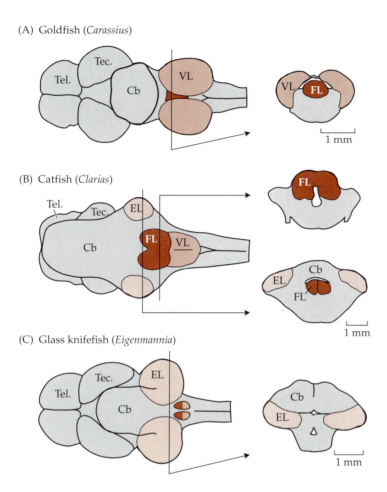

(A) Goldfish (*Carassius*)

(B) Catfish (*Clarias*)

(C) Glass knifefish (*Eigenmannia*)

Figure 5.10 Mosaic Evolution of the Medullary Lobes in Teleosts
(A) Goldfishes and carps have large facial lobes and huge vagal lobes (FL and
VL, respectively) that serve as the main medullary centers for processing gustatory information (see Chapter 6). (B) In catfishes, the facial lobes are larger
than they are in goldfishes (relative to the rest of the brain), but the vagal
lobes are smaller. Catfishes also have an electrosensory lateral line lobe (EL)
that is not present in goldfishes or other nonelectroreceptive teleosts. (C) The
glass knifefish *Eigenmannia* and other gymnotoid electric fishes have small
vagal and facial lobes (one may not even want to call them lobes), but their
electrosensory lateral line lobes are enormous. In A–C, dorsal views of the
brain are on the left, transverse sections on the right. Section levels are indicated by arrows and lines. Other abbreviations: Cb = cerebellum; Tel. = telencephalon; Tec. = Tectum. (After Nieuwenhuys et al., 1998; Bullock and
Heiligenberg, 1986.)

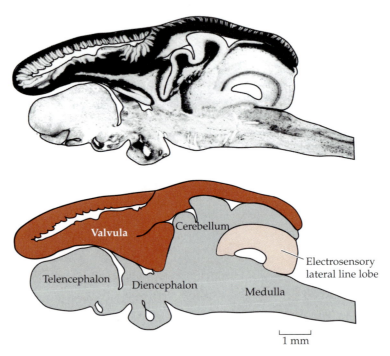

Figure 5.11 The Cerebellar Valvula of Mormyrids The best example of mosaic evolution is the cerebellar valvula of mormyrid electric fishes. It lies just rostral to the main body of the cerebellum in all teleosts but is enlarged beyond all allometric expectations in mormyrids. These fishes also have large electrosensory lateral line lobes that are structurally and functionally similar to those of catfishes and gymnotoids (see Figure 5.10) but evolved independently of them. (From Bell and Szabo, 1986.)

vertebrate brains is the "valvula" of mormyrid electric fishes (Nieuwenhuys and Nicholson, 1969; Meek et al., 1992). This valvula is a relatively small rostral extension of the cerebellum in most teleosts, but covers virtually the entire brain of mormyrids (Figure 5.11); its size has not been measured but surely it exceeds all allometric expectations.

Thus, there clearly are some spectacular examples of mosaic evolution in nonmammals. Unfortunately, we cannot say anything about the relative frequency of mosaic evolution in nonmammals because no comprehensive analyses of either mosaic evolution or concerted evolution have been published on nonmammals. A recent brief report by Iwaniuk et al. (2004) suggests that avian brains evolved mosaically, but my own look at the avian data (especially Portmann, 1947) indicates that, to some extent at least, bird brains also evolved concertedly (Figure 5.12). Similarly, cartilaginous fish brains show some concerted evolution (see Demski and Northcutt, 1996) even though the Torpedo's

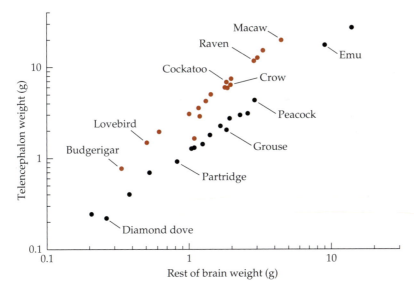

Figure 5.12 Concerted and Mosaic Evolution in Bird Brains   Within any one order of birds, telencephalon size correlates tightly with the size of the remaining brain, indicating concerted evolution. However, in parrots and corvids (red data points) the telencephalon is 2–3 times larger, relative to the remaining brain, than one would expect on the basis of data from ratites, galliformes, and columbiform birds (black data points). Since parrots and corvids (e.g., jays and crows) are not close relatives of one another, these data suggest that the proportional size of the telencephalon increased at least twice among birds. Curiously, the allometric line for parrots and corvids has the same slope as the data for other birds. (Data from Portmann, 1947.)

electric lobes clearly evolved mosaically. Far more detailed studies are required, however, to determine the relative frequencies of concerted and mosaic evolution in these and other nonmammalian brains. Also interesting to contemplate is that in most nonmammals neurogenesis is not limited to the embryonic period, as it is in most mammalian brain regions, but continues into adulthood (see Zupanc, 2001; Doetsch, 2001; Reh, 2001). This species difference in development raises the possibility that nonmammalian brains may sometimes evolve mosaically by modifying rates of adult neurogenesis (or death). To date, we have few data relevant to that hypothesis.

Toward a synthesis

Based on the data discussed in the previous section, it seems obvious that mosaic evolution of individual brain regions has, at least sometimes, occurred. But does it follow that the developmental constraint hypothesis is falsified, and that concerted evolution is not a real phenomenon? Surely not, for despite the examples of mosaic evolution we have discussed, the great majority of species differences in region size do seem to fall well within the 2- to 3-fold limits proposed by Finlay and Darlington (1995), especially if we restrict our comparisons to species within the same order (e.g., primates, bats, or parrots; see Figure 5.12). Therefore, concerted evolution is a general principle that holds most of the time but not always (see Chapter 1). This is not to say that mosaic evolution was not important in brain evolution. As the vast literature on key innovations and mass extinctions exemplifies (Lauder and Liem, 1989; Courtillot, 1999), even rare events can be of profound significance in evolution. Indeed, I suspect that mosaic evolution is more likely than concerted evolution to cause major changes in brain function and, therefore, more likely to open up new ecological niches and possibilities for further change. If this is true, then mosaic evolution should be more common between classes than between orders, more common between orders than between families, and so forth. Because the frequency of mosaic evolution seems, indeed, to increase with taxonomic level (see Figure 5.7), mosaic evolution was probably at least as important as concerted evolution when we consider vertebrate brain evolution overall. In any case, the preceding argument reveals that the debate about whether brain evolution is concerted *or* mosaic essentially dissolves into an empirical question about the relative frequencies and importance of both.

This probabilistic view of mosaic and concerted evolution is a step forward, but it still tends to reinforce the notion that brain evolution is either concerted or mosaic, never both. This is somewhat misleading, since the existence of mosaic evolution does not imply the total absence of constraints. Instead, it is better to think of mosaic evolution as occurring against a background of at least some conserved constraints. This implies that mosaic evolution interacts with the constraints that cause concerted evolution but what, precisely, is the nature of those interactions? To answer this difficult question, we must think more deeply about the mechanisms that underlie both mosaic evolution and concerted evolution. According to Finlay and Darlington (1995), concerted evolution occurs when developmental schedules are lengthened, but what developmental changes drive mosaic evolution?

This question has rarely been addressed explicitly and few relevant data exist. However, if we recall the conditions under which our model of late equals large is falsified, we can make some educated guesses about the kinds of evolutionary changes in development that would lead to mosaic evolution. For example, any regions that are born earlier than expected in some species should end up being smaller than expected in adults; evidence to that effect was recently provided by Clancy et al. (2000) in their analysis of human brain development. In addition, any evolutionary changes in the initial size of a precursor region (i.e., shifts in the expression boundaries of genes that give the region its identity) should lead to deviations from concerted evolution, since the model assumes that initial size remains constant. Indeed, I suspect that such a change in initial precursor size underlies the observed grade shift in relative telencephalon size between parrots (or songbirds) and most other birds (see Figure 5.12). Unfortunately, we currently have no data relevant to this hypothesis, and alternative hypotheses are possible. Still, such hypotheses are testable and merely contemplating them reveals that evolutionary changes in a single developmental parameter can yield mosaic evolution even as other developmental constraints remain in place. Ultimately, further work along these lines should help to clarify the complex interplay between constraints and change, between concerted and mosaic evolution.

Functional Correlates of Brain Region Size

Thus far, I have discussed how the size of individual brain regions changed during the course of vertebrate evolution; now, let us ask the complementary functional question: how does changing a region's size affect brain function and/or animal behavior? As we saw in this chapter's introduction, the early phrenologists argued that a region's size relative to other regions (its proportional size) reflects the organism's capacity to perform whatever function is localized to the region, all else being equal. Later studies confirmed that different brain regions do tend to perform different functions, which means that it is indeed reasonable to say that, in general, different functions are localized differentially within the brain (Young, 1990). Moreover, physiological investigations showed that in species with highly developed sensory or motor abilities, the corresponding neocortical regions do tend to be enlarged (relative to their size in less specialized species). Edgar Adrian, for example, showed that in the somatosensory cortex of pigs the highly sensitive snout is dramatically overrepresented, relative to its representation in most other mammals (see Finger, 2000). Similarly Welker and Campos (1963) showed that raccoons, which have unusually agile paws, also have an unusually large representation of the paws in their sensory and motor cortices. As this kind of evidence accumulated, an increasing number of neuroscientists became convinced that, somehow, bigger is better in the brain. Yet, hardly anyone tried to explain why bigger should be better and what, precisely, those terms mean. Only Harry Jerison (1973) tried to formalize the idea that bigger is better with his "principle of proper mass." Therefore, let us begin our discussion of functional correlates with a review of how Jerison defined and used the principle of proper mass.

The principle of proper mass

Jerison originally defined his principle of proper mass thus: "The mass of neural tissue controlling a particular function is appropriate to the amount of information processing involved in performing the function" (Jerison, 1973, p. 8). This implies that, given a region's mass, we should be able to calculate how much information processing is required to perform its function or, conversely, that we can predict a region's size from its function. But making such calculations is terribly difficult even for relatively simple organs (e.g., the lung; see Taylor and Weibel, 1981) and, to date, no one knows how to make such calculations for the brain. Therefore, I suspect that Jerison intended something else. Perhaps he meant that *differences* in neural mass correlate (at least approximately) with *differences* in information processing capacity. If that is true, then who and what is being compared? Jerison himself provided an answer when he wrote:

> This [principle of proper mass] implies that in comparisons among species the importance of a function in the life of each species will be reflected by the absolute amount of neural tissue for that function in each species. It also implies that, within a species, the relative masses of neural tissue associated with different functions are related to the relative importance of the functions in the species. Among the mammals, "visual" species have enlarged superior colliculi and an enlarged visual cortex, and "auditory" species have enlarged inferior colliculi and an enlarged auditory cortex. (Jerison, 1973, pp. 8–9)

According to this passage, the principle of proper mass can be used in both inter- and intraspecific comparisons, and the quantity being compared is absolute region size (which amounts to proportional size in intraindividual comparisons). That much is consistent with Jerison's original definition. Curiously, however, Jerison no longer links region size to information processing demands, as he did in his original definition, but to "the importance of a function in the life of a species." Are those two variables equivalent? I think not. Consider, for example, the suprachiasmatic nucleus. It is one of the brain's smallest cell groups, relative to overall brain size, and performs very little information processing, since its neurons measure mainly diffuse light intensity. Yet, this nucleus is very important in the life of virtually all vertebrates, since it is the central component of the "biological clock" that times many vital activities. Indeed, the suprachiasmatic nucleus is among the most conservative of all brain structures: even blind mole rats retain circadian rhythms and a sizeable suprachiasmatic nucleus (Cooper et al., 1993a). Thus, small size may be linked to small processing capacity (as per Jerison's definition), but even small brain regions may be quite important in an organism's life. Therefore Jerison's statement that "the importance of a function in the life of each species will be reflected by the absolute amount of neural tissue for that function" simply does not hold reliably for intraspecific (intraindividual) comparisons. What about comparisons among species? Are interspecies *differences* in absolute size linked to *differences* in "importance"? Again, the answer is: not reliably. Consider the mammalian olfactory system, and let us begin with Jerison's own words:

> It is often stated, probably with some justice, that primitive mammals tended to have enlarged olfactory bulbs and were, therefore, "olfactory" animals. . . . A quantitative analysis, however, reveals that the absolute size of the olfactory bulbs in mammals has increased throughout evolution. . . Thus, it may be surmised that living mammals are generally more efficient in their ability to use olfactory information or in the amount of olfactory information that they use compared to their fossil ancestors. However, the elaboration of other systems in the brain has been even greater in the evolution of the mammals, so much so that the olfactory system has been overshadowed in size. (Jerison, 1973, p. 8)

Here, Jerison refers to the generally accepted idea that the earliest mammals were small, with small brains and correspondingly small olfactory bulbs, whereas many recent mammals, especially the primates, have larger brains with larger olfactory bulbs (see Chapters 8 and 9 for more details on early mammals and primates). Because primate olfactory bulbs are small in proportion to the rest of the brain, we tend to think of primates as having a feeble sense of smell. However, in terms of absolute size, most primate olfactory bulbs (and certainly most primate olfactory cortices) are at least as large as those of the small-brained early mammals (see Figure 5.8A). Therefore, Jerison used his principle of proper mass to predict that most primates should have a fairly good sense of smell. This prediction has, by and large, been confirmed: even humans have a better sense of smell than we used to think (see Laing et al., 1991; Laska and Seibt, 2002). But what about the *importance* of olfaction in life? Is it also correlated with absolute olfactory bulb size? Most evolutionary biologists would probably agree with Jerison that early mammals were "olfactory" and that in later mammals, the olfactory system was overshadowed by the other sensory systems (see Chapter 9). Therefore, unless "importance in life" is not a zero-sum game, the importance of olfaction must have *decreased* as the absolute size of the olfactory bulbs *increased*. This conclusion seems inescapable. But did Jerison not write that "in comparisons among species the importance of a function in the life of each species will be reflected by the absolute amount of neural tissue for that function in each species"? Where did our reasoning go astray?

In my opinion, the chief difficulty with Jerison's principle of proper mass is that it mixes together several concepts that are best treated separately. If we try simultaneously to think about absolute, proportional, and relative size, as well as both intra- and interspecific comparisons, our heads are bound to spin. Therefore, I recommend that we decompose the problem of how region size relates to function into more manageable units and restrict our analysis mainly to interspecific (or interindividual) comparisons. This will not solve all difficulties, but it is helpful. Let us begin with the relationship between absolute size and functional capacity.

Absolute size and functional capacity

The idea that a region's absolute size reflects its functional capacity is perhaps not "obviously right" (Jerison, 2001), but it makes a great deal of sense. The larger a brain region is, the more neurons it contains (see Chapter 4) and, collectively, those neurons should be able to "do more" than the homologous neurons in another species. Evolutionary increases in the size of a motor region, for example, should increase the region's ability to control muscle contractions. On the sensory side, the situation is similar.

In a larger retina, for example, each neuron has a smaller visual receptive field (since larger retinas generally contain more neurons rather than just larger neurons), thereby increasing the retina's ability to resolve small stimulus objects. More centrally in the brain, receptive fields may be more difficult to define, but even there, greater neuron number generally implies a more finely tuned response. Moreover, when neuron number is increased, the neurons that receive their inputs from the enlarged neuron pool should be able to integrate over a greater number of inputs and, thereby, achieve better sensitivity and/or accuracy (Kawasaki, 1997). Finally, any increase in the number of neurons implies an increase in the number of synapses and, if we accept that learning and memory depend on synapse modification, then a region's ability to "learn" new input or output patterns might well increase with absolute region size. All of these ideas are intuitively obvious, but they are *necessarily* true only when other things are equal, when no other regions have changed in size and when the system's overall structural organization remains the same.

Unfortunately, in brain evolution "other things" are rarely "equal." As we discussed above, brain regions rarely change size in a purely mosaic manner, which means that changes in the absolute size of one region are almost always associated with changes in the size of other regions. Many aspects other than size are likewise subject to evolutionary change. Earlier in the chapter when I compared olfactory capacity across mammals, for example, I did not mention that humans have a much higher fraction of nonfunctional pseudogenes in their olfactory receptor repertoire than New World monkeys or mice (Rouquier et al., 2000). This evolutionary decrease in functional receptor diversity almost certainly explains (at least in part) why humans have a relatively poor sense of smell, despite the considerable absolute size of their olfactory bulb (and olfactory cortex). Another complication is that different mammals are probably specialized to detect different odorants, which makes it difficult to compare "olfactory capacity" across species (see Laska and Seibt, 2002). Additionally, there is the possibility that homologous brain regions may differ in their connections (see Chapter 7). The cerebellar valvula, for example, is not only larger in mormyrid electric fish than in other teleosts (see Figure 5.11), it also receives novel inputs from electrosensory brain regions (Finger et al., 1981). We may say that, therefore, it does more than other valvulas, but qualitative differences certainly complicate quantitative comparisons. Finally, we must recognize that homologous brain regions may differ in internal structure. They may, for example, contain different numbers of subdivisions (subnuclei or laminae; see Chapter 6) and that is likely to impact their functional capacity.

For all these reasons, it is difficult to establish a tight linkage between absolute size (or neuron number) and functional capacity. As we saw in Chapter 4, there is some correlation between absolute brain size and various kinds of intelligence, but this correlation may also (or instead) be due to the increased "neocorticalization" of larger brains; that is, to an increase in the proportional size of neocortex. Similarly, when neurobiologists look for correlations between the size of individual brain regions and behavior, they almost always focus on a region's proportional and relative size, not on its absolute size. In part, they do this to factor out differences in the size of other regions (i.e., to equalize other things mathematically), but they also do it because general experience has shown that correlations with absolute region size tend to be weaker than correlations with proportional or relative brain size (Figure 5.13). So, if absolute size is so difficult to work with, why should we continue to think about it? Mainly because it is undeniable that changes in the number of neurons within a region should have at least some effect on the region's function. I shall come back to that later. At this point, let us ask a different question: How can a small nucleus in a small brain be better than a slightly larger nucleus in a much larger brain? Could it be that the ability of a nucleus to influence behavior is determined, at least in part, by how large it is in relation to the rest of the brain? Let us explore that possibility.

Proportional size and influence

Jerison argued that the relative size of different brain regions within a species is indicative of how important their respective functions are in the life of the species. This amounts to saying that a region's proportional size (i.e., its size relative to the rest of the brain) reflects its importance in the life of an organism. We have already seen that this proposition cannot be generally true since the suprachiasmatic nucleus, for example, is small but important. Yet, if we look at the proportional size of the various cranial nerves, we see that species that depend heavily on vision have disproportionately large optic nerves, whereas those that depend more heavily on hearing have disproportionately large auditory nerves (Combe, 1836; see quotation above). More generally, it seems that if we know the principal function of the various cranial nerves, then we can use their sizes *relative to one another* to predict a great deal about an animal's sensory and motor capabilities. We can also, it seems, use the pro-

(A) Absolute volume correlation

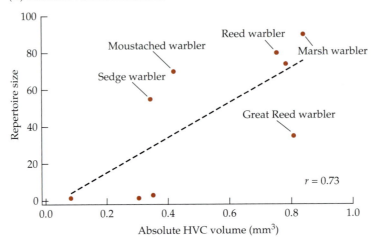

(B) Proportional volume correlation

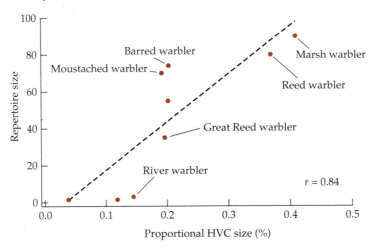

(C) Relative volume correlation

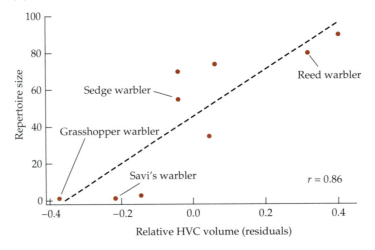

◀ **Figure 5.13 The Size of Nucleus HVC in Songbirds**   The size of nucleus HVC, which is involved in the control of avian song, correlates with song repertoire size across various species of European warblers. The strength of this correlation increases as we consider (A) absolute HVC volume, (B) proportional HVC volume, and (C) relative HVC volume. The Pearson's correlation coefficients (r) increase steadily as we go from absolute to proportional to relative size, which is indicative of the increase in correlation strength. These data illustrate that a structure's relative size is generally the best predictor of functional capacity, at least when the comparisons involve diverse species. Proportional and residual HVC sizes were calculated using telencephalon volume rather than overall brain size because the latter was not measured. (Data from Szekely et al., 1996.)

portional size of the main auditory and visual regions in the mammalian midbrain (the inferior and superior colliculi, respectively) to predict which sensory modality is most important in the life of a species (Figure 5.14). But why do these predictions work when others don't? Mainly, they work because we surreptitiously switched from intraindividual to interindividual comparisons.

Contrary to what you might at first surmise, comparing regional proportions *within a brain* is relatively uninformative. Yes, the fact that our olfactory cortex is proportionately smaller than our visual cortex suggests that olfaction is generally less important than vision in our lives, but what if you have a natural gas leak in your house? Then olfaction becomes critical. Alternatively, consider trying to determine whether vision or audition is more important in our lives. Would you rather grow up blind or deaf? The more we try to rank the importance of an individual's various senses, abilities, brain regions, or systems, the more dubious those efforts seem. Interindividual comparisons, on the other hand, are relatively straightforward and instructive. For instance, the finding that the hippocampus tends to be larger in London cabbies than in average Englishmen (Maguire, 2000) is consistent with the notion that taxi drivers tend to have superior navigation skills (since the hippocampus is involved in spatial memory; discussed in more detail later in this chapter). Similarly, the finding that the visual cortex tends to be considerably larger in Australian aborigines than in Caucasians (Klekamp, 1994) is consistent with the finding that the former tend to have superior visual abilities. In both instances, we do not know whether the differences in region size are heritable (i.e., due to differences in genes rather than experience) but that is immaterial for our purposes. What is important is that the interindividual differences in brain region size correlate with differences in behavior. If a region is proportionately larger in one individual than in another, then it is likely to be more important in the former's life (other things being equal). This rule is fairly general, holding for comparisons between members of the same species (conspecifics) as well as for comparisons between species, but why does the rule itself exist? Why should comparing the proportional sizes

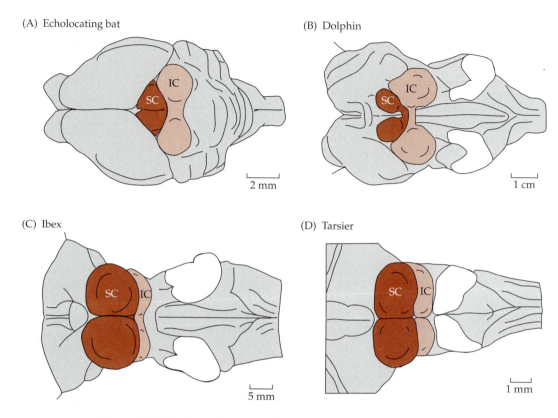

(A) Echolocating bat

(B) Dolphin

2 mm

1 cm

(C) Ibex

(D) Tarsier

5 mm

1 mm

Figure 5.14 The Superior and Inferior Colliculi in Various Mammals If an animal's superior colliculus (SC) is larger than its inferior colliculus (IC), then it is likely to use vision more than audition in its normal life; if its inferior colliculus is larger, then it is likely to depend heavily on audition. This rule is consistent with data from (A) an echolocating tomb bat, (B) a bottlenose dolphin, (C) an ibex, and (D) a tarsier (a prosimian). While echolocating bats and dolphins rely heavily on their auditory systems for communication and orientation, ibexes and tarsiers are predominantly visual. Accordingly, the inferior colliculi are larger than the superior colliculi in bats and dolphins, while the reverse is true for ibexes and tarsiers. In B–D, the telencephalon and cerebellum are removed to expose the midbrain colliculi; the cerebellar peduncles are cut and colored white. (A after Baron et al., 1996; B after Langworthy, 1967; C after Schober and Brauer, 1974; D after Tilney, 1927.)

of homologous brain regions among individuals or among species be more informative than comparing the proportional size of structures within an individual?

Before we try to answer that, let us ask a slightly different question: How do individual differences in the proportional size of homologous brain regions affect brain function? If we assume that the number of projection neurons (the number of neurons that project out of a brain region) increases with absolute region size, then any increase in a region's proportional size should, other things being equal, lead to an increase in the number of axons that project from that region to the rest of the brain. This implies that differences in a region's proportional size should correlate with differences in the region's ability to influence or "control" its target regions. The neocortex, for example, is proportionately larger in primates than in other mammals (as we discussed earlier) and ought, therefore, to send more axons to the medulla (one of its targets). The medulla itself gets inputs from a variety of different regions but, if the above considerations are correct, then the proportion of those inputs that arise from neocortex must be larger in primates than in other mammals; this, in turn, would allow the neocortex to exert more control over the medulla in primates. This principle of disproportionately large areas being disproportionately well-connected is discussed at length in Chapter 7. The evidence for it is limited but generally supportive. Moreover, if we accept the idea of large equals well-connected, then we would expect differences in a region's proportional size to correlate with differences in its importance to normal brain function (since those other regions would probably be more dependent on the enlarged region's influence). We might predict, for example, that neocortical lesions in large-brained primates should have more severe functional and behavioral consequences than corresponding lesions in a species with proportionately smaller neocortices. This idea, which is generally known as "functional neocorticalization," is old and controversial (see Chapter 7) but its potential explanatory power is significant, especially when we consider human brain evolution (see Chapter 9). For now, suffice it to say that the rule of large equals well-connected predicts that phylogenetic increases in a region's proportional size should increase its influence within the brain, and that influence within the brain tends to translate into importance in life.

Thus, we can conclude that interindividual *differences* in region size tend to correlate with *differences* in influence and importance. This rule is not foolproof, but it works fairly well. For instance, if you sorted the brains of many different bat species according to the proportional size of their auditory regions, the top few species in your list would most likely belong to the echolocating, highly auditory microbats, rather than the more visual megabats (see Figure 4.13). Similarly, the fact that catfishes and goldfishes have much larger gustatory brain regions than most other teleosts (see Chapter 6) almost certainly reflects that the sense of taste is more important in their lives. Thus, comparisons of proportional brain region size across different individuals can yield some useful information about behavioral differences. What we have not yet considered, however, is that differences in proportional brain region size are often accompanied by major differences in absolute brain size. In those cases, some of the differences in regional proportions might be automatic consequences of the differences in overall brain size (Finlay and Darlington, 1995). If that is true, do differences in proportional brain region size still predict differences in behavior? I think they do, but as you will see in the following section, the matter is complex.

Relative size and adaptation

Most comparative studies focus on a brain region's relative size, which is best defined as any statistically significant deviation from a region's expected size, given the species' overall brain size. Determining statistical significance in such comparisons is not a simple matter, especially if we have small sample sizes and nonindependent data points, but these details must be hashed out in the technical literature. For our purposes, we can simply say that all instances of mosaic evolution, such as those mentioned earlier, qualify as significant changes in a region's relative size (since concerted evolution is defined as change that is in accord with expectations). But what do these mosaic deviations from expected region size tell us about brain function and/or animal behavior? The most widely accepted answer is that changes in a region's relative size reflect adaptive evolution. That is, if a region is larger than expected, we tend to assume that the function(s) performed by that region were "selected for." But why should that be so? And does the converse also hold: Are evolutionary changes that are in line with expectations never adaptive? To answer these questions, let us start with the origins of the idea that relative size is linked to adaptation.

Because evolutionary changes in absolute brain size consistently entail allometrically predictable changes in the absolute and proportional size of individual brain regions, we might say that evolutionary changes in absolute brain size *must* lead to changes in the absolute and proportional size of many individual regions. This is an important point because, according to many evolutionary biologists (e.g., Gould and Lewontin, 1979), any changes that are necessary consequences of other changes in an organism cannot be adaptations (i.e., they cannot be due to natural selection). Moreover, according to some authors, the converse is also true: Any deviations from allometric expectations *must* be adaptive. Julian Huxley (1932), for example, wrote the following:

> If, however, a way could be found of taking account of the changes in absolute size which so frequently accompany directive evolution, and automatically induce changes in proportions of limbs, etc., a further analysis might be possible, which would enable us to distinguish what we might call the consequential changes in form from the strictly adaptive, meaning by the former those changes in proportion which, unless counter-acting growth-mechanisms are evolved, automatically accompany changes in size, and by the latter such changes as are specifically related to mode of life, and presumably brought about by natural selection. . . . (Huxley, 1972 p. 110)

Essentially, Huxley argued here that mosaic evolution is adaptive, concerted evolution is not. But if we accept this argument and if, as we concluded above, most brain evolution is of the concerted kind, then we are forced to conclude that, most of the time, brain evolution is not adaptive. A rather counterintuitive conclusion (at least to many neuroscientists)!

One way to avoid this inference is to challenge Huxley's proposed dichotomy and to argue, instead, that changes in the size of an individual brain region may well be adaptive even if they are predictable from overall brain size. This argument was pioneered by Finlay and Darlington (1995) who argued, for example, that if an organism's reproductive success were enhanced by having a larger inferior colliculus (because this would allow them to locate their prey more accurately), then an increase in the absolute size of the inferior colliculus would be selected for even if it entailed, as a necessary consequence, increases in the absolute size of many other brain regions. In this hypothetical case, the increased absolute size of the inferior colliculus would be adaptive, even though its size would be in line with allometric expectations. The consequential changes in all the other regions, in contrast, would be non-adaptive (although they might turn out to be useful later on in evolution). This is a radical hypothesis because it challenges Huxley's, and perhaps most biologists', view of how natural selection "works." For that reason alone, Finlay and Darlington's argument deserves serious consideration. But is the argument correct?

To answer that, we must consider costs and benefits. In the hypothetical example we discussed above, we presumed that only the increase in the size of the inferior colliculus is beneficial to the animal and that no benefits accrue from increasing the size of other regions that would grow in concert with the inferior colliculus (e.g., the neocortex). But those other, consequential size increases would be associated with some costs because neural tissue is metabolically expensive to build and maintain (see Chapter 4) (Aiello and Wheeler, 1995). Those metabolic costs are probably substantial, since neural tissue is lost quite readily whenever it is not needed, as in domesticated animals (Kruska, 1988; Ebinger and Röhrs, 1995) or queen ants (Julian and Gronenberg, 2002). Therefore, in our hypothetical example, the benefits of increasing inferior colliculus size would have to offset the significant costs of increasing the size of all the other regions (which are, by definition, not beneficial to the animal). This seems unlikely. But what if we imagined a different scenario? What if the benefits derived not from changes in the size of the inferior colliculus, which is a relatively small brain region, but from changes in a structure that is already large, such as the neocortex? Then Finlay and Darlington's proposition becomes more plausible, for the benefits associated with increasing the size of the already large structure would be more likely to outweigh the costs associated with the consequential increases in the size of all the other, relatively small structures (that, by themselves, may not be beneficial). In general, we may conclude that the kind of focused adaptation by concerted evolution that Finlay and Darlington proposed should occur more frequently when selection acts on relatively large structures, such as the neocortex, than when it acts on relatively small structures, such as the inferior colliculus. Thus, Finlay and Darlington's argument is neither totally correct nor totally false; its validity depends on the details of the case.

So, where does that leave us with regard to the presumed relationship between relative brain region size and adaptation? Clearly, Huxley's claim that *only* evolutionary changes in relative size can be adaptive must be wrong because, at least for some large (late-born) brain regions, the benefits of increased size might outweigh the costs of all consequential changes. We might say that evolutionary changes in relative brain region size are more likely to be adaptive than changes that are in line with allometric expectations, since the former are likely to be larger than the latter in absolute magnitude, but this only underscores that Huxley's dichotomy is probabilistic rather than hard-and-fast. In addition, we must question whether it is proper to assume that Huxley's "consequential changes in form" are always devoid of benefit. Just because a change in size is part of a concerted pattern does not mean it must be functionally insignificant. If changes in a region's proportional size lead to changes in the region's influence and importance, should they not do so regardless of whether the change is mosaic or concerted? I think so, and argue that point extensively in Chapter 9, when we discuss the human brain. Finally, we must admit that even changes in a region's relative size need not be due to natural selection; they could be due to chance or, more technically, genetic drift (Suzuki et al., 1989). Given all these complications, can we ever show that any given evolutionary change in brain region size is (or was) adaptive?

As I mentioned in Chapter 1, there are no simple answers to this question. We might try to manipulate a brain region's size experimentally and then measure the animal's reproductive success, but this is difficult. Alternatively, we can try to determine whether similar changes in brain region size are consistently associated with particular ecological niches and/or behaviors, even when those changes occurred independently of one another. A nice example of the latter comparative method is provided by Healy and Guilford (1990), who showed that, among birds, those species that evolved diurnal habits consistently have smaller olfactory bulbs, relative to allometric expectations, than birds that became crepuscular or nocturnal. This finding considerably strengthens the hypothesis that increases in relative olfactory bulb size were an adaptive response to nocturnality in birds (most likely, early birds were diurnal). The comparative method cannot be used to test all hypotheses of adaptation, however, because not all evolutionary changes have occurred more than once. Indeed, most adaptive changes are likely, a priori, to have occurred only once or a few times independently, and all of those instances would be impossible or difficult to detect by means of the comparative method. Thus, the linkage between a region's relative size and adaptation is neither tight in general nor easy to demonstrate in any particular case.

Despite the problems associated with showing that any particular change in relative size is due to natural selection, it seems intuitively obvious that changes in a region's relative size must have important consequences for brain function and animal behavior. Even this assertion entails some complexities, however, since changes in a region's relative size always entail changes in its absolute and/or proportional size that have, as we have seen, functional consequences of their own. In fact, it seems to me that any functional consequences of species differences in relative region size *must*, somehow, be due to differences in absolute and/or proportional region size (if we disregard, for the moment, the possibility of changes in internal organization or connectivity; see Chapters 6 and 7). Therefore, if we want to understand how species differences in behavior are related to differences in relative brain region size, we must *also* think about species differences in absolute and proportional brain region size. In other words, now that we have become clear about the distinctions between absolute, proportional, and relative brain region size, we must try to synthesize what we have learned about all three kinds of change. We must put Humpty-Dumpty back together again. As an illustration of this synthetic approach, consider the following test case.

Synthesis: The avian hippocampus

One of the most widely discussed examples of a behaviorally relevant species difference in brain region size involves the hippocampus of birds. The avian hippocampus looks quite different from its mammalian counterpart but is probably homologous to it (see Chapter 8). A variety of behavioral studies further suggest that it, like the mammalian hippocampus, is required for some aspects of spatial learning and memory (e.g., Papadimitriou and Wynne, 1999). Armed with this information, John Krebs and his collaborators (1989) asked whether, among birds, species that regularly store food items for future consumption (storers) have larger hippocampi than species that do not store food (nonstorers). In order to answer this question, they measured hippocampal size in 35 species of songbirds from 9 different families. Their principal finding was "that the volume of the hippocampal complex relative to brain and body size is significantly larger in species that store food than in species that do not" (p. 1388). By now, it should be clear that this statement, taken literally, entails some ambiguity because hippocampus volume "relative to brain and body size" might refer to the hippocampus's proportional size, its relative size, or both. Nor does it exclude the possibility that absolute hippocampus volume is also causally related to food storing capacity. In order to resolve these uncertainties, let us examine the data for ourselves.

If we plot hippocampus volume against telencephalon volume (using the latter as a substitute for overall brain size, which was not measured), we find that the two variables are highly correlated with one another and that, in general, absolute hippocampus volume increases with telencephalon volume (Figure 5.15A). The proportional size of the hippocampus decreases, however,

Figure 5.15 Hippocampus Size and Food Storing in Songbirds Correlations between hippocampus size and food storing capacity in diverse songbird families. Data points from species that store their food (red) tend to lie above the best-fit allometric lines (dashed lines), whereas data points for nonstorers (black) tend to lie below those lines, regardless of whether we consider (A) absolute or (B) proportional hippocampus size. Particularly interesting are comparisons between storing and nonstoring species within the same family (i.e., within titmice and within corvids). In those comparisons, the storers' hippocampi are larger in absolute, proportional, and relative size. (After Krebs et al., 1989.)

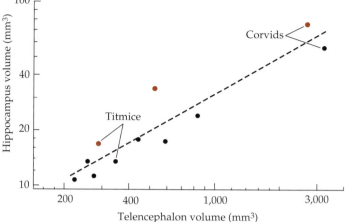

(A) Absolute hippocampus size

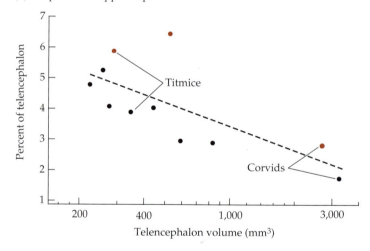

(B) Proportional hippocampus size

as overall telencephalon size increases (see Figure 5.15B). In other words, the songbird hippocampus scales against the songbird telencephalon with negative allometry. The deviations from allometric expectations tend to be less than 2-fold, which means that these data conform to Finlay and Darlington's standard for concerted evolution. Even those relatively small variations seem to be behaviorally significant, however, since the data points for storers consistently lie well above the allometric line, while those for nonstorers tend to lie below the line ($p < 0.01$; Krebs et al., 1989). Therefore, we can conclude that the *relative* size of the hippocampus is greater in songbirds that are storers than in those that are not. This is basically the conclusion that Krebs et al.

came to, even though they used somewhat ambiguous language to report it. Moreover, this basic finding has now been confirmed and extended in further studies (Basil et al., 1996; Healy and Krebs, 1996; Volman et al., 1997). Because relative hippocampal size correlates with food storing behavior in so many different species, many of which evolved this ability independently of one another, application of the comparative method suggests that those increases in relative hippocampus size were probably due to natural selection—that is, they were adaptive.

This is wonderful, considering how rarely evolutionary changes in brain anatomy can be linked to natural selection, but how do the differences in relative hippocampus volume *explain* the differences in food storing ability? As I argued earlier, changes in relative region size can bring about functional changes only because they change the region's absolute and/or proportional size. So, we might ask, does the capacity for food storing emerge whenever the hippocampus exceeds some absolute size threshold? Apparently not, since some nonstorers have hippocampi that are larger in absolute size than the hippocampi of many storers. Next, we might ask whether the avian hippocampus must be of a certain proportional size before it has enough "influence over other brain regions" to support food storing behavior? Again, the answer is negative, since several nonstorers have proportional hippocampal sizes that exceed those of food-storing corvids. This leaves us with a paradox: only evolutionary changes in functional or proportional region size can have behavioral consequences, yet neither of those variables correlates with food storing behavior.

Well, as you might have figured out, the paradox is only an apparent one, which arose because we compared so many different species at once (which we did originally in order to make inferences about adaptation). If, instead, we compare only a few closely related species, such as storing and nonstoring species within the same family, then a very different picture emerges. Both in corvids and in titmice, the storing species have larger hippocampi than the nonstoring species, in terms of both absolute and proportional size. Marsh tits, for example, store food, and their hippocampus is 31% larger than it is in the great tit, a nonstorer, even though the entire telencephalon is 21% smaller in the latter species (Krebs et al., 1989). Thus, when we compare closely related species, the differences in relative hippocampal size are consistently accompanied by corresponding differences in absolute and proportional size.

That allows us to construct the following hypothesis about how evolutionary changes in hippocampus size are linked to food storing ability. If we assume that storers evolved from nonstorers, then food storing probably evolved at least in part because the absolute size of the hippocampus increased. This hippocampal size increase was not coupled to a significant increase in overall brain size, which implies that the proportional size of the hippocampus likewise increased. If we further accept (1) that the hippocampus is associated with spatial memory even in birds that do not store food, and (2) that increases in absolute and proportional region size correlate, all else being equal, with functional capacity and influence over other brain regions, respectively, then the hypothesized increase in absolute and proportional hippocampal volume would have led to an increase in spatial memory capacity and accessibility (i.e., in how well spatial information is remembered and in how well the stored information is passed on to other brain regions that act on it). This enhancement of spatial memory would have contributed causally to the emergence of food storing behavior, which, in turn, would have allowed individual birds to retrieve food when others starved; to the extent that this benefit outweighed the costs of maintaining a larger hippocampus, it would have been selected for. This hypothesis would apply to both storers whose nonstoring ancestors had small brains (e.g., ancestral titmice) and to species whose ancestors were large-brained (e.g., corvids). However, if we simultaneously examine many different species that differ widely in overall brain size, then the correlations between absolute and proportional hippocampus size and food storing disappear, and only the observed correlation between relative hippocampus size and food storing remains robust (see Figure 5.15).

I consider this explanation to be plausible, but two caveats remain. The first is that no one has ever done a proper cladistic reconstruction of food storing behavior in birds, so that we actually do not know whether storers evolved from nonstorers or *vice versa*. Should such a reconstruction reveal that nonstorers evolved from storers, then the hypothesis would have to be inverted: one would have to argue that the metabolic savings associated with reducing absolute hippocampal volume outweighed the benefits of having good spatial memory, leading to a reduction in absolute hippocampal volume without a major change in overall brain size. The second caveat is that hippocampal growth may be modulated by experience: If young birds from at least some storing species are deprived of food storing opportunities, then hippocampal neurogenesis is reduced and cell death increased (Clayton, 1998). This observation suggests that the phylogenetic increases in hippocampus size might be *due* to food storing, rather than *responsible* for it. This possibility is interesting, but "it is of no evolutionary consequence whether a trait is sensitive to environmental variation, as long as the actual historical environment regularly provides the required input" (Griffiths and Gray, 1994, p. 279), which in this case is the opportunity to store food. Putting it differently, the capacity of a storer's hippocampus to respond to storing experience with increased growth may itself be a heritable feature that is unique to food-storing birds (see Oyama, 1985). If this is true, then the mechanism by which storers increased the absolute size of their hippocampus would not be a change in early embryogenesis (as we had assumed in our discussion of late equals large), but a modification of juvenile development that increased the ability of environmental factors to modulate cell death and postembryonic neurogenesis (which is common in the avian telencephalon). This would be interesting because it would show that (contrary to what is generally assumed) not all evolutionary change in brain anatomy is due to changes in embryogenesis. It would not, however, invalidate the hypothesis that phylogenetic increases in the absolute and proportional size of the *adult* hippocampus are causally linked to the evolution of food storing capacity.

Thus, it is possible to construct a synthetic hypothesis about how evolutionary changes in the size of the avian hippocampus are linked to evolutionary changes in behavior. An important element of this hypothesis is that absolute brain region size was altered in the absence of major changes in overall brain size. Since the magnitude of the hypothesized size increase was less than 2-fold, it does not satisfy Finlay and Darlington's criterion for mosaic evolution but, since overall brain size did not change significantly, the hippocampal size increase is not an instance of concerted evolution either. I suggest we think of it as "mildly mosaic" evolution. Indeed, I suspect that many behaviorally significant changes in brain region size occurred in this mildly mosaic manner. Often, however, it is difficult to figure out precisely what has changed. The temporal lobe of the neocortex, for example, is larger in humans than in other primates when we look at its absolute or relative size, but its proportional size is smaller (Rilling and Seligman, 2002). In this case, absolute and proportional region size have changed in opposite directions, making it difficult to discern how brain function was altered. Such complexities notwithstanding, the avian hippocampus and food storing data illustrate that a careful analysis of how individual brain regions have changed in absolute, proportional, and relative size, combined with a detailed functional analysis of what the region does, can provide a solid foundation on which to build more elaborate explanations of how evolutionary changes in brain structure relate causally to evolutionary changes in behavior.

Conclusions

We have seen that evolutionary changes in brain region size come in several flavors. Probably the most common kind of change in size follows the allometric rule of late equals large (Finlay and Darlington, 1995); this kind of change is concerted because individual brain regions evolve in concert with one another. The second major kind of change in size is called mosaic evolution (Barton and Harvey, 2000) because, in this mode, individual brain regions seem to evolve independently of one another, defying allometric expectations. We may debate how large the deviations from allometry have to be before we recognize a change as being mosaic, but it seems fairly clear that severely mosaic evolution (deviations that are greater than 2- to 3-fold) is less frequent than concerted evolution, at least for mammalian brains. This does not mean that mosaic evolution has been unimportant in vertebrate brain evolution, and it would certainly be wrong to think of all vertebrate brains as varying according to a single, unchanging set of allometric rules. Putting it crudely, the brain of a goldfish may be smaller than the brain of a rat, but the size of its individual brain regions is not determined solely, or even predominantly, by the general scaling rules that apply to mammals. Scaling rules may be ubiquitous in the sense that they are unavoidable, but they are not immutable. We have also seen that both mosaic and concerted evolution can lead to changes in brain function and behavior. It is difficult to determine whether any particular change in region size is adaptive but its functional consequences are, in principle, identifiable as long as we have good data on what the region does and distinguish carefully between changes in absolute, proportional, and relative brain region size.

Thus, we have come a long way from the old phrenologists' ideas about how brain regions vary in size. Franz Gall was right to note that, across species, different brain regions may vary in size independently of one another. He was oblivious, however, to the idea that differences in the proportional size of various brain regions could be related to differences in overall brain size by some law-like, allometric relationship. That was a more recent development, and the next few years are likely to see even further progress on this front, as more data become available on the mechanisms that can cause and modify allometries. Gall was also right to argue that differences in brain region size are causally related to differences in behavior. He had been rather vague about why that relationship should exist, but we now have much better ideas about how differences in absolute and proportional brain region size affect brain function and behavior. Particularly prescient, though rarely appreciated, was the early phrenologists' admonition that correlations between a brain region's size and an animal's behavior are best made when other things are equal.

The importance of this *ceteris paribus* ("all else being equal") clause becomes apparent when we try to compare brains that differ dramatically in absolute brain size. In such cases, so many regions vary in both absolute and proportional size that it becomes quite difficult to decipher which changes in behavior are due to which changes in brain region size. Gall had no comparative data on the size of the avian hippocampus—at the time, no one was even sure whether birds had a hippocampus (Smith, 1910)—but if he had seen the data gathered by Krebs et al. (1989), he also would have emphasized the comparisons between species that differ relatively little in terms of absolute brain size. He knew that the *ceteris paribus* clause plays a central, if often underappreciated, role in virtually all comparative analyses. And that brings us to a set of other factors that we have largely ignored thus far, namely the brain's internal structure and connectivity. Gall and his contemporaries knew next to nothing about the brain's structural complexity, and we still know relatively little about how that complexity evolves. Some data and ideas are, however, starting to emerge. We will discuss them in the following two chapters.

6 Evolutionary Changes in Brain Region Structure

We have now ...a large collection of detailed and reliable data upon the structure, neurone connections, and embryology of the anterior regions of the brain. The farther this ...work has proceeded the more clear has become the continuity of phylogenetic development of brain structure. New structures have not appeared, and although old structures have been modified, the process of evolution leading to the human brain has been one of gradual development of the elements inherent in the structures or areas laid down in very early vertebrate ancestors with gradual shifting of emphasis, growth, or decline of this or that element, all under the influence of environment and changing habits.

—J. B. Johnston, 1923

In Chapter 5, we focused on evolutionary changes in the size of individual brain regions and assumed that "other things" were "equal." This approach is popular among evolutionary neuroscientists because data on brain region size are relatively easy to obtain and can, at least sometimes, be related to an animal's behavior. We also know enough about brain development to help us think about how changes in development can lead to evolutionary changes in brain region size. All that is useful, but how often are those other things really equal? Surely brain regions sometimes change in connectivity and function (changes discussed in Chapter 7). In addition, brain regions may change in their internal organization or "structure," which is what shall occupy us in this chapter. Although a region's structure may change in myriad ways, I here focus only on the most radical changes, for that is where controversy reigns. As our starting point, let us take J. B. Johnston's dictum that "new structures have not appeared." Since Johnston was one of evolutionary neurobiology's founding fathers (see Chapter 2), his views are widely respected. Why, then, do so many neurobiologists, including me, continue to refer to some brain regions as "new" (e.g., *neo*cortex)? Was J. B. Johnston wrong? Or are many neurobiologists simply confused about what counts as novelty in evolution? Let us begin to explore these questions by considering their historical context.

Near the dawn of evolutionary neuroscience, Ludwig Edinger (1908) proposed that as vertebrate brains evolved, a novel "neencephalon" was added to an older "palaeencephalon" (see Figure 2.8). This idea became widely accepted in the first half of the twentieth century and transformed into the more general view that brains generally evolve by the accretion of novel parts—hence the proliferation of terms such as "neopallium," "neocortex," "neocerebellum," and "neostriatum" (see Chapter 2). This view is appealing because it implies that brains became more complex over time: If evolution tends to add new parts to old brains, then the number of different parts within a brain must increase with time, and that constitutes an increase in complexity. Since the image of evolution as striving tirelessly toward increased complexity was widespread in Edinger's day, the idea of "evolution by accretion" captivated many minds. So why did Johnston object to it so forcefully? Most likely, Johnston simply became convinced, on account of his own investigations and the accumulating literature, that Edinger's supposedly new brain regions did, in fact, have homologues in "lower" vertebrates. For example, Johnston wrote extensively about fishes and reptiles having a "general pallium" that differs in appearance from mammalian neocortex but is nonetheless homologous to it and, therefore, not new (Johnston, 1923). This conclusion was radical at the time and failed to deter most of Johnston's contemporaries from using terms like "neopallium" and "neocortex." In the last twenty years, however, Johnston's dictum has become more widely accepted, and calls to replace the term "neocortex" with "isocortex" are now commonplace (see Chapter 2). Indeed, today's evolutionary neuroscientists often pride themselves on having found homologues for structures that were once thought to be new in (unique to) one lineage or another (see Butler and Hodos, 1996). This presents us with an apparent paradox: if "new structures have not appeared," how can brains have become more complex over evolutionary time?

One way to resolve this paradox is to argue that the supposed increase in brain complexity is illusory, that brains did *not* become more complex with time (see Lashley and Clark, 1946). That position is untenable. Although complexity is notoriously difficult to quantify, we can use the number of distinct regions within a brain as a proxy for complexity (McShea, 2000; Bullock, 2002). Of course, counting brain regions is likewise problematic because some brains have been studied more intensively than others (which generally means that more of their cell groups have been described) and because some neuroanatomists may identify multiple cell groups where others recognize just one. However, comparative neuroanatomists do have reasonably well-established standards for deciding what constitutes a distinct cell group (Kaas, 1982), and they have consistently described more cell groups in the forebrain of amniotes than in the forebrain of hagfishes, lampreys, or amphibians (Wicht and Northcutt, 1992). Even naïve observers are likely to agree that, across the major vertebrate lineages, significant differences in the number of distinct brain regions exist (Figure 6.1). If we subject these differences to a parsimony analysis (Figure 6.2), we can conclude that brain complexity probably increased and decreased several times during the course of vertebrate brain evolution. It almost certainly increased in teleosts, with the origin of amniotes and again in mammals, and it probably decreased in lungfishes and salamanders. Moreover, if we step back and look at the overall trend, we see that brains have more often increased in complexity than decreased. Therefore, evolutionary changes in brain complexity are not illusory but real. Brain complexity has at least sometimes, and on average, increased. Which returns us to our paradox: if "new structures have not appeared," how can

(A) Catfish

(B) Macaque

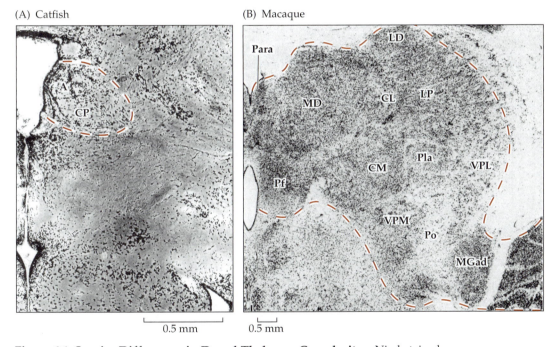

0.5 mm   0.5 mm

Figure 6.1 Species Differences in Dorsal Thalamus Complexity Nissl-stained transverse sections through the dorsal thalamus (dashed line) of (A) a bullhead catfish and (B) a crab-eating macaque. The latter is much larger and is divisible into many more distinct cell groups. In both A and B, the brain's midline is on the left and dorsal is at the top. Abbreviations: A = nucleus anterior; CL = central lateral nucleus; CM = centre médian; CP = central posterior nucleus; LD = lateral dorsal nucleus; LP = lateral posterior nucleus; MD = mediodorsal nucleus; MGad = medial geniculate nucleus; Para = anterior paraventricular nucleus; Pf = parafascicular nucleus; Pla = anterior pulvinar; Po = posterior nucleus; VPL = ventral posterior lateral nucleus. (A from Striedter, 1990a; B from Jones, 1985.)

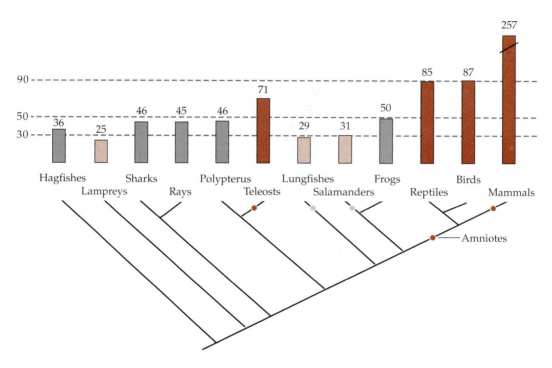

Figure 6.2 A Cladogram of Forebrain Complexity Shown across the top of this cladogram are counts of how many different cell groups experts have described in the forebrains of representative species from each major group of vertebrates. References to those studies are given in Wicht and Northcutt (1992). Differently colored bars represent different grades of forebrain complexity. The colored circles on the cladogram indicate approximately when forebrain complexity is likely to have increased (dark red circles) or decreased (light red circles).

the number of brain regions have increased? If new parts were not added to old parts, how did the parts proliferate?

One possibility is that brain regions proliferate by "phylogenetic segregation," (Ramón-Moliner and Nauta, 1966). This simply means that an ancestral brain region becomes subdivided, or "segregated," into two or more brain regions in a descendant. If brain regions indeed proliferate by phylogenetic segregation, it resolves the paradox, for phylogenetic segregation does not involve any *addition* of new regions: Individual subdivisions may be "new" insofar as they are nonexistent in the ancestor, but they are not added to an old complement of structures. Thus, phylogenetic segregation is a clear alternative to Edinger's old notion of evolution by accretion. Because Edinger's ideas about brain evolution fell out of favor after 1950 (or thereabouts), the segregation hypothesis has become widely accepted as *the* major process underlying the phylogenetic "complexification" of vertebrate brains. Indeed, there is now considerable evidence for phylogenetic segregation. We will get to that shortly, but let us step back first and ask an important theoretical question: Is phylogenetic segregation the *only* process by which brain complexity may increase? Some evolutionary neuroscientists have argued that it is (Ebbesson, 1984), but that position is likewise untenable. Think about the human neocortex and its multitude of distinct areas! Did all those areas evolve by phylogenetic segregation from a few primitive areas in early mammals? According to most students of mammalian brain evolution, the answer is no (discussed in detail later in the chapter). But how can new parts simply be added to old brains? They cannot evolve from nothing, can they?

Any attempts to answer these questions must begin with a clarification of what is meant by "new" and "old." Since the use of these terms in evolutionary biology is inextricably linked to the concept of homology, the next section explores how homology relates to novelty. In particular, I will argue that we can distinguish between two kinds of novelty in evolution, namely: (1) novelties that result from the "phylogenetic conversion" of one character into another; and (2) novelties that result from the "phylogenetic proliferation" of characters. Included in the first category is, for example, the evolution of laminated structures from unlaminated ones. Included in the second category are instances of both "phylogenetic segregation" and "addition" (to be defined later). As you shall see, the notion that new brain regions are sometimes added to old ones (evolving, as it were, "from nothing") is not as impossible as it might at first appear.

Homology and Novelty

Although the concept of homology is sometimes derided as a quagmire of vague and indistinct ideas (e.g., Spemann, 1915), this reputation applies only to its borderland, where novelty dwells. Within the framework of cladistics (see Chapter 2), homologues are fairly easy to define: They are the features or characters that evolved just once, rather than multiple times independently. In practice, this means that comparative biologists establish homologues by identifying comparable characters in multiple species and then mapping their distribution onto a well-established phylogeny to see if the characters are likely to have evolved just once or more than once among the taxa under consideration. Most of the time, the answer is clear; there is no doubt, for example, that brains are homologous across all vertebrates, or forebrains, or dorsal thalami, and so forth. Precisely because homology is so ubiquitous, it tends to be taken for granted. Controversy erupts only when we are unsure about how the species under investigation are related to one another (i.e., when the phylogeny is not "well established"), or when the characters that are being compared are quite dissimilar. The first of these problems can be solved by systematists and need not concern us here. The second problem is more difficult and relates directly to the issue of novelty in evolution. Consider the following thought experiment:

Imagine that you are comparing two sister species, A and B, and that any similarities shared between them also are found in the next most closely related species C (Figure 6.3). In that case, any character shared by species A and B would, according to cladistic methodology, be homologous between them. But how do you decide which character in species A is comparable to which character in species B in the first place? Assuming, for example, that

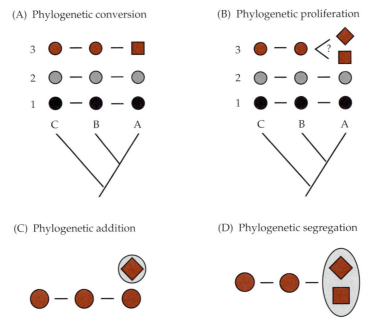

Figure 6.3 The Evolution of New Structures   Truly novel structures may evolve in several different ways. (A) Novelty by "phylogenetic conversion" occurs when clear one-to-one homologies exist for all characters in all species, but one character (red square) is so different from its homologues (red circles) that it deserves to be called "new." (B) Novelty by "phylogenetic proliferation" occurs when one species has more characters than another and that condition is derived (not primitive). There are two conceptually different ways in which phylogenetic proliferation may occur. (C) If one of the characters in the species with "too many" characters cannot be readily homologized, but one-to-one homologies can be found for all other nearby characters, then proliferation by "phylogenetic addition" has probably occurred. (D) If two characters in the derived condition *together* exhibit all or most of the attributes of a single character in the primitive condition, then proliferation by "phylogenetic segregation" has probably occurred. In reality, of course, the comparisons are rarely this simple.

you have identified the character "lateral geniculate nucleus" in species A, what criteria can you use to find the "same" character in species B? This is not a question of homology per se but of character identification, of constructing a putative hypothesis of homology that you can later test by means of parsimony analysis (Kitching et al., 1998; Striedter, 1999). This first step of any comparative analysis is rarely made explicit, but it is crucial. Character identification is always based on similarities between species, even though similarity is not part of our definition of homology. Thus, to find the putative "lateral geniculate" in species B, you would look for a cell group that is similar to species A's lateral geniculate nucleus in position, cytoarchitectural appearance, histochemistry, connectivity, physiological properties, and/or function. If the similarities are readily forthcoming, then the character identities are obvious, the putative homologies clear, and we can proceed to test whether our putative homologies are bona fide homologies or due to independent evolution (see Chapter 2). In our thought experiment, with its assumptions about character distribution and phylogenetic relationships, any putative homologies will be bona fide homologies, but what happens when you can only identify a "somewhat similar" character in species B? What if it exhibits only 7 of the 10 similarities you were looking for? Or 5, or 3? Somewhere along that road, comparative biologists begin to sweat; they start to disagree about what is comparable to what and, hence, to argue about homologies.

One way of dealing with the problem of marginally similar characters is to avoid it altogether, to declare up front that you are interested only in comparing highly similar characters. Systematists, for instance, may exclude from their analyses of phylogenetic relationships any characters that are not unambiguously "the same" across species. Most evolutionary neuroscientists, however, are more interested in evolutionary change than in phylogenetic relationships per se and are more willing, therefore, to compare marginally similar characters. But how do they decide that two or more characters are "the same" when they are only marginally similar? Most comparative neurobiologists do this by weighting different kinds of similarities differently, assigning greater significance to those attributes that they consider most conservative in evolution. Rudolf Nieuwenhuys (1994), for example, has argued that a region's topological position (its position relative to other cell groups) tends to be highly conserved in evolution and is therefore the most valuable criterion of character identification. This view is widely shared (e.g., Puelles and Medina, 2002), but some authors have argued that neuronal connections and histochemistry are at least as conserved as topology and, therefore, at least as important for character identification and homology (e.g., Campbell and Hodos, 1970). Most likely, this debate will continue for some time because decisions about how to weight different kinds of character similarities must be made a posteriori rather than a priori. That is, the debate will cease only when we have reached consensus about which aspects of brain structure (or function) are most conserved in evolution (Striedter, 1999).

Nonetheless, most authors nowadays seem to agree that topological position—of both the adult structures and their embryonic precursors—is the most useful criterion of character identification for brain regions.

Now, if we are willing to weight topological similarity more than other kinds of similarity, then we may homologize some characters that differ so much from one another in connections (or some other attribute) that it makes sense for us to call one of them "new." Let us return to our thought experiment: If a structure in species A has the same topological position as a structure in species B but differs radically from it in cytoarchitecture, then we may (given our assumptions) consider the two structures to be homologous between the two species, but we may also want to indicate that one of them was radically reorganized during phylogeny. Therefore, we may want to call the derived structure in species A "novel by phylogenetic conversion" (meaning it has been converted from an old character; Figure 6.3A). This is the sense in which I consider mammalian neocortex to be new; it has a homologue in nonmammalian vertebrates but has changed so much that it deserves to be called new (see Chapter 8). Similarly, when ancestrally homogeneous structures become laminated during the course of evolution, their structure and function may change so much that the designation "new" may well be warranted (see the discussion later in this chapter). Obviously, the identification of "novelty by conversion" involves some subjectivity, since different observers may well have different thresholds for deciding that a character is "new," but why should novelty not lie in the eyes of the beholder? In a way, it always does, since everything in evolution can be viewed as either old or new, depending on your point of view and/or level of analysis (see Catania et al., 1999b). Therefore, I suggest that we use attributions of novelty simply to indicate when and where radical change occurred. That is, in fact, how most comparative biologists use the descriptor "new." Having settled that, let us expand the thought experiment.

If species A, B, and C each exhibit the same total number of different characters, then comparing these characters across species is relatively simple. For example, if each species has three different characters, and if two of them are already identified as being the same across species, then the third character is also likely to be the same across the three species (since the other characters are already spoken for). But what if you have an unequal number of characters in the different species? What if species A has four characters, while B and C have three? Then you have what I call "phylogenetic proliferation" (Figure 6.3B; or its opposite, phylogenetic reduction, but let us keep things simple). Thinking about how characters might proliferate, I recognize two possibilities: either one of the two remaining characters in species A is due to "phylogenetic addition," or both of the remaining characters in species A arose by "phylogenetic segregation" from the single remaining character in B and C (Figure 6.3C,D). How are we to decide between those two alternatives? This question has not been debated extensively but, in practice, comparative neurobiologists use the following rule of thumb. If, of the two remaining characters in species A, one is far more similar than the other to the remaining characters in species B and C, then the other, nonmatching character was probably added in species A. On the other hand, if both of the remaining characters in species A are equally similar to the remaining characters in B and C, then segregation most likely occurred. In this case, we might say that two characters in species A are "collectively homologous" to a single character in B and C. Many neurobiologists refer to such one-to-many homologies as a "field homology" (Smith, 1967), but I prefer the term "collective homology" because it allows for the possibility that homologous adult structures may develop from nonhomologous embryonic fields (see Chapter 3).

Thus, we can logically distinguish phylogenetic conversion from phylogenetic proliferation and, within the latter category, segregation from addition. In practice, these distinctions can be more difficult to draw because different kinds of change may prevail at different levels of the brain's structural hierarchy. For instance, the phylogenetic conversion of one brain region into another may, at a lower level of analysis, entail the phylogenetic addition of some new cell types (which probably occurred in neocortex evolution; see Chapter 8). Nonetheless, as long as we specify our level of analysis and compare closely related animals, the categories of phylogenetic conversion and proliferation, as well as segregation and addition, are useful. In the next section I use these categories to discuss some specific examples of evolutionary change in brain structure.

Phylogenetic Conversion: Lamination

The phylogenetic conversion of one brain region into a structurally different, new brain region may entail changes in a variety of different attributes, such as neuronal connectivity, histochemistry, or cytoarchitecture. Particularly interesting is the conversion of an ancestrally unlaminated brain region into a laminated one (let us define laminated as consisting of cellular and/or fibrous layers, or laminae, that lie parallel to one another). The paradigm case of phylogenetic lamination is the evolution of mammalian neocortex from an unlaminated, or at least poorly layered, homologue (see Chapter 8), but the neocortex is only one of many laminated structures in vertebrate brains. Accordingly, let us here consider several other brain regions that became highly laminated during the course of evolution, namely the vagal lobe of cyprinid teleosts, the torus semicircularis of gymnotoid electric fishes, and the dorsal lateral geniculate (LGN) of mammals. All of these structures are widely acknowledged to be homologous to unlaminated structures in related species, which is why I consider them to be instances of phylogenetic conversion. If we were to consider each lamina a separate brain region, we could also interpret phylogenetic lamination to be an instance of phylogenetic pro-

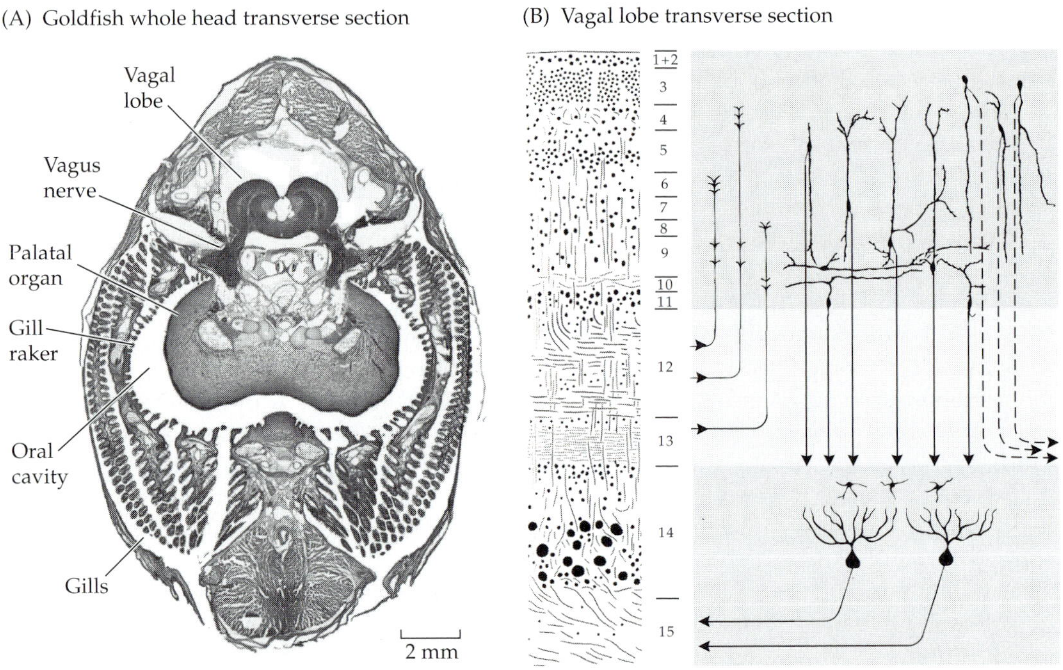

(A) Goldfish whole head transverse section

(B) Vagal lobe transverse section

Figure 6.4 The Vagal Lobe of Cyprinids   The vagal lobe of cyprinid teleosts is perhaps the most spectacularly laminated brain region in vertebrates. (A) A transverse section through the head of a goldfish reveals the vagal lobe, the palatal organ, which lines the roof of the oral cavity, and the gill rakers. (B) A schematic cross section through the vagal lobe reveals 15 laminae that are divisible into a sensory zone (laminae 1–11) and a motor zone (laminae 14 and 15), separated by a zone that contains predominantly axons (laminae 12 and 13). The left side of B shows neuronal cell bodies and axons; the right side depicts various cell types in their entirety, with the axons drawn schematically. (A from Farrell et al., 2002; B after Morita et al., 1983; Morita and Finger, 1985; Nieuwenhuys et al., 1998.)

liferation (with individual laminae resulting from phylogenetic segregation or addition), but evolutionary neuroscientists have traditionally considered laminae to be *parts* of a larger cell group, rather than cell groups themselves (e.g., Kaas, 1982). This is the convention I have followed here.

One of the most beautifully laminated brain regions, the vagal lobe of cyprinid teleosts (Figure 6.4A), is located in the medulla and controls the "lowly" process of feeding. In a typical cyprinid (e.g., a goldfish or a carp), the vagal lobe comprises roughly 20% of the brain (Brandstätter and Kotrschal, 1990) and consists of 15 separate laminae (Figure 6.4B) (Morita and Finger, 1985). Its 11 most superficial layers constitute a "sensory zone" that receives gustatory inputs from inside the mouth and is homologous to part of the nucleus of the solitary tract in other vertebrates. Its two deepest layers constitute a "motor zone" that innervates intraoral muscles and is homologous to part of nucleus ambiguus in other vertebrates. Within the sensory zone, different kinds of inputs tend to be confined to different laminae: information from taste buds in the palatal organ comes in to layers 6 and 9, while afferents from taste buds on the gill rakers terminate in layers 4 and 9 (Morita and Finger, 1985). Both inputs are topographically organized such that nearby taste buds project to nearby sites within a lamina, and both maps are in register (i.e. aligned) with one another (though the palatal organ map is more precise). Individual neurons within the sensory zone typically have dendrites that extend radially (i.e., orthogonally) across several laminae, and only a few neurons have dendrites that are oriented parallel to the laminae (Morita et al., 1983). The axons of sensory zone neurons either ascend to the diencephalon or descend to neurons that lie deep to them within the motor zone. Since the dendrites of these motor neurons have rather limited tangential spreads, we can say that information within the cyprinid vagal lobe tends to flow radially from the superficial layers to the deeper ones. This nearly rectilinear design of the vagal lobe circuitry suggests that some rather special computation is being performed. In order to discover what that special something is, we have to consider the vagal lobe's functions.

One function of the cyprinid vagal lobe is to pass gustatory information to the diencephalon. This is not surprising, since the noncyprinid homologues of the vagal lobe's sensory zone generally perform this kind of gustatory "relay" function (Kanwal and Finger, 1992). In cyprinids, however, the vagal lobes are also involved in a highly derived process of sorting good food from bad. If you have ever watched a goldfish eat, you may have noticed that it regularly spits out much of what it takes into its mouth. This odd behavior is part of a food-sorting routine that is unique to cyprinids (though some birds, notably parrots and ducks, exhibit similar food-sorting skills). Briefly, cyprinids have taste buds located throughout the inside of their mouth, especially on the gill rakers and on a special "palatal organ" (see Figure 6.4A). When these taste buds contact a tasty morsel, they pass that information on to neurons in the vagal lobe's sensory zone, which then project to motor neurons that innervate muscles in the palatal organ. These muscles are relatively small and organized in such a way that local contractions cause small "bumps" to form. Remarkably, gustatory stimulation at one location in the oral cavity causes a palatal organ bump to form at that same site (Sibbing et al., 1986), so that tasty bits of food are squeezed between the palatal organ and the gill rakers. Thus, the tasty food is held in place while the remaining items are spit out. Now, the principal function of the vagal lobe in this food-sorting process is that it houses the "reflex arcs" that connect taste buds to muscle fibers in a point-to-point, highly topographical manner (Finger, 1997). Given these observations, it seems likely that the vagal lobe of cyprinids evolved its laminar structure because lamination somehow allowed for better food sorting, but can we be more specific about the benefits of lamination?

One likely benefit of lamination is that it tends to minimize neuronal connection lengths. If multiple sensory and/or motor maps are distributed across different layers but in register with one another, then corresponding points in different maps can be connected by dendrites that extend across the laminae and by short, radially coursing axons. If the maps were located in different brain regions, much longer connections would be needed to connect the corresponding points. Thus, the rectangular arrangement of cellular laminae, dendrites, and axons in the cyprinid vagal lobe minimizes the region's overall volume, much like the hexagonal pattern of honeycombs minimizes their space requirements (Peterson, 1999). This is important because space is limited within the skull and because neural tissue is metabolically expensive to build and operate (see Chapter 4). In addition, minimizing connection lengths should minimize conduction times and, therefore, increase processing speed. Collectively, these savings in space, energy, and conduction time are likely to represent a significant benefit, particularly when neuron number is as large as it is in the cyprinid vagal lobe. Another likely benefit of lamination is that it probably facilitates the emergence of some functions that are more complex than the reflexes I just described. For example, short axonal connections within one of the vagal lobe's motor laminae may be all that is needed to generate "peristaltic waves" that spread across the palatal organ and move retained food into the esophagus. Similarly, descending axons might modulate the flow of information through a radial "column" of the vagal lobe simply by extending radially across the laminae. Presently, we have only hints that such waves and modulatory influences actually exist in the cyprinid vagal lobe (Finger, 1988), but I suspect that further studies will reap rich rewards.

Another likely reason for lamination to evolve is that it may be "easy" to develop. To see why this is so, consider first the mechanisms involved in generating topographical maps within the brain. As Roger Sperry first proposed, topographical projections probably develop mainly because axon outgrowth and termination are, to a large degree, chemotactic functions (Sperry, 1963). Briefly, the idea is that embryonic nervous systems are replete with molecules that are expressed in crisscrossing concentration gradients (if those molecules diffuse through extracellular space, they are called "morphogens"). Developing axons supposedly "read" those molecular gradients, grow to specific locations within them, and then terminate there. If neighboring neurons have similar "instructions" about where to terminate (which is likely because neighboring cell bodies are exposed to very similar molecular environments during their own early development), then topographical projections or "maps" result. These projections may not be terribly accurate, but they can later be honed by activity-dependent mechanisms that reduce the overlap between different sets of terminals (Figure 6.5A). For instance, if adjacent projection neurons have similar firing patterns (as they probably do in most sen-

(A) Topographical map formation

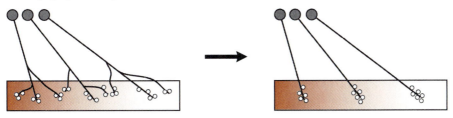

(B) Combined map and laminae formation

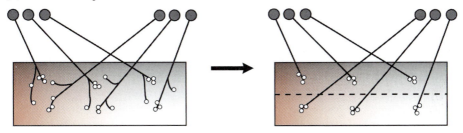

Figure 6.5 Development of Topographical Maps and Laminae   Lamination might be relatively "easy" to develop because the same basic mechanisms that are already used to build topographical maps could also be used to create laminar structures. (A) The first step in topographical map formation is that axons from adjacent input neurons grow to different but overlapping locations along a molecular gradient (here represented by a color gradient). The second step involves an activity-dependent reduction of this overlap, yielding a more refined topography (right). (B) In order to develop lamination, all that is needed is to add a second, orthogonal gradient (gray shading) and a second set of axons that terminates preferentially at a different point along that gradient. Activity-dependent terminal segregation would then "pull" those terminal zones apart to generate two separate topographical maps that are in register with one another (right). Some evidence supports this model of map and lamina formation, but it is still largely hypothetical.

sory systems), and if neurons with similar firing patterns compete against neurons with different firing patterns for synaptic sites, then any initially rough topography should become more refined over time (Willshaw and von der Malsburg, 1976). Many mapped projections within the nervous system are thought to be generated in this manner, though concrete evidence exists for only a few systems, and some puzzles remain (Goodhill and Richards, 1999; Thanos and Mey, 2001).

Of interest for our present discussion is that the same basic mechanisms that are apparently used to build topographical maps may also be used to build laminar structures with multiple maps. Imagine, for example, that two different sets of neurons project topographically to the same target, using the same molecular gradient to find their appropriate termination sites. In that case, two overlapping topographical maps would form (see Figure 6.5B). If the target structure contains an additional molecular gradient that is oriented orthogonally to the topography-determining gradient, and if the two sets of axons preferentially grow to different locations within that additional gradient, then the two topographical maps would become at least partially segregated along the radial dimension. That is, the presence of an additional "biasing" gradient would cause one of the two maps to lie above the other one, at least on average. If this partial segregation were subsequently enhanced by activity-dependent terminal segregation, a distinctly laminar termination pattern would result and the two maps would "automatically" be in register with one another (since they were based on the same topography-determining molecular gradient). This model of laminar development has never been tested explicitly, but it is supported by computer simulations (Elliott and Shadbolt, 1999) and is consistent with most of the available data on laminar development in the lateral geniculate nucleus (see Brunso-Bechtold and Casagrande, 1981; Rakic, 1981; Penn et al., 1998; Huberman et al., 2002). Therefore, we can conclude that laminar brain regions probably evolved at least in part because they are built readily.

Before proceeding further with this notion of "developmental ease," let me introduce another remarkable example of phylogenetic lamination, the electrosensory midbrain of gymnotoid electric fishes (Figure 6.6). As we saw in Chapter 5, gymnotoids are related to cyprinids but do not have extensive vagal lobes. Instead, they have a well-developed electrosensory system that is used mainly for locating objects in the environment (Bastian, 1986). Basically, these fishes generate a weak electric field with an electric organ in their tail and then use electroreceptors on their body surface to detect how that electric field is distorted by nearby objects. One structure that is involved in processing this information is the midbrain torus semicircularis (TS), which contains several distinct zones (one of which is probably homologous to the mammalian inferior colliculus). The electrosensory zone of this gymnotoid TS consists of 12 different laminae and 48 anatomically distinct cell types (Figure 6.6)(Carr and Maler, 1986). A parsimony analysis of toral lamination indicates that it is a derived feature of gymnotoids, since the TS is not laminated (or is at best poorly laminated) in other teleosts (Cuadrado, 1987). The electrosensory mormyrids, for example, have a well-developed TS, but it contains no obvious laminae (Bell and Szabo, 1986). Even the catfishes, which are closely related to gymnotoids and have a "passive" electrosensory system (Finger, 1986), do not have obvious toral laminae. Therefore, we can conclude that the laminated TS of gymnotoids almost certainly evolved from an unlaminated (or poorly laminated) homologue. Which raises the same question we asked of the cyprinid vagal lobe: What is the benefit of lamination in this brain region?

(A) The gymnotoid midbrian

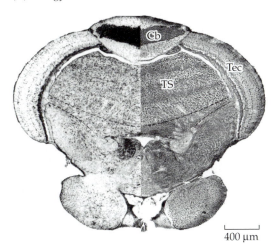

400 μm

Figure 6.6 The Electrosensory Torus Semicircularis of Gymnotoids Gymnotoid electric fishes have an enormous, highly laminated midbrain torus semicircularis. (A) In a cross section through the midbrain, you can see that the electrosensory portion of the torus semicircularis (TS) lies ventral to the cerebellum (Cb) and medial to the optic tectum (Tec). The left half of this section was stained to reveal cell bodies, the right half to show axons. Twelve distinct laminae are apparent. (B) In Golgi stains, 48 different neuron types can be identified within the electrosensory TS. Most neuronal processes course either parallel to the laminae or orthogonal to them. Adjacent laminae tend to receive inputs from different brain regions and most of those inputs are topographically organized. Other abbreviations: EL(P) and EL(T) = P- and T-type afferents from the electrosensory lateral line lobe; Vdesc = descending trigeminal nucleus; Tec = optic tectum; Tl = torus longitudinalis. (A from Nieuwenhuys et al., 1998; B from Carr et al., 1981; Carr and Maler, 1986; Nieuwenhuys et al., 1998.)

(B) Laminar circuits in the gymnotoid midbrain

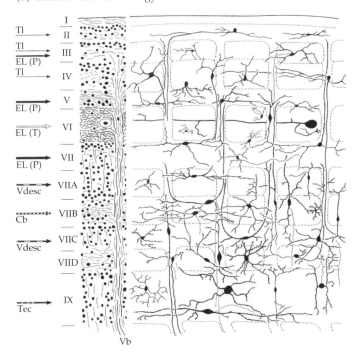

Again, we must consider circuitry and function. The principal inputs to the TS come from the electrosensory lateral line lobe (EL; see Figure 5.10 and Figure 6.6B). These inputs are topographically organized such that information from the fish's head is represented rostrally and information from the fish's dorsal aspect medially. Information from two different types of electrosensory receptors is conveyed to different sets of laminae: Information about the timing of an electrical signal is sent to layer 6, whereas information about a signal's strength is sent mainly to layers 3, 5, and 7 (Carr et al., 1981). Connectivity within the TS is quite complex but tends to be either parallel or orthogonal to the layers (see Figure 6.6B). The large cells in layer 6, for example, project almost exclusively to other cells in the same layer. Because these intralaminar connections extend for considerable distances within layer 6, individual cells in that layer integrate electrosensory information from widely separated parts of the body surface (Heiligenberg, 1986). Neurons in adjacent layers, which receive different afferents, then "tap" into that computed information by extending dendrites radially into layer 6 (Rose and Heiligenberg, 1985). Other neurons are also likely to integrate information across toral layers, but these have not been studied in detail. Turning to the outputs of the gymnotoid TS, we see projections to several targets. One prominent target is the optic tectum, which is itself laminated and topographically organized. Another target is the diencephalic nucleus electrosensorius, which is involved in controlling the electric organ. Finally, the TS is known to project topographically to a rhombencephalic nucleus preeminentialis, which in turn projects back to the EL. The latter projection appears to modulate how EL cells respond to electrosensory inputs (Bastian, 1998; Berman and Maler, 1999).

Given these data, we can draw some comparisons between the gymnotoid TS and the cyprinid vagal lobe. In both structures, different types of inputs project to different layers and are in register with one another topographically. This suggests that in both systems, the radial alignment of multiple topographical maps is a crucial element of how the circuits develop and/or function. An intriguing difference between the two structures is that the TS has not only *inter*laminar connections but also many long *intra*laminar connections (especially in layer 6; discussed earlier). These intralaminar connections allow gymnotoids to compare information from different locations of the body surface, which is useful in various behavioral contexts (Heiligenberg, 1986). So, why does the vagal lobe of cyprinids contain so few intralaminar connections? Perhaps they simply have not been looked for, or perhaps they would simply not be useful for food sorting, which is a highly localized function. The question is difficult. It is interesting to note, however, that both interlaminar and intralaminar connections are also, like laminae themselves, relatively easy to develop. To form interlaminar connections, growing axons merely need to remain confined to regions that have the same molecular "address" in the topography-determining gradient as their site of origin. Intralaminar connections are even easier to grow, for axons need only to remain within their lamina of origin. From this perspective, we would expect laminar structures to exhibit mainly intralaminar connections, interlaminar connections, or both. We would expect few other kinds of connections (e.g., connections that course diagonally across laminae) because they would require more complex developmental "instructions." Indeed, virtually all laminar structures are dominated by intralaminar and/or radial interlaminar connections. A fascinating exception is the dorsal lateral geniculate nucleus, which we will now consider in detail.

The mammalian dorsal lateral geniculate nucleus (LGN) receives visual inputs mainly from the retina and is reciprocally interconnected with the visual cortex. It is cytoarchitecturally homogeneous in monotremes and in many marsupials, poorly laminated in many small mammals (e.g., hedgehogs, echolocating bats, and rabbits), and divisible into 4 or more distinct laminae in primates, carnivores (i.e., cats, dogs, seals, and their close relatives), and some marsupials (Sanderson, 1974; Kaas et al., 1978; Kahn and Krubitzer, 2002b). Given this phylogenetic distribution, it is most parsimonious to conclude that a highly laminated LGN evolved at least three times independently, namely, in the lineage that led to primates and tree shrews (see Kaas, 2002), in carnivores, and in some marsupials (e.g., kangaroos). That much is widely accepted. Less clear is *why* the LGN became laminated in these diverse lineages. The observation that different LGN laminae receive their visual inputs from different eyes (Figure 6.7) suggests that lamination in the LGN might have evolved to prevent the convergence of visual information from the two eyes within the LGN. This hypothesis seems reasonable, since binocular convergence within the LGN might be associated with a loss of information that the visual cortex could use for other purposes. However, it can be at best a partial explanation for why lamination evolved, because many mammals have more LGN laminae than eyes. That is, if the LGN became laminated *merely* to segregate information from the two eyes, we would expect to see only two or three laminae, not four or more (three layers would result if the information from one eye is sandwiched between regions that receive inputs from the other eye, as it is in rats; see Figure 6.7A).

The most influential hypothesis to explain why those "extra" laminae evolved was proposed by Gordon Walls (1953). Essentially, Walls proposed that the mammalian retinocortical pathway comprises multiple "channels" that carry different kinds of visual information, project to different LGN laminae, and remain separate until they reach the neocortex. According to Walls, the LGN laminae evolved to keep the different channels from mixing in the LGN. This hypothesis is supported by the observation that physiologically and morphologically different kinds of retinal cells indeed project to different LGN laminae. In macaque monkeys, for instance, color-sensitive retinal neurons project predominantly to the outer, parvocellular layers of the LGN, whereas color-insensitive cells project mainly to the deeper, magnocellular layers (Derrington et al., 1984). Although these data support Wall's hypothe-

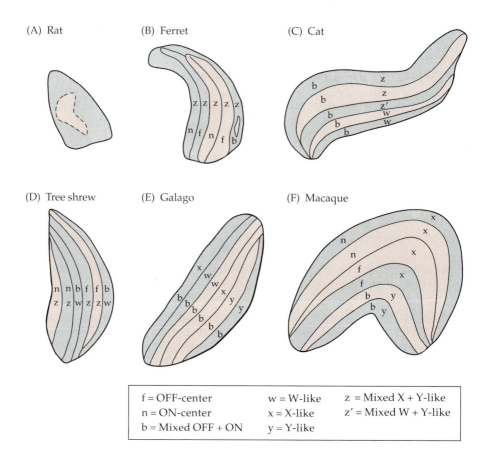

(A) Rat

(B) Ferret

(C) Cat

(D) Tree shrew

(E) Galago

(F) Macaque

f = OFF-center	w = W-like	z = Mixed X + Y-like
n = ON-center	x = X-like	z′ = Mixed W + Y-like
b = Mixed OFF + ON	y = Y-like	

Figure 6.7 Lamination and Functional Segregation in the LGN   The dorsal lateral geniculate nucleus (LGN) is relatively small and not obviously laminated in (A) rats and many other mammals, but it contains multiple laminae in various other mammals, including (B) ferrets, (C) cats, (D) tree shrews, (E) galagos, and (F) macaques. Laminae that receive their inputs mainly from the ipsilateral retina are colored light red; those with contralateral inputs are shown in gray. The dashed line in panel A indicates that the boundary between contralateral and ipsilateral retinal input regions is not cytoarchitectually distinct in rats. The lowercase letters in B–F indicate the kinds of physiological response properties most commonly found in the various laminae. Obviously, different species exhibit different patterns of functional segregation within the LGN. (After Jones, 1985; Strykyer and Zahs, 1983; Conley, 1988; Holdefer and Norton, 1995.)

sis, there are considerable disagreements in the literature about exactly what kind of information passes through which LGN laminae (see Levitt et al., 2001). Particularly interesting is that different species exhibit different patterns of functional segregation within the LGN (see Figure 6.7). In galagos, for example, X- and Y-like retinal cells (i.e., retinal cells with different morphologies and different physiological response properties) project to different LGN laminae, but this is not the case in minks, ferrets, or tree shrews, where individual laminae receive inputs from both X- and Y-like cells; conversely, ON- and OFF-center cells are segregated into different laminae in ferrets and tree shrews, but intermingled in prosimians (see Casagrande and Norton, 1991). In light of these species differences, we cannot argue, as Walls did, that information from different functional classes of retinal cells *must* be segregated into different LGN laminae for effective processing.

In fact, functional segregation can be accomplished without laminae as long as inputs project to restricted target regions (e.g., single cells or "patches" of cells) and those targets are not interconnected with one another. That kind of punctate connectivity is precisely what we see in the LGN of large primates, for retinal neurons in those species typically project to only a few adjacent LGN neurons, and those target cells typically have few intra- or interlaminar connections within the LGN (Guillery, 1966; Wong-Riley, 1972). Even the dendrites of most LGN neurons are relatively short (Figure 6.8). Therefore, different retinal input channels would mix minimally within the LGN even if there were no laminae. In other words, Walls was correct in pointing out that maintaining channel segregation is a crucial aspect of what the LGN does, but he was wrong to suggest that lamination evolved specifically to achieve that segregation. Why then did laminated LGNs evolve? Part of the

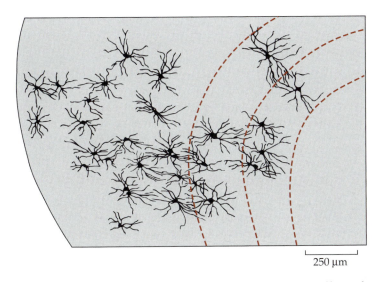

250 µm

Figure 6.8 Neurons in the LGN of a Squirrel Monkey   Shown here are Golgi-stained neurons in a section through the LGN of a squirrel monkey. The dendrites of most neurons in the LGN are not predominantly parallel or orthogonal to the laminar boundaries (dashed lines). This is unusual for such a highly laminated structure (contrast this figure with Figures 6.4B and 6.6B). (After Wong-Riley, 1972.)

Figure 6.9 Extrinsic Connections of the LGN   The external inputs to the LGN are, in contrast to its internal connections (see Figure 6.8), radially organized. (A) If a small lesion is made in the visual cortex of a mandrill (Kaas et al., 1972), neurons degenerate within a cone-shaped "projection column" (light red area) in the LGN. This shows that the projection from LGN to visual cortex is topographically organized and that the maps in different laminae are in register with one another. (B) Tracer injections into the visual cortex of an owl monkey (Ichida and Casagrande, 2002) reveal that individual corticogeniculate axons traverse and terminate in several LGN laminae like "toothpicks in a club sandwich" (Walls, 1953). (C) Tracer injections into the thalamic reticular nucleus of a cat (Wang et al., 2001) yield a swath of labeled neurons (red dots) and terminal arborizations (light red zone) that extends radially through all LGN laminae. These data demonstrate that the mammalian LGN has the same kind of radial translaminar connections that are typical of all laminar brain regions. (A after Kaas et al., 1972; B after Wang et al., 2001; C after Ichida and Casagrande, 2002.)

answer is most likely that, as I discussed earlier, lamination is relatively easy to evolve; but surely there must also have been some functional benefit to a laminated LGN. What was that benefit? In the case of the cyprinid vagal lobe and the gymnotoid TS, we argued that the short interlaminar connections afforded by a laminar organization were functionally advantageous, but does this explanation work also for the LGN? At first glance, the answer seems to be no, for LGN neurons are virtually devoid of interlaminar connections (be they axonal or dendritic). The LGN seems to be an exception to the rule.

A more detailed look reveals, however, that the exception is more apparent than real, for some neurons that lie outside the LGN do take advantage of its laminar organization by making strikingly radial and interlaminar connections (Figure 6.9). These extrinsic connections, which account for more than 80% of all synapses within the LGN (Sherman and Koch, 1986), come mainly from the visual cortex and the thalamic reticular nucleus. Both projections are topographically organized and extend across all LGN laminae (Robson, 1983; Wang et al., 2001; Ichida and Casagrande, 2002). Moreover, the axons of individual neurons in the visual cortex often terminate in radial "projection columns" that transect multiple laminae "like one of the toothpicks in a club sandwich" (Walls, 1953, p. 1). Therefore, although neurons intrinsic to the LGN have relatively few radial interlaminar connections, neurons extrinsic to the LGN have axons that do extend across multiple laminae in a distinctly radial fashion. Functionally, the projections from the visual cortex and the thalamic reticular nucleus probably serve to regulate or "gate" the flow of information through the LGN (Singer, 1977; Sherman and Koch, 1986). If this is true, then the radial interlaminar nature of those projections would allow for spatially selective visual attention—that is, the modulation of all visual information that originates from a particular location in the visual field (Desimone and Duncan, 1995; O'Connor et al., 2002). Alternatively, the extrinsic interlaminar projections might mediate some aspects of binocular depth

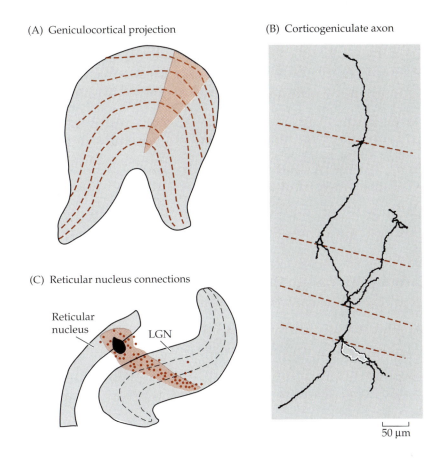

(A) Geniculocortical projection

(B) Corticogeniculate axon

(C) Reticular nucleus connections

Reticular nucleus

LGN

50 μm

perception (McIlwain, 1995). Either way, the rectilinear design of the visual cortical and thalamic reticular projections to the LGN is likely to improve some aspect(s) of vision with a minimum of axonal wiring.

Therefore, we can conclude that lamination is fairly easy to develop and functionally advantageous, because corresponding neurons in multiple maps can be interconnected with a minimum of dendritic or axonal wiring. This happy confluence of developmental ease and functional benefit probably explains why lamination evolved so frequently among the vertebrates (as well as in invertebrates; see Bullock and Horridge, 1965; Strausfeld et al., 2000).

Phylogenetic Proliferation: Segregation

The term "phylogenetic segregation" was coined by Ramón-Moliner and Nauta (1966) to describe the evolution of clearly differentiated nuclei from a diffusely organized brain stem core. I here use the term to denote instances where an ancestrally homogeneous brain region evolved into multiple distinct regions, none of which is homologous in a one-to-one manner with the old, ancestral region (see Figure 6.3D). As in the previous section, let us confine our analysis to the level of distinct cell groups, which we may define as "delineable clusters of neuronal somata" (Nieuwenhuys et al., 1998, p. 42) with similar connections and/or functions (Kaas, 1987; Passingham et al., 2002). And, to illustrate the principle, let us focus on two regions that differ enormously between the major groups of vertebrates, namely the dorsal thalamus and the posterior tuberculum.

The dorsal thalamus is well known to contain far more nuclei in amniotes than in anamniotes (Figure 6.10A; see also Figure 6.1). Roughly 10 years ago, Ann Butler (1994, 1995) explained this taxonomic difference in cell group number by arguing that the dorsal thalamus of all vertebrates is divisible into a "lemnothalamus" that receives mainly retinal inputs and a "collothalamus" that receives its dominant input from the midbrain roof (the superior and inferior colliculi in mammals). In fishes and amphibians, Butler argued, the lemnothalamus consists of just a single nucleus, called nucleus anterior, which lies rostrally within the dorsal thalamus and is closely apposed to the ventricular zone, where its neurons were born. In reptiles, birds, and mammals, on the other hand, the lemnothalamus comprises numerous distinct cell groups, most of which are located at some distance from the ventricle (see Figure 6.10A). According to Butler, none of these lemnothalamic nuclei in amniotes are strictly homologous (i.e., homologous in a one-to-one manner) to the nucleus anterior of anamniotes. Instead, Butler hypothesized that all of

Figure 6.10 Proliferation by Phylogenetic Segregation   Two examples of phylogenetic proliferation by segregation of an ancestrally simple structure into several discrete nuclei. (A) According to Ann Butler (1994), the dorsal thalamus is divisible into two sets of nuclei, namely the lemnothalamus (light red) and the collothalamus (gray), that develop from two distinct patches of the embryonic ventricular zone (remnants of which are shown in red and black at the bottom of each rectangle). The lemnothalamus consists of a single nucleus in anamniotes, but it consists of numerous nuclei in amniotes, particularly in mammals, which implies that the number of lemnothalamic nuclei probably increased during amniote phylogeny. Since none of the lemnothalamic nuclei in amniotes are homologous in a one-to-one manner to the single lemnothalamic nucleus in anamniotes, this proliferation of lemnothalamic nuclei was probably due to phylogenetic segregation rather than addition. (B) Phylogenetic segregation also occurred in the preglomerular complex (PGC, light red), which is part of the posterior tuberculum (see Figure 3.8). In basal ray-finned fishes, such as polypterus, sturgeons, or gars, the PGC is either absent or represented by only a few, relatively indistinct cell groups. In teleosts, however, the PGC is divisible into at least 6 distinct cell groups. Since these cell groups in teleosts are not readily divisible into "old" and "new" cell groups, they probably evolved by phylogenetic segregation. The gray cell groups in this figure represent other components of the posterior tuberculum. (A after Butler, 1994; B after Braford and Northcutt, 1983; Striedter, 1990a; Rupp and Northcutt, 1998.)

the lemnothalamic nuclei in amniotes develop from a single embryonic region that is homologous "as a field" to the embryonic precursor of nucleus anterior in anamniotes (Butler, 1994). Translated into the terms of this discourse, Butler proposed that all lemnothalamic nuclei in amniotes are "collectively homologous" to nucleus anterior and were derived from it by phylogenetic segregation.

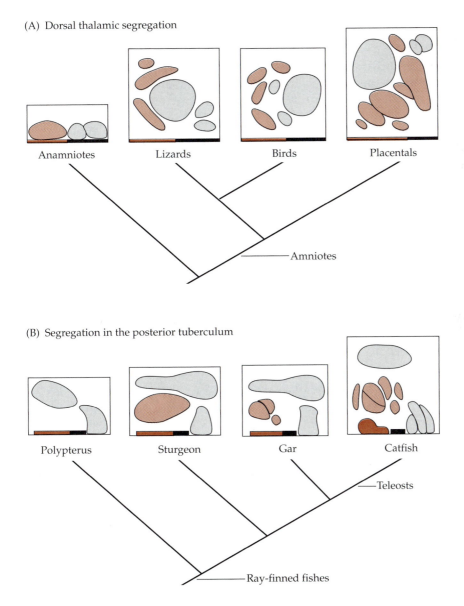

(A) Dorsal thalamic segregation

Anamniotes   Lizards   Birds   Placentals

Amniotes

(B) Segregation in the posterior tuberculum

Polypterus   Sturgeon   Gar   Catfish

Teleosts

Ray-finned fishes

Butler's hypothesis of dorsal thalamic segregation is reasonable because there really is no single structure in the dorsal thalamus of amniotes that can be homologized to the nucleus anterior of anamniotes. However, Butler's hypothesis also raises new questions. One of these concerns development. Over 50 years ago, Jerzy Rose had described five "pronuclei" in the rabbit's embryonic dorsal thalamus (Rose, 1942), but do these five pronuclei develop from two distinct embryonic "fields," as Butler has proposed? Recent gene expression data suggest that the embryonic dorsal thalamus is divisible into three separate zones rather than two. This finding could be interpreted as contradicting Butler's hypothesis, but perhaps two of these three zones developed from a single zone at an even earlier developmental stage (Redies et al., 2000; Puelles, 2001). The data needed to resolve this issue are not yet available. Another important question that remains to be addressed is the extent to which the dorsal thalamus of amniotes evolved new attributes that are not present at all in nucleus anterior. As far as I can tell, reciprocal connections between dorsal thalamus and dorsal pallium evolved only in amniotes, but how many other "new" features are there in the dorsal thalamus of amniotes? At this point, the question remains open, mainly because the dorsal thalamus of anamniotes has not yet been studied in detail. However, given that amniotes diverged from anamniotes roughly 300 million years ago, we would actually expect the dorsal thalamus of amniotes to exhibit numerous attributes that are not typical of nucleus anterior. That expectation does not contradict the hypothesis that the amniote dorsal thalamus evolved by phylogenetic segregation, for it is rarely the case that phylogenetic segregation manifests itself *solely* as a segregation of preexisting attributes (or cell types) into distinct cell groups. Most of the time, it is accompanied by the evolution of at least some novel attributes.

A second major brain region that is likely to have evolved by phylogenetic segregation is the preglomerular complex of ray-finned fishes (see Figure 6.10B). This preglomerular complex probably constitutes a migrated portion of the posterior tuberculum (see Figure 3.8) and has no obvious homologue outside of ray-finned fishes (Braford and Northcutt, 1983; Northcutt, 1995; Rupp and Northcutt, 1998). Most interesting for present purposes is that the number of distinct preglomerular nuclei varies considerably across species. In the oldest ray-finned fishes (i.e., polypterus; see Figure 3.1), the preglomerular complex can hardly be identified, and in sturgeons, which are the next oldest lineage of ray-finned fishes, it consists of a single, fairly homogeneous cell group (see Figure 6.10B) (Braford and Northcutt, 1983; Striedter, 1990a). In gars and amia, which evolved soon after the sturgeons, the preglomerular complex consists of three distinct cell groups (Mueller and Wullimann, 2002), and in teleosts, which are the newest and most species-rich group of ray-finned fishes, that number doubles again (see Wullimann, 1997). Thus, we here have a clear case of phylogenetic proliferation. But was the increase in cell groups due to phylogenetic segregation or phylogenetic addition? That question has never been addressed in print, but I think addition seems unlikely since it is not the case that some preglomerular nuclei are clearly "old" across the ray-finned fishes, while others are "new." Instead, the teleostean preglomerular nuclei all share at least some similarities with the more homogeneous preglomerular complex of the "basal" ray-finned fishes; they all seem similarly "new." This implies that the preglomerular nuclei of teleosts probably proliferated by means of phylogenetic segregation rather than addition.

Again, the segregation in this case was accompanied by the evolution of new attributes. For instance, some preglomerular neurons in teleosts are apparently born far away from the ventricle, which is highly unusual (Mueller and Wullimann, 2002). Moreover, many preglomerular nuclei have evolved novel sensory inputs and telencephalic projections that make them similar to the dorsal thalamic nuclei of amniotes, even though the two brain regions are clearly not homologous (see Striedter, 1990b). One preglomerular nucleus, the "nucleus prethalamicus" of squirrel fishes, even evolved a laminar structure and reciprocal connections with the telencephalon (Figure 6.11) (Ito and Vanegas, 1983). These attributes are not part of the posterior tuber-

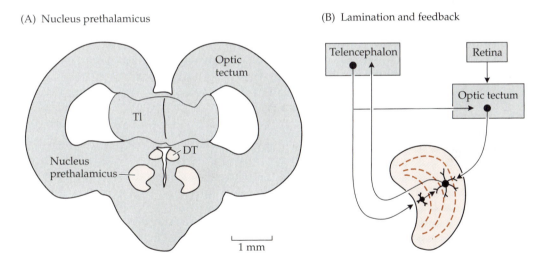

Figure 6.11 Nucleus Prethalamicus in Squirrel Fishes The nucleus prethalamicus is a preglomerular nucleus that is remarkably similar, but not homologous, to the pulvinar nucleus of primates (see Chapter 9). (A) In this cross section through the midbrain and diencephalon of a squirrel fish, the nucleus prethalamicus lies ventrolateral to the dorsal thalamus (DT; see also Figure 6.1). (B) The nucleus prethalamicus contains several laminae (dashed red lines) and resembles the mammalian pulvinar insofar as it receives major inputs from the optic tectum and has reciprocal connections with the telencephalon. These similarities must be due to convergent evolution because the pulvinar is a dorsal thalamic nucleus (and because basal ray-finned fishes lack anything that resembles the nucleus prethalamicus; see Figure 6.10). Other abbreviations: DT = dorsal thalamus; Tl = torus longitudinalis. (After Ito and Vanegas, 1983.)

cular region in nonteleosts, which means that they must have been added as teleosts evolved. As in the case of dorsal thalamic evolution, we do not yet know the full spectrum of innovations that the preglomerular complex of teleosts evolved. However, given that teleosts comprise roughly half of all vertebrate species and occupy an enormous variety of ecological niches (Kotrschal et al., 1998), we can expect the preglomerular complex to have evolved numerous new attributes. Teleost brains seem odd in many ways (at least if we look at them through our primate eyes), but their preglomerular region is undeniably one of their most peculiar "new" structures.

Phylogenetic Proliferation: Addition

Ideally, phylogenetic addition is invoked only when the alternative hypotheses of phylogenetic segregation and conversion have been eliminated, when all potential homologues in other species are accounted for and the region of interest is clearly "left over." This prerequisite is rarely satisfied, but some clear-cut examples of phylogenetic addition exist in all major brain divisions. In the medulla, for example, the electrosensory lateral line lobe (EL; see Figure 5.10) is best viewed as a phylogenetic addition to the "old" complement of sensory brainstem nuclei (see Finger et al., 1986; McCormick, 1989). Similarly, the cerebellar valvula, which is so enormous in the mormyrids (see Figure 5.11), is unique to teleosts and is most reasonably interpreted as a rostral addition to the cerebellar "corpus," which is homologous to the cerebellum of nonteleosts (Nieuwenhuys et al., 1998). In the midbrain, the electrosensory portion of the TS in gymnotoids and catfishes (see Figure 6.6) evolved as a phylogenetic addition to the old complement of toral nuclei because it does not have a discrete homologue in nonelectroreceptive teleosts (see Striedter, 1991). Yet another example of phylogenetic addition is the torus longitudinalis, which lies along the medial edge of the optic tectum in all ray-finned fishes (see Figure 6.11A)(Northmore, 1984; Ito et al., 2003), but in no other vertebrates. As these examples illustrate, brain evolution does at least occasionally add new structures to ancestral brains. Edinger may have been wrong to suppose that the entire dorsal telencephalon was added in amniotes, but his idea of evolution by accretion is not to be dismissed outright.

But wait, you might object, how can "something evolve from nothing," as the idea of phylogenetic addition seemingly implies? This concern is natural, but saying that something *evolved* from nothing does not amount to saying that it *developed* from nothing (ex nihilo). Clearly, every adult structure has a developmental precursor. However, not every adult structure has to have an evolutionary precursor in ancestral adults. To see why this is true, imagine a developmental precursor region that gives rise to a single structure X in some ancestral species A (Figure 6.12A). Imagine further that some part of that same homologous precursor region in a descendant species B becomes sub-

(A) Phylogenetic addition by de novo gene expression

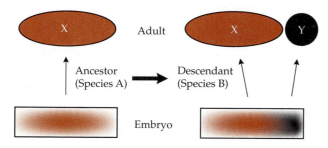

(B) Phylogenetic addition by precursor expansion

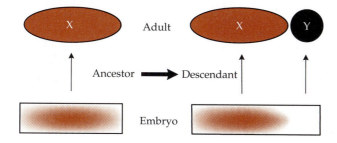

Figure 6.12 Developmental Mechanisms for Phylogenetic Addition This figure schematically illustrates how new adult brain regions may be added to phylogenetically old structures. (A) If a new gene expression pattern (gray gradient) appears at the edge of a conserved embryonic precursor region (rectangles), then this modified precursor region may develop into a new adult structure (region Y). Meanwhile the remaining, unmodified precursor region is likely to autoregulate, meaning that it will still develop into its "normal" adult derivative (region X). If this occurs, then X is homologous between ancestor and descendant, but Y is new by phylogenetic addition. (B) Phylogenetic addition may also occur when an embryonic precursor expands so that some of its parts are no longer exposed to their old developmental stimuli, leaving an area (here represented by the "blank" region in the right rectangle) free to embark on a novel developmental trajectory. This second mechanism probably accounts for the addition of new cortical areas in large-brained mammals (see Figure 6.17).

jected to a novel developmental gene (or combination of genes) and embarks upon a novel developmental trajectory, leading to some new adult structure Y. The crucial question is: What happens to the remainder of the developmental precursor, the portion that does not exhibit the altered gene expression pattern? Most likely, the unmodified portion of the precursor will follow its ancestral trajectory and develop into a perfectly good adult structure X (see Figure 6.12A). After all, we already know from numerous experimental embryological studies that partial ablations of a precursor region tend not to disturb subsequent development (see Huxley and de Beer, 1934; Serluca and Fishman, 2001). Now, if X develops normally in species B, then it must be homologous to structure X in species A. Structure Y, in contrast, has no homologue in species A. It is a phylogenetic addition, even though it clearly developed from a phylogenetically old developmental precursor. This thought experiment reveals that phylogenetic addition is not a theoretical impossibility (see also Striedter, 1998b). But does it actually occur? And, if so, does it result from the kinds of developmental changes I have described?

These questions are difficult to answer, but the hypothesis is plausible because all of the added structures I mentioned lie adjacent to regions that are very similar to them, as we would expect if they develop from the same "old" precursor region (Figure 6.12A). The cerebellar valvula, for instance, is structurally quite similar to the caudally adjacent corpus cerebelli (Nieuwenhuys and Nicholson, 1969; Meek et al., 1986a,b). Similarly, the main auditory nucleus in the brain stem of frogs, which has no homologue in other amphibians or in anamniotes, lies immediately lateral to the "old" vestibular nuclei, with which it shares some similarities (Fritzsch et al., 1984). This topological adjacency suggests that the frog's auditory nucleus evolved because some new developmental stimulus (e.g., incoming auditory axons) impinged upon the margins of an old precursor region. Unfortunately, we still know relatively few details about how the frog's brain stem or the teleosts' valvula develops. Therefore, let us now turn to the most thoroughly studied, and most heavily debated, example of phylogenetic addition in vertebrate brains, namely, the addition of new neocortical areas.

The notion that mammalian neocortex is divisible into numerous distinct areas that span the cortical depth dates back at least to Brodmann (1909), who based his conclusions on comparative cytoarchitectural data (see Chapter 2). Later studies added connectional and physiological criteria to determine which cortical subdivisions deserve the status of "area" and proposed numerous modifications to Brodmann's initial scheme (Kaas, 1982, 1987). Although debates continue about the status of some cortical regions (e.g., Rosa and Krubitzer, 1999; Zeki, 2003), many cortical areas are now firmly established. Especially important for our present discussion is that Brodmann was fundamentally correct when he claimed that the number of cortical areas differs between species, being larger, for example, in monkeys than in hedgehogs or rats (Figure 6.13). According to Jon Kaas and his collaborators, who have examined the cortex of more than two dozen different mammalian species, most small mammals have less than 20 (or so) distinct neocortical areas (Northcutt and Kaas, 1995; Catania et al., 1999a). Macaques, in contrast, have over 30 distinct areas in their visual system alone (Figure 6.14) and probably at least 50 areas overall (van Essen et al., 1992). For humans, the data are extremely limited, but there is no reason to believe that they have fewer neocortical areas than a macaque (Killackey, 1994; Disbrow et al., 2000; Wallace et al., 2002) (see Chapter 9). If we subject these data to a parsimony analysis, the

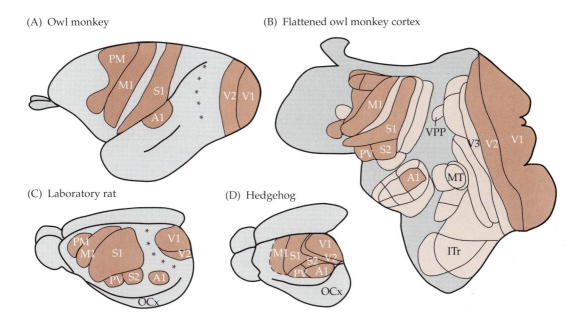

Figure 6.13 "Old" and "New" Neocortical Areas A comparative analysis of mammalian neocortical areas reveals that some areas are "old" whereas others are "new." The old, highly conserved neocortical areas (here shown in red) include the primary and secondary sensory and motor areas. The new areas are interposed between the old areas and generally consist of higher-order sensory or "association" cortices. Crucially, the territory devoted to new areas is much larger in monkeys (A and B) than in (C) rats. This is most evident when we look at the band of cortex lying between the visual, somatosensory, and auditory cortices (here denoted by asterisks). (D) In hedgehogs, which have small neocortices that probably resemble those of early mammals (see Chapter 8), virtually the entire neocortex consists of "old" neocortical regions. This implies that the new neocortical areas were probably added in phylogeny. Unfortunately, it is not yet clear whether hedgehogs have a premotor cortex. Panel B shows some of the new neocortical areas (light red) that have been identified in owl monkeys. Abbreviations: A1 = primary auditory cortex; FEF = frontal eye field; ITr = rostral inferior temporal cortex; M1 = primary motor cortex; MT = middle temporal area; OCx = olfactory cortex; PM = premotor cortex; PV = ventral parietal cortex; S1 = primary somatosensory cortex; S2 = secondary somatosensory cortex; V1 = primary visual cortex; V2 = secondary visual cortex; V3 = tertiary visual cortex; VPP = ventral posterior parietal cortex. (A–C after Northcutt and Kaas, 1995; D after Catania et al., 2000; B modified to show V3 after personal communication with Jon Kaas.)

answer is clear: the number of cortical areas has increased during primate phylogeny. A similar analysis reveals that neocortical areas also proliferated in at least some carnivores, for domestic cats reportedly have at least 30 different cortical areas, including 15 within the visual system (Figure 6.14). Additional increases in the degree of neocortical subdivision may have occurred

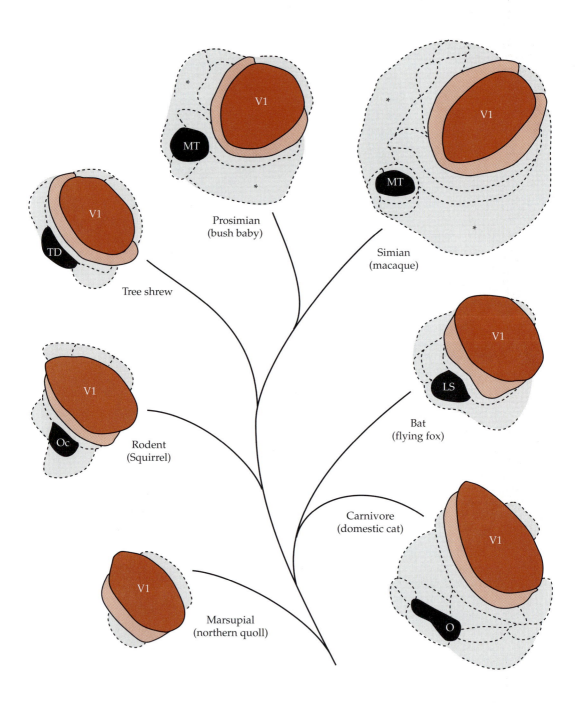

Tree shrew

Prosimian
(bush baby)

Simian
(macaque)

Rodent
(Squirrel)

Bat
(flying fox)

Carnivore
(domestic cat)

Marsupial
(northern quoll)

◀ **Figure 6.14 A Phylogenetic Tree of Visual Cortical Areas**   The evolutionary history of the visual areas in the mammalian neocortex remains the subject of debate. Clearly, all mammals have a primary visual cortex (V1; dark red areas). According to Marcello Rosa (1999), all placental mammals also have a more anterior visual cortical area (black areas) that is called the middle temporal area (MT) in primates but goes by various different names in nonprimates. Most mammals also have a secondary visual cortex (V2; light red areas), but whether any other visual cortical areas can likewise be homologized across all mammals remains uncertain. The so-called tertiary visual area (V3), for instance, can be identified in both prosimians and simians, as well as cats, but not in tree shrews; this suggests that V3 evolved independently in primates and carnivores (Lyon and Kaas, 2002), but the issue is not settled yet (see Rosa, 1999). Despite such uncertainties, what seems fairly clear is that the number of visual cortical areas has increased with neocortex size (see also Figure 6.16) and that "new" visual areas typically appeared at the margins of the "older" areas (see also Figure 6.17). In this figure, regions marked by asterisks comprise multiple areas whose borders are not shown, and the shapes of individual areas are distorted. Other abbreviations: LS = lateral suprasylvian area; O = occipitotemporal area; Oc = caudal occipitotemporal area; TD = dorsal temporal area. (After Rosa, 1999.)

in cetaceans and elephants (or other large-brained mammals; see Chapter 4), but too little is known about the neocortex of these species to allow firm conclusions on that point. In any case, neocortical areas certainly proliferated at least twice among mammals.

So, did this phylogenetic proliferation of neocortical areas occur by segregation or addition? Much of the older literature is unclear on this question, since terms such as "progressive differentiation" (von Economo, 1929) include either possibility. An unambiguous argument in favor of segregation was made by Richard Lende (1969), who argued that primary motor and somatosensory areas overlapped in early mammals and later segregated into distinct areas. More generally, Lende argued that "the present patterns of cortical localization evolved from a polymodal pallium containing completely superimposed representational [i.e., sensory and motor] areas" (Lende, 1969, p. 273). As you shall see in Chapter 7, some aspects of Lende's hypothesis are probably correct. However, much of its supporting evidence has been called into question, and detailed physiological mapping studies have shown that even in hedgehogs, which have a very small and probably primitive neocortex (see Chapter 8), the principal auditory, visual, and somatosensory areas exhibit sharp boundaries and virtually no overlap (Kaas et al., 1970; Catania et al., 2000). Indeed, Jon Kaas and his collaborators have now mapped the neocortex in a wide variety of mammals and have found that all of them have the same basic and well-segregated complement of primary and secondary sensory areas (see Kaas, 1987; Krubitzer, 1995; Northcutt and Kaas, 1995; Catania et al., 1999a). On the motor side the situation is less clear, but most placental mammals do have a primary motor cortex and some premotor regions (see Chapters 7 and 8). Thus, the data indicate that most "lower-order" sensory and motor areas evolved early in mammalian evolution and were retained in most lineages (see Catania et al., 1999a for a possible exception). If that is true, what are we to make of all those higher-order sensory and "association" areas (Kaas, 1999) in primates and carnivores? When and how did they evolve?

Looking across mammals, it quickly becomes apparent that the higher-order areas are far less conserved than the lower-order ones. For instance, if we compare the visual cortex between macaques and cats, we find that, although some higher-order areas are superficially similar between the two species (Payne, 1993), most of them are not found in other mammals, which implies that they probably evolved independently in primates and carnivores (Rosa, 1999; Kaas and Lyon, 2001). Moreover, if we extend our analysis to rats (see Figure 6.13), we find that the lower-order visual, somatosensory, and auditory areas lie closer together than they do in primates or cats, leaving much less territory in which higher-order areas might lurk. This "interposed cortex" (indicated by asterisks in Figure 6.13) might contain one-to-one homologues of *some* higher-order visual areas in primates or cats (e.g., V2 and MT; see Figure 6.14) but is unlikely to harbor strict homologues for *all* of them (Rosa, 1999; Kaas and Lyon, 2001). Perhaps, you might argue, the interposed cortex in rats and other small-brained mammals is an "amalgam" of all the higher-order visual areas in primates or cats and, therefore, homologous to them collectively. This would imply that the higher-order areas of monkeys and cats evolved out of that ancestral amalgam by phylogenetic segregation. There is, however, no experimental support for this hypothesis, and all of the really small mammals have virtually no interposed cortex (see Figure 6.13D)(Catania et al., 1999a; 2000). This is important because the earliest mammals were almost certainly quite small (see Chapter 8), and if they had no interposed cortex, then the higher-order visual cortical areas could not all have evolved from it by phylogenetic segregation. Some of them, it seems, must have evolved by phylogenetic addition.

But how, precisely, were those new cortical areas added in phylogeny? One answer to this question was proposed by Leah Krubitzer (1995), who argued that new cortical areas evolve by the selective aggregation of modular heterogeneities within "old" areas. Specifically, she proposed that cortical area addition begins with the invasion of an old cortical area by novel inputs, which then causes the old area to become studded with novel patches or modules that receive the new inputs. Later in phylogeny, those new modules may aggregate into a coherent area, which amounts to having a new cortical area added in phylogeny. This "module aggregation hypothesis" is consistent with the observation that many cortical areas indeed contain various modules, such as "somatosensory barrels" and "ocular dominance stripes" (see LeVay and Nelson, 1991; Purves et al., 1992). The hypothesis is also supported by the observation that thalamic inputs to the neocortex seem to play a major role in the developmental specification of cortical areas (see Rakic et al., 1991;

Kahn and Krubitzer, 2002a). However, the module aggregation hypothesis fails to specify what phylogenetic changes in development would lead to the selective aggregation of discontinuous modules into a coherent area. Conceivably, module aggregation might result when the module-inducing axons express more molecules that cause them to fasciculate (form bundles) as they approach the neocortex, causing modules to be induced right next to one another. More data is sorely needed to support or falsify this hypothesis.

A second, very different explanation for the phylogenetic addition of neocortical areas was proposed by John Allman and Jon Kaas (1971), who argued that cortical areas might duplicate in evolution just as genes or other body parts are known to duplicate from one generation to the next (Gregory, 1935; Ohno, 1970; Kaas, 1984). An appealing aspect of this "area duplication hypothesis" is that the functional redundancy created by the duplication would allow one of the duplicates to diverge in structure and function (during subsequent generations) while the original function is maintained. However, that appeal is shared by the modular aggregation hypothesis, for modules might likewise diverge functionally from the matrix in which they are embedded. Moreover, the mechanisms that could cause area duplication remain largely mysterious. It is unlikely, for instance, that the appearance of a new cortical area is due *merely* to the duplication of some gene. More likely, area duplication is due to changes in gene *expression*. But how, in detail, might that work? A partial answer to that question comes from the work of Tomomi Fukuchi-Shimogori and Elizabeth Grove (2001), who showed that when FGF8, a molecule that is normally expressed at the rostral pole of the developing telencephalon (Figure 6.15A), is artificially expressed in the caudal telencephalon as well, an additional somatosensory area develops just caudal to the normal one (Figure 6.15B). Intriguingly, this new cortical area seems to be a mirror image of the "old" somatosensory map, just as most phylogenetically added brain regions seem to be mirror images of their older neighbors (e.g., Rosa, 1999). Overall, these data suggest that cortical areas might duplicate whenever genes that are involved in their developmental specification are expressed in additional areas. However, this mechanistic explanation still fails to specify what might normally cause "old" genes to be expressed in novel areas.

Thus, the two major hypotheses that have been proposed to explain area addition in the neocortex both remain unsatisfactory. Most important, neither hypothesis takes absolute brain size into account. This omission is important because primates and carnivores differ from other mammals not only in possessing additional neocortical areas, but also in possessing larger brains (see Chapter 4). Indeed, the number of cortical areas in a given species correlates rather tightly with the total amount of neocortex that is available for area formation (Figure 6.16)(Changizi, 2001). This suggests that cortical expansion is causally related to area addition. The question is: how? One possible answer emerges from the observation that many of the molecules important for neo-

(A) Telencephalic gradients

(B) An added cortical area

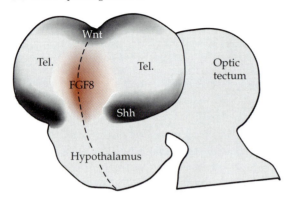
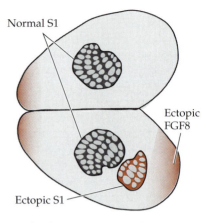

Figure 6.15 Gradients and Duplication in the Developing Neocortex Recent studies have revealed the existence of several molecular gradients in the developing telencephalon and researchers have begun to manipulate those gradients experimentally. (A) In this schematic rostrolateral view of an early embryonic mouse brain, several molecular gradients are indicated by red and black shading. Note that fibroblast growth factor 8 (FGF8) is expressed most heavily at the rostral pole of the developing telencephalon (Tel.). (B) If FGF8 is also experimentally expressed at the caudal pole of the telencephalon, an additional neocortical area can be induced (Fukuchi-Shimogori and Grove, 2001). This additional "ectopic" area is essentially a mirror-duplication of the primary somatosensory cortex (here shown in gray with "barrel fields"). Other abbreviations: Shh = sonic hedgehog; Wnt = wingless type. (After O'Leary and Nakagawa, 2002.)

cortical subdivision are expressed at the margins of the embryonic neocortex (see Figure 6.15) (see O'Leary and Nakagawa, 2002 for review). If these molecules are "morphogens" that can diffuse no more than a few hundred microns into the neocortical precursor tissue, then a phylogenetic expansion of the neocortical precursor tissue should cause some central areas within the precursor to remain unexposed to those morphogens. Such an absence of ancestrally present morphogens might lead to a re-routing of developmental trajectories (Striedter, 1998b) and, hence, to the appearance of a new cortical area (see Figure 6.12B). According to this hypothesis, we would expect new cortical areas to be added near the center of the enlarged neocortex. As Friedrich Sanides (1967) put it, new neocortical areas should be added like "growth rings," from the neocortical periphery toward its center. This hypothesis is plausible but not supported by the evidence, for new cortical areas do not generally appear in the center of the neocortex (Kaas, 1987). Instead, they are added in between the older areas, which are dispersed across the neocortical

(A) Double logarithmic coordinates

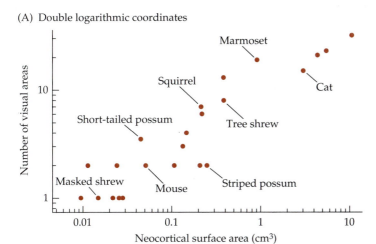

(B) Linear coordinates

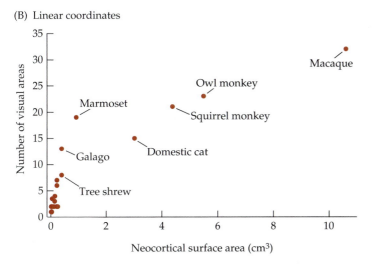

Figure 6.16 Neocortical Size and Area Number   The number of visual cortical areas correlates rather well with the total neocortical surface area in a species. (A) In double-logarithmic coordinates, the data points tend to fall on a straight line. (B) In linear coordinates, the number of visual areas increases steeply with cortical area as long as cortical area is less than about 2 cm^3. After that, the scaling relationship becomes less steep and individual cortical areas become larger in absolute size. This increase in the absolute size of cortical areas may be associated with an increase in the "modularization" of individual areas (i.e., with the appearance of ocular dominance columns, blobs, etc.) in large-brained species. (Data from Jon Kaas, Leah Krubitzer, and their many collaborators; data compiled by Cheung et al., in press.)

terrain (see Figure 6.13). Therefore, we need a different hypothesis to explain how neocortical expansion causes the addition of new cortical areas.

Currently, no such hypothesis has been offered, but Marcello Rosa (2002) recently proposed the outlines of one. In essence, Rosa suggests that some of the old neocortical areas, specifically V1 and area MT (see Figure 6.14), act as intracortical organizing centers that emerge early in neocortical development, emit some sort of morphogen, and interact with activity-dependent mechanisms to control area differentiation in the surrounding neocortex (Figure 6.17). Loosely speaking, they act like the "seeds" in crystal formation. Rosa further proposes that if these intracortical organization centers lie close to one another, as they would in a species with a relatively small amount of neocortex, then only a single cortical area (V2) can be induced between them. However, if the organizing centers are pulled apart because of a phylogenetic increase in the size of embryonic neocortex, then additional cortical areas can be induced within the "extra space" (see Figure 6.17A). This hypothesis incorporates many important aspects of cortical development and evolution, but it requires further elaboration. It is not yet clear, for instance, whether area MT really has a homologue in nonprimates (Allman and Kaas, 1971; Northcutt and Kaas, 1995). We also do not know which morphogens, if any, originate from V1 or MT, or how the borders between the additional areas are specified. Finally, it remains unclear how the putative organizing centers (i.e., V1 and MT) are themselves set up. Perhaps they are induced by the aforementioned signaling centers at the margins of the developing neocortex (see Figure 6.15)

Figure 6.17 A Developmental Hypothesis of Cortical Area Addition   An intriguing hypothesis about how visual cortical areas develop was proposed by Marcello Rosa (2002). This hypothesis is based on the observation that the topographical maps of adjacent areas are generally continuous with one another and it presumes that area MT (the medial temporal area) is primitive for placental mammals (see Figure 6.15). (A) Lateral view of the left hemisphere of a macaque's telencephalon, with the medial portions of the visual cortex "unfolded" to make them visible from this perspective. (B) Each cortical area contains a complete topographical representation of the retina, and these topographical maps are generally continuous across adjacent areas. Thus, visual stimuli that impinge upon adjacent parts of the retina are represented in adjacent portions of the visual cortex, even as we cross area boundaries. The gray "rings" in this figure indicate cortical regions that represent stimuli at various degrees of retinal eccentricity (given in degrees). Pluses and minuses indicate regions that encode stimuli from above or below the retinal horizon, respectively. (C) Rosa proposed that V1 and MT develop early and then emit unspecified molecular signals (arrows) that organize the surrounding regions, causing them to develop into the higher-order visual cortices. An obvious extension of this model is that, as V1 and MT are pulled apart phylogenetically (by increases in early neocortical precursor size), additional, new areas appear between them. (After Rosa, 2002.)

(A) Visual cortex unfolded

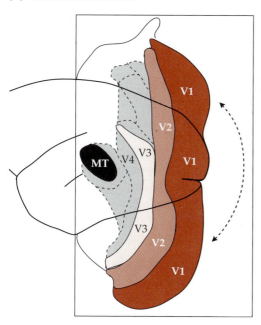

(B) Global retinotopy

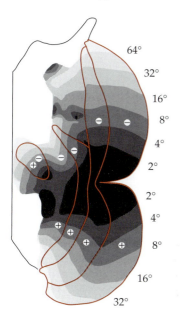

(C) Rosa's hypothesis

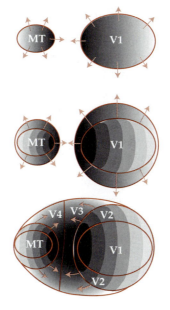

but thalamic inputs could be involved as well (Huffman et al., 1999). Thus, many questions remain. Nonetheless, Rosa's hypothesis is the kind of explanatory model that will, in the long run, allow us to comprehend at the mechanistic level of analysis how cortical areas both develop and evolve.

Conclusions

J. B. Johnston was probably correct to question Edinger's hypothesis that the entire dorsal telencephalon was simply "added on" in evolution, but his claim that "new structures have not appeared" was excessive. Clearly, some brain regions have changed so much in evolution that we may call them "new." If those new brain regions have strict (one-to-one) homologues in other species, then they are "novel by conversion"; if no such homologues can be identified, they are "novel by proliferation." We have also seen that the proliferation of brain regions may result from either phylogenetic segregation or addition. Either way, brain complexity increases. Naturally, brain complexity may also decrease in evolution, as it has in salamanders and lungfishes (see Northcutt, 1986). In those instances, we speak of phylogenetic reduction rather than proliferation, of coalescence and subtraction rather than segregation and addition. However, phylogenetic increases in brain complexity clearly outnumber the decreases, just as increases in absolute brain size outnumber decreases (see Chapter 4). This implies that brain complexity has, on average, increased over the course of vertebrate phylogeny—a conclusion that resonates with our intuitive notions of how brains have changed with time. It is important to point out, however, that brain complexity increased independently in different lineages, with largely divergent results. For instance, both mammals and teleosts evolved a large and complex diencephalon, but in mammals it was the dorsal thalamus that segregated into many nuclei, whereas in teleosts the posterior tuberculum became unusually complex. Similarly, both primates and carnivores evolved complex visual cortices, but most of their "new" cortical areas are probably not homologous between them. To put it simply, complex brains evolved repeatedly among the vertebrates but the details of that complexity tend to vary between clades.

The question of how and why novel brain structures evolved remains fascinating but unsettled. We have seen that laminated brain regions probably evolved so frequently because they are both easy to develop and functionally advantageous. That same argument can also be applied to phylogenetic proliferation by segregation and/or addition. For instance, we discussed that phylogenetic increases in the size of an embryonic precursor region may actually "force" its adult derivatives to become more heterogeneous and, hence, complex (see Figures 6.12 and 6.17). At the same time, increases in the number of subdivisions within a brain region are probably functionally advantageous because they allow individual subdivisions to become specialized for different functions. After all, division of labor is generally a good idea. Excessive specialization may reduce a system's flexibility (e.g., in response to brain damage; see Chapter 9), but there likely is no other way to make brains really big (see Chapter 7). Therefore, it seems reasonable to conclude that brains generally became more complex in evolution because more complex brains were (1) easy to develop, (2) functionally useful, and (3) the only way to increase neuron number without sacrificing functionality. This conclusion is perhaps daring, since we still know very little about how and why brain structure sometimes changes radically, but it constitutes a useful working hypothesis.

In the last two chapters, we have focused on the size and structure of individual brain regions. We have dealt with brain function only in passing and as if it were an attribute of individual brain regions. This was expedient but simplistic, for no brain region is an island. As you probably realize, all brain regions are connected with at least some other brain regions—many of them massively so. Consequently, it must be the case that all brain regions perform their normal function through their interactions with those other brain regions. For present purposes, this means that if we want to understand how evolutionary changes in brain structure are linked to evolutionary changes in behavior, we must look beyond individual brain regions and also examine how neural pathways and circuits evolve. This topic shall occupy us in the following chapter.

7 Evolution of Neuronal Connectivity

Evolutionary morphology has been unable to deal with problems concerning transformations of complex systems. It has basically restricted itself to the analysis of changes in metric traits, such as cusp height in teeth, or very simple meristic traits, such as number of spots on butterfly wings. . . . The problem of how complex systems with many interacting component parts can evolve is a major challenge in evolutionary biology.

—Pere Alberch, 1984

In Chapter 6, I argued that a brain's complexity can be measured by counting how many structurally and functionally distinct brain regions are contained within it. This is a reasonable approach, but it fails to capture the most intriguing aspect of the brain's complexity, namely its connectivity. It is that connectivity that gives the brain its special function, that makes it a complex system *par excellence*. In order to get a feeling for the brain's connectional complexity, consider that an average neuron in the neocortex of a mouse is covered by about 8,000 synapses (Schüz and Palm, 1989). Since neurons may synapse multiply onto any given postsynaptic neuron (as well as onto themselves), we cannot say that an average neuron receives inputs from 8,000 other neurons. However, detailed examination of connected neuron pairs in the mammalian forebrain have revealed that each neuron tends to have less than 10 functional synapses on each of its postsynaptic cells (see Pavlidis and Madison, 1999; Alonso et al., 2001; Silver et al., 2003). Venturing beyond the forebrain, neuronal connectivity becomes more difficult to estimate, but cerebellar granule cells are extremely numerous and each contact several hundred postsynaptic cells (Llinás et al., 2004). Therefore, we can guesstimate that an average neuron in the brain of a "typical" mammal, such as the laboratory mouse, is directly connected to at least 500 other neurons. If this is true, and if human brains are wired similarly (discussed later in the chapter), then we can surmise that the roughly 100 billion neurons of a typical human brain (Lange, 1975; Pakkenberg and Gundersen, 1997) are interconnected by at least

50 trillion different connections. This estimate is crude but, any way you look at it, brains do contain an enormous amount of "wiring"! Moreover, if you have ever tried to memorize a neuronal circuit diagram, you probably know that the circuitry is neither perfectly regular nor totally random, which means that you cannot use any single, simple rule to remember its design. In other words, the brain's circuitry hovers in that fascinating realm between regularity and randomness, order and disorder, where most truly complex systems dwell (Kauffman, 1993).

One of the most fascinating questions we can ask about any complex system is how it changes over time. If you add or eliminate some components, how does that affect the rest of the system? Do changes in one component cascade throughout the system or are they quickly dampened by some sort of buffering mechanisms? Are all kinds of change equally likely, or do the system's internal dynamics favor some trajectories? And do complex systems always become more complex with time? Although these questions are difficult to answer for any kind of complex system (e.g., Jeong et al., 2001; Thelen, 2003), interest in them is growing fast. Back in 1984, the pioneering morphologist Pere Alberch admitted that "evolutionary morphology has been unable to deal with problems concerning transformations of complex systems," but that is beginning to change. An ever-growing number of morphologists is becoming interested in how complex morphologies evolve, and these scientists are making good headway (e.g., Müller and Newman, 2003). Even in neurobiology, progress is being made. It is still the case, however, that evolutionary neurobiologists generally prefer to study "metric traits," such as overall brain size, or "simple meristic traits," such as species differences in the number of distinct brain regions. Few have grappled with the more difficult question of how changes in one part of the brain relate to changes in other, interconnected parts. Using the words of D'Arcy Thompson, a great theoretical biologist, we might say that the typical neurobiologist "when comparing one organism with another, describes the differences between them point by point . . . [and]. . . falls readily into the habit of thinking and talking of evolution as though it had proceeded on the lines of his own descriptions, point by point, and character by character" (Thompson, 1959, p. 1036). Or, in the case at hand, brain region by brain region.

Historically, some neurobiologists did write about how neural circuitry evolves, but their efforts were generally based on inaccurate data. For instance, C. J. Herrick, whom we encountered briefly in Chapter 2, had proposed that sensory inputs from the dorsal thalamus "invaded" the telencephalon in early amniotes and displaced the olfactory inputs to the telencephalon. Herrick further proposed that this invasion created an environment "favorable for cortical differentiation" (Herrick, 1948, p. 105) and was critical for the evolution of the mammalian neocortex. Although Herrick's "invasion hypothesis" was supported by a variety of circumstantial evidence, such as the relatively small size of both dorsal thalamus and telencephalon in all anamniotes (see Chapter 6), it was falsified by subsequent discoveries. Once experimental methods were developed to actually trace specific axons through the brain, as opposed to inferring their course from fiber-stained sections (see Chapter 2), it became clear that large parts of the telencephalon are devoid of direct olfactory projections not only in amniotes, but also in amphibians, sharks, and ray-finned fishes (see Figure 2.10)(Scalia et al., 1968; Ebbesson and Heimer, 1970; Scalia and Ebbesson, 1971). In addition, those new methods revealed that the dorsal thalamus projects to the telencephalon not just in amniotes, but in all major groups of vertebrates (see Northcutt, 1981). Collectively, these findings proved that the invasion hypothesis was wrong, at least in the form that Herrick had proposed (see Northcutt and Puzdrowski, 1988).

After the demise of Herrick's invasion hypothesis, most evolutionary neurobiologists began to stress that neuronal connections are highly conserved across the vertebrates. The most extreme proponent of this view was Sven Ebbesson, who had been among the first to apply the new tracing techniques to nonmammalian vertebrates. He argued that "projections vary primarily in quantity and degrees of differentiation, but are principally to the same targets and *never* to an unusual target" (Ebbesson, 1980, p. 185, emphasis in original). This view was extreme in its ostracism of "unusual" connections (see the discussion later in this chapter) but most of Ebbesson's colleagues agreed that neuronal connections are far more conserved across vertebrates than Herrick and his contemporaries had believed (e.g., Campbell and Hodos, 1970; Nauta and Karten, 1970). As Glenn Northcutt, another pioneer of the new axon tracing methods, put it: "Many, if not most, neural pathways appear to be very stable phylogenetically, and the majority of these pathways appear to have arisen with the origin of vertebrates or, shortly after, with the origin of jawed vertebrates" (Northcutt, 1984, p. 70). This conclusion soon gained widespread acceptance because subsequent studies continued to report many unexpected similarities in neuronal connectivity between major vertebrate groups (e.g., Karten and Shimizu, 1989; Reiner et al., 1998). William Hodos and Ann Butler clearly expressed this view in their recent discussion of how sensory systems evolve: "To be sure, certain senses wax and wane with increasing and decreasing specialization; in general, however, patterns of consistency can be observed. The central nervous system appears to have been far more conservative in its evolution than our scientific predecessors of fifty years ago would have suspected" (Hodos and Butler, 1997, p. 190).

Indeed, one can fill many pages with descriptions of how connections in different groups of vertebrates resemble one another (see Butler and Hodos, 1996; Nieuwenhuys et al., 1998). The existence of those similarities is very important because, without them, it would be difficult to extrapolate research results across species. Finding similarities is also satisfying because similarities are fairly easy to explain: if they are found in closely related species, then they probably are due to descent from a common ancestor that already had the feature in question. But what about species differences in neuronal connectivity? Clearly, they are also commonplace, especially if we compare distantly related groups of animals (see Butler and Hodos, 1996; Nieuwenhuys

Connectivity ———>behavior

et al., 1998). Indeed, it is probably fair to say that species differences in neuronal connectivity account for most of the behavioral differences between species. This raises a question that is central to this chapter: How can we explain species differences in neuronal connectivity?

One way to answer this question is to seek functional explanations for each species difference in connectivity. We might, for instance, hypothesize that reciprocal connections between diencephalon and telencephalon evolved in mammals (and some teleosts; see Figure 6.11) in order to provide a substrate for physiological feedback loops and rhythmicity (Steriade, 2003). Unfortunately this approach is arduous, because it is generally difficult to demonstrate how changes in brain physiology contribute to an animal's fitness (its ability to produce offspring or help kin) and because the functional effects of changes in neuronal connectivity generally remain far from clear. Neuroscience is simply not mature enough to allow for many hypotheses about how specific changes in neuronal connections affect overall brain function. This does not, however, mean that the phylogenetic variation in neuronal connections must remain incomprehensible. Instead, it means that we must strive to improve our understanding of how neural circuits "work" and seek out additional kinds of explanations for how and why species differences exist. For instance, we can try to explain at least some species differences in terms of developmental "rules" or "constraints," as Finlay and Darlington did for variation in brain region size (see Chapter 5). As you will see, our current attempts to provide developmental explanations for species differences in connectivity remain incomplete, but they are a promising beginning. In addition, we can look at the variation in neuronal connectivity from a more abstract, global perspective. To illustrate this point of view I use this chapter's final section to review the logic and evidence behind the fascinating claim that neuronal connections have generally evolved to minimize their length. Finally I shall argue, just as I did in Chapters 5 and 6, that no single rule or principle can provide us with a realistic understanding of how brains evolve. Only a synthesis of several different "theories" and perspectives enables us to comprehend how neural circuitry and other "complex systems with many interacting component parts" (Alberch, 1984) evolve.

Epigenetic Population Matching and Cascades

The most likely mechanism by which evolutionary changes in the size of one brain region can affect the size of other brain regions is the axon-mediated modulation of cell death. Strange as it may seem, most brain regions contain 20–80% more neurons at the end of neurogenesis than in adulthood, the excess being eliminated by naturally occurring cell death (Oppenheim, 1985). In the lumbar region of the developing spinal cord, for instance, roughly 40% of the motor neurons degenerate during the period when those neurons first make contact with the hind limb musculature (Hamburger, 1975). The most fascinating aspect of this motor neuron loss is that it is regulated, at least in part (see Bennet et al., 2002), by the amount of target tissue that is available for innervation. For instance, when the developing hind limb is removed in a young chick embryo, virtually all lumbar motor neurons die; conversely, when a supernumerary limb is surgically implanted, the amount of naturally occurring motor neuron death is generally reduced (Hollyday and Hamburger, 1976; Tanaka and Landmesser, 1986). Apparently, developing motor neurons compete with one another for some "trophic factor" that is produced by muscle fibers, taken up by the axon, and required for the neuron's survival. As a result of this competition for trophic support, the number of projection neurons is effectively matched to the number of available target cells. Such "epigenetic population matching" (Katz and Lasek, 1978) is useful because it reduces or eliminates the need to specify genetically how many neurons are needed to innervate each muscle. We can think about it thus: If epigenetic population matching did not exist, then any genetic variation in muscle size would require matching mutations to adjust the size of the associated motor neuron pool. Such matching mutations would be rare, and any variation that is due to other mechanisms, such as chance or nutritional factors during development, could not be matched at all. Therefore, it makes sense that epigenetic population matching would have evolved. More generally, we can say that epigenetic population matching probably evolved as a mechanism to ensure that all brain (and muscle) cells receive at least some innervation and that excess neurons are eliminated "automatically"—without requiring additional genetic mutations.

Most important from an evolutionary point of view is that epigenetic population matching can facilitate the evolution of morphologically diverse brains by reducing the number of matching mutations required to create viable new phenotypes. For instance, if some genetic mutation causes an individual to develop unusually large leg muscles, then epigenetic population matching would ensure that those muscles are innervated by a greater number of motor neurons; no additional mutations would be required (as long as the increase in leg size is not too large). In support of this hypothesis, a variety of circumstantial evidence may be adduced. If you look at the spinal cord of a typical tetrapod, for instance, you immediately notice that it contains two prominent bulges: one at the low cervical level, containing the motor neurons that innervate the forelimb, and another at the lower lumbar level, where the hind limb motor neurons sit (Figure 7.1). As the population matching hypothesis predicts, snakes and other limbless tetrapods do not possess these limb-related spinal enlargements. Similarly, cetaceans and manatees lack hind limbs and possess either severely reduced lumbar enlargements or none at all (Dexler and Eger, 1911; Breathnach, 1960; Morgane and Jacobs, 1972). Especially interesting is the turtle spinal cord. It contains virtually no thoracic motor neurons, just as one would expect from the fact that turtles have virtually no thoracic musculature (see Figure 7.1B). These correlations between muscle size and motor neuron number are consistent with the population-

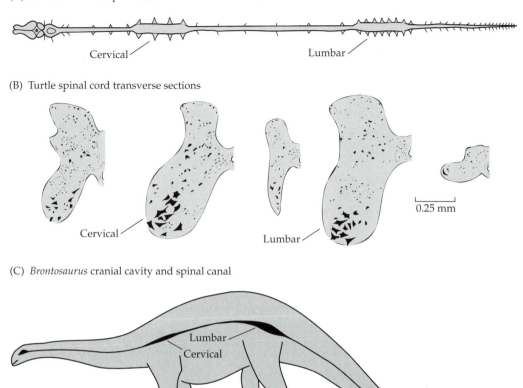

(A) Turtle brain and spinal cord

Cervical

Lumbar

(B) Turtle spinal cord transverse sections

Cervical

Lumbar

0.25 mm

(C) *Brontosaurus* cranial cavity and spinal canal

Lumbar

Cervical

Figure 7.1 Differential Elaboration of the Spinal Cord The observation that vertebrate spinal cords generally contain more motor neurons in those body segments that are "muscular" supports the epigenetic population-matching hypothesis. (A) In turtles and most other tetrapods, the spinal cord exhibits two major enlargements; one innervates the forelimb, the other the hind limb. (B) Cross sections through the spinal cord at various rostrocaudal levels (i.e., upper cervical, lower cervical, mid-thoracic, lumbar, and sacral) reveal that the limb-related enlargements reflect, at least in part, an increase in the number of motor neurons (i.e., the large cells in the ventral horn of the spinal cord). Turtle spinal cords have virtually no motor neurons at mid-thoracic levels, which makes sense because turtles have almost no thoracic (i.e., rib-related) musculature. (C) The brain and spinal cord of *Brontosaurus*, as reconstructed from fossil specimens. In this species the hind limb enlargement of the spinal cord probably exceeded the brain in size. (A and B from Kusama et al., 1979; C after Moodie, 1915.)

matching hypothesis. Unfortunately, however, we do not yet know the extent to which the population matching is epigenetic. That is, we do not yet know whether these species differences in motor neuron number result only from species differences in the amount of epigenetically induced cell death or involve heritable differences in neurogenesis as well. Therefore, we can cur-

rently conclude only that the phylogenetic diversification of spinal cord morphologies probably involved at least some epigenetic population matching.

Moving out of the spinal cord and into the brain, we find more evidence
consistent with the epigenetic population-matching hypothesis. First of all,
numerous studies have now shown that many neurons in developing brains
do depend on trophic factors for their survival. Often these trophic factors are
target-derived, as they are for the motor neurons described above, but they
may also come from presynaptic cells (Linden, 1994). In addition, interconnected brain regions do sometimes vary in a coordinated fashion, just as the
hypothesis predicts. A nice example of such coordinated circuit evolution is
provided by the series of parallel "loops" that pass through the cerebellar cortex and the inferior olive in mammals (Figure 7.2). Apparently, when one element in one of these loops becomes hyper- or hypotrophied, the others follow

(A) Primates   (B) Dolphins and whales

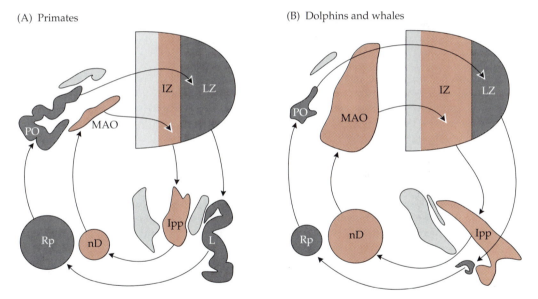

Figure 7.2 Population Matching in Trans-cerebellar Circuits   Species differences
in the size of two looplike circuits through the cerebellum provide additional evidence
for the population-matching hypothesis. The outermost circuit (gray) runs from the
principal division of the inferior olive (PO) to the lateral zone of the cerebellar cortex
(LZ), to the lateral deep cerebellar nucleus (L, also known as the dentate nucleus in
some species), to the parvocellular red nucleus (Rp), and back to the inferior olive. All
elements in this circuit are significantly larger in (A) primates than in (B) cetaceans (relative to other brain regions). A second, parallel loop (red) courses through the main
accessory olive (MAO), the intermediate zone of the cerebellar cortex (IZ), the posterior
interposed nucleus (Ipp), and the nucleus Darkschewitsch (nD). All elements in this second circuit are unusually large in cetaceans, but small in primates. (After Kooy, 1917;
Ogawa, 1935; Korneliussen, 1968; Voogd et al., 1998; Onodera and Hicks, 1999.)

Figure 7.3 The Visual System of the Blind Mole Rat   Most retinal targets are ▶
unusually small in blind mole rats, but the suprachiasmatic nucleus (SCN) is about as
large as we would expect it to be if mole rats were not "blind." (A) If we chart retinal
axons (red lines) and terminal zones (red areas) onto schematic transverse sections
through the midbrain and diencephalon of a blind mole rat (*Spalax ehrenbergi*), we can
see that most retinal targets are relatively small. (B) Quantitative comparisons confirm
that most retinal target regions are significantly smaller in blind mole rats than in hamsters, which have a similar absolute brain size. Remarkably, the mole rat's SCN is similar in size to that of hamsters. This is consistent with the observation that blind mole
rates have well-developed circadian rhythms, for those rhythms are controlled by the
SCN in other vertebrates. It also indicates that the phylogenetic reduction of retina size
in blind mole rats did not entail a corresponding reduction in the size of *all* retinal targets. Abbreviations: dLGN = dorsal lateral geniculate nucleus; Ha = habenula; LP = lat-
eroposterior nucleus; MG = medial geniculate nucleus; nOT = nucleus of the optic tract;
PO = posterior nucleus; SCN = suprachiasmatic nucleus; Sup Co = upper layers of the
superior colliculus; vLGN = ventral lateral geniculate nucleus; VM = ventromedial
nucleus; VPL = lateral ventroposterior nucleus; VPM = medial ventroposterior nucleus.
(After Cooper et al., 1993b.)

suit (Korneliussen, 1968; Voogd, 2003). This pattern of interspecies variation is
consistent with the population matching hypothesis but, again, no developmental data are available to determine if the population matching in these
loops is *epigenetically* controlled, as the hypothesis requires. Such evidence is,
however, available for some circuits in the mammalian visual system. Specifically, we know that experimental removal of the retina during development
leads to a dramatic reduction in the size of most retinal targets (Cullen and
Kaiserman-Abramof, 1976; Finlay et al., 1986). We also know that in naturally
blind (or nearly blind) species, such as the blind mole rat, most retinal target
regions are considerably smaller than they are in species with well-developed
eyes (Figure 7.3) (Cooper et al., 1993a,b). Collectively, these data strongly suggest that at least some of the interspecies variation in retinal target size is an
epigenetic consequence of variation in retina size. Curiously, there is one retinal target structure that is not reduced in blind mole rats, namely the suprachiasmatic nucleus (SCN). Even this exception to the rule remains consistent with
the epigenetic population-matching hypothesis, however, because SCN neurons can survive without trophic support from the retina (Lehman et al., 1995).

Given that changes in the size of one brain region can epigenetically induce
matching changes in the size of regions that are interconnected with it, we
might expect changes in the size of one region to propagate like an avalanche
throughout a neural system or, possibly, the entire brain. The possibility of
such "epigenetic cascades" (Wilczynski, 1984) is intriguing because it would
allow evolution to "buy system-wide changes cheaply in terms of the number of genetic changes required to effect a functional change" (Finlay et al.,
1987, p. 103). Evidence for such cascades can indeed be found in the mam-

(A) Blind mole rat retinal projections

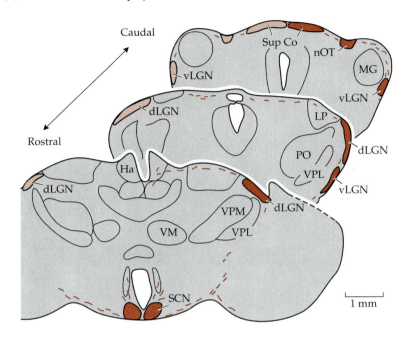

(B) Blind mole rat versus hamster

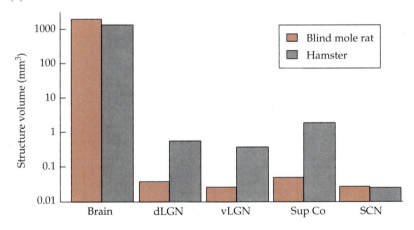

malian visual system. As I mentioned earlier, mammals that are blinded during early development have abnormally small retinal targets. They also have abnormally small visual cortices (Rakic et al., 1991; Dehay et al., 1996b). Since the retina connects to the visual cortex only indirectly, via the LGN (see Chapter 6), the cortical size reduction must be an epigenetic cascade effect (it may not be a "trophic" effect, but that is not vital to this argument). Given these data, it is likely that the unusually small size of the visual cortex in blind mole rats (Cooper et al., 1993b) is also, at least in part, an epigenetic cascade effect. On the other hand, we also find some evidence against the epigenetic cascade hypothesis. Primates, for instance, have large retinas and a generally well-developed visual system, but some of their visual circuits, particularly those through the pretectum, are much smaller than they are in other mammals (Kaas and Huerta, 1988). In fact, whenever you have multiple pathways diverging from a common source, different species tend to elaborate different pathways to different degrees (e.g., Northcutt and Wullimann, 1988). This is clearly not what one would expect if changes in the size of one brain region propagate like falling dominos throughout the brain. So, what is going on?

One way to explain why epigenetic cascades do not percolate along all possible pathways is that individual brain regions vary in the degree to which their size is influenced by trophic factors or other developmental stimuli. We have already seen, for instance, that SCN neurons are unusually resistant to the loss of synaptic inputs during development. Another factor counteracting epigenetic cascades is that brain regions with multiple afferent and/or efferent connections may derive axon-mediated trophic support from multiple sources (Finlay et al., 1987). This would reduce the influence of variations in a single source and, thus, dampen the cascade effects. Finally, there is the possibility that changes in the size of one connection lead to compensatory changes in the size of other connections. Such compensatory effects may involve changes in the strength of already existing connections or in the formation of connections that are, to use Ebbesson's term, "unusual" (discussed later in this chapter). Compensation of the latter kind is well known from embryological studies showing that when the LGN is deprived of its normal retinal input during early development, it comes to receive abnormal auditory and/or somatosensory inputs, particularly if those inputs are deprived of their normal targets (Schneider, 1973; Asanuma and Stanfield, 1990; Bhide and Frost, 1992). Such compensatory "re-routing" of axonal connections also occurs in some other systems (Kitzes et al., 1995), but the extent to which it accounts for natural species differences in connectivity remains unclear. Some authors have argued that the apparent invasion of visual cortex by auditory and/or somatosensory inputs in blind mole rats (Figure 7.4A) represents such an epigenetic cascade effect (Bronchti et al., 2002), and this suggestion is supported by the discovery of similar results in experimentally blinded opossums (Figure 7.4B) (Kahn and Krubitzer, 2002). Still, the issue remains controversial, (Cooper et al., 1993b) and further studies, particularly on other species and systems, are needed to determine how frequently compensatory innervation occurs in evolution and how effective it is at dampening or buffering epigenetic cascades (Katz and Lasek, 1978).

Therefore, we can say that epigenetic cascades do occur sometimes but are unlikely to propagate far in any systems that exhibit a significant amount of

(A) Blind mole rat

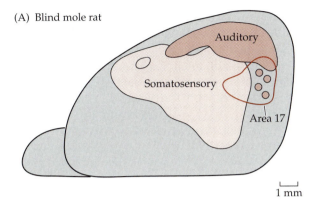

(B) Blinded opossum

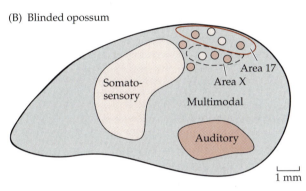

(C) Normal opossum

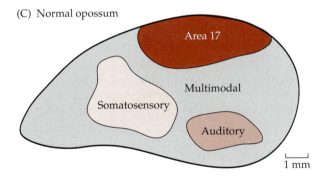

Figure 7.4 Compensatory Innervation in Blind Animals
Naturally blind and experimentally blinded animals have expanded auditory and somatosensory representations in their neocortex. (A) In blind mole rats the primary visual cortex (Area 17, outlined in red) is relatively small but identifiable on the basis of cytoarchitectonic criteria. Its cells do not respond to visual stimuli (Cooper et al., 1993b). They do, however, respond to auditory and/or somatosensory stimuli (Bronchti et al., 2002), which is not the case in other rodents. These data suggest that auditory and somatosensory inputs have "invaded" the blind mole rat's visual system at some juncture. (B) A very similar kind of compensatory innervation is seen in short-tailed opossums (*Monodelphis*) that are experimentally blinded before their visual system has matured. Again, the primary visual cortex (Area 17) is smaller than it is in normal opossums (C), and cells within it respond to auditory and/or somatosensory stimuli. Curiously, blinded opossums also exhibit an entirely novel neocortical area (Area X) (see also Rakic et al., 1991; Dehay et al., 1996a; Striedter, 1998). (A after the work of various authors, principally Bronchti et al., 2002; B and C after Kahn and Krubitzer, 2002.)

connectional divergence and/or convergence. In other words, we would expect to see cascades in fairly linear circuits, such as the transcerebellar loops mentioned earlier (see Figure 7.2), but not in the more reticulate circuits that are so typical of brains. This makes sense in evolutionary terms because, if the cascades were not dampened so frequently, then it would be almost impossible for evolution to change the size of one brain region without also changing the size of all other regions. That is, if epigenetic cascades were commonplace, mosaic evolution of the kind I discussed in Chapter 5 would be virtually impossible. It would also be difficult to evolve brains that differed in anything other than absolute size. Thinking about it in more abstract terms, we can say that the mechanism of epigenetic population matching does provide for links between the brain's various interacting parts, but these links are weak, on average, rather than strong. This is interesting because weak linkages are precisely what one would expect in complex systems that have evolved to maximize their own "evolvability"—that is, their own ability to evolve further (Conrad, 1990; Volkert and Conrad, 1998). It has been argued, for instance, that gene networks cannot evolve if all gene linkages are pleiotropic (Wright, 1986) and that those networks became "modular" specifically to overcome this pleiotropy limitation (Wagner and Altenberg, 1996). If this is true, then neural circuit evolution may be quite similar to gene network evolution, but this analogy remains largely unexplored. Therefore, let us now deal with several specific hypotheses about how neurobiological networks evolve.

The Parcellation Hypothesis

Epigenetic population matching involves mainly *quantitative* changes in the size of various neuronal connections but, as I mentioned earlier in the discussion of compensatory innervation, *qualitative* changes in neural connectivity could, at least in theory, occur as well. So, how frequently do qualitatively new connections appear in evolution? And, when brain regions proliferate (see Chapter 6), how are those extra regions connected to the remaining brain? The most explicit and ambitious answer to these questions was proposed by Ebbesson in 1980, when he stated that "nervous systems become more complex, not by one system invading another, but by a process of parcellation that involves the selective loss of connections of the newly formed daughter aggregates and subsystems" (Ebbesson, 1980, p. 213).

In giving this answer, Ebbesson explicitly rejected both Herrick's invasion hypothesis and the idea that brain regions can ever be added to ancestral brains. Instead, he proposed that brains become more complex only by means of phylogenetic segregation (see Chapter 6) and that this segregation is caused by the loss of old connections from some of the "daughter aggregates" (Figure 7.5). This parcellation theory was criticized by some of Ebbesson's colleagues who noted that this "theory" is really just a hypothesis, that it did not originate with Ebbesson, that it describes a pattern rather than a "cause," and that it disregards some evidence for invasion (see Ebbesson, 1984; Rehkämper, 1984). In response to these critiques, Ebbesson acknowledged that parcellation is not a hard rule but a "trend" (Ebbesson, 1984). Nonetheless, he continued to assert that connectional invasion is rare and insignificant. By now, it is quite clear that Ebbesson underestimated the role of invasion in brain evolution, but let us come back to that. First, let us examine whether the

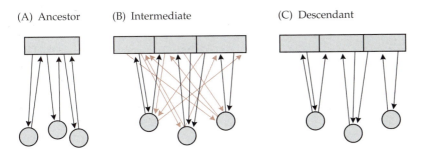

(A) Ancestor   (B) Intermediate   (C) Descendant

Figure 7.5 Ebbesson's Parcellation Theory   In order to grasp the essence of Ebbesson's parcellation "theory," imagine (A) an ancestral rectangular brain region (gray) that has reciprocal connections with three other regions (gray circles). (B) Now imagine that the rectangular brain region segregated phylogenetically into three distinct brain regions, all of which retained the same set of three reciprocal connections (red and black arrows combined). According to Ebbesson, some of those connections would be lost (red arrows), leaving a derived condition (C) in which each of the three rectangular brain regions retains only a subset of their ancestral connections). (After Ebbesson, 1984.)

core of Ebbesson's idea, namely that phylogenetic segregation is associated with a selective loss of connections, is supported by the evidence.

According to Ebbesson himself, a prime example of the parcellation process is the evolution of multiple visual pathways to the telencephalon. Specifically, Ebbesson claimed that in sharks and other anamniotes, projections from the retina and from the optic tectum converge on a single target in the dorsal thalamus, which then sends axons to the telencephalon (Figure 7.6). This single thalamic target, Ebbesson argued, segregated into two separate regions as amniotes evolved and, as part of that process, each of the "daughter aggregates" lost one of its visual inputs: The cell group commonly referred to as the dorsal lateral geniculate (LGN) lost its tectal input, while the cell group known in primates as the pulvinar lost the direct input from the retina (Ebbesson, 1984). This scenario would be a nice illustration of Ebbessonian parcellation if its underlying data base were valid but, as Glenn Northcutt (1991) has pointed out, the data tell a somewhat different story. All anamniotes do possess an "anterior nucleus" (see Chapter 6) that projects to the telencephalon and receives both retinal and tectal inputs, just as Ebbesson had claimed. However, all anamniotes also have a second dorsal thalamic nucleus that receives tectal inputs and projects to the telencephalon, but receives no direct inputs from the retina (see Northcutt, 1991). This "dorsal posterior nucleus" clearly resembles the primate pulvinar in terms of connectivity (see Figure 7.6). Therefore, multiple visual pathways through the dorsal thalamus have existed all along in vertebrate evolution. In addition, we now know that the LGN receives at least some inputs from the optic tec-

(A) Postulated anamniote condition

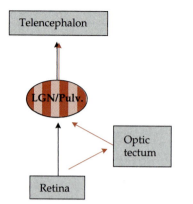

(B) Mammalian condition

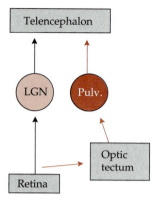

(C) Cartilaginous fish condition

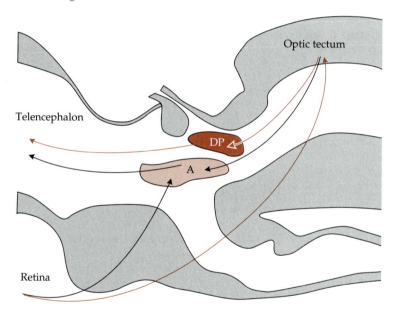

tum (or superior colliculus) in many amniotes (Kaas and Huerta, 1988; Wild, 1989). Therefore, we cannot say that the two major visual nuclei of the dorsal thalamus have selectively "lost" some of their inputs during vertebrate phylogeny. In other words, the data on visual pathway evolution do not provide good evidence for Ebbesson's parcellation theory.

Another frequently cited example of Ebbessonian parcellation is the evolution of distinct somatosensory and motor cortices from a "sensorimotor amalgam." This idea originated with Richard Lende (1963), who electrically

◀ **Figure 7.6 Theory and Data in the Visual System**   Ebbesson's hypothesis about how the visual circuits through the dorsal thalamus evolved is not consistent with the data now available. (A) According to Ebbesson (1980, 1984), anamniotes have only a single dorsal thalamic nucleus that conveys visual information to the telencephalon. (B) In contrast, Ebbesson considered amniotes to have at least two such nuclei: one receives visual inputs directly from the retina and is called the lateral geniculate nucleus (LGN) in mammals; the other receives its visual input from the optic tectum and is called the pulvinar (Pulv.) in primates. According to this scenario, the amniote condition would have evolved from the anamniote condition by parcellation. The data indicate, however, that Ebbesson's scenario is probably incorrect. (C) In this schematic diagram of a shark's visual circuits, as viewed from a medial perspective with the relevant cell groups in their anatomically correct positions, you can see that anamniotes also have two visual pathways that pass through the dorsal thalamus to the telencephalon. One pathway courses through nucleus anterior (A) and is probably homologous to the LGN pathway in mammals; the other courses through the dorsal posterior nucleus (DP) and is probably homologous to the mammalian pulvinar pathway. Therefore, the condition of having "two visual systems" (Schneider, 1969) is probably primitive for vertebrates and not the result of Ebbessonian parcellation. (A and B after Ebbesson, 1980; C after Northcutt, 1991.)

stimulated the cerebral cortex of various marsupials and found that all regions from which body movements could be elicited also responded to somatosensory stimuli (Figure 7.7A). In contrast, similar experiments in placental mammals had revealed little or no overlap between motor and somatosensory cortex (Figure 7.7B). From this Lende concluded that the condition of having separate motor and somatosensory cortices evolved from a more primitive condition where those regions are "completely superimposed" (Lende, 1969). One problem with this inference is that Lende's technique for localizing cortical areas was rather crude. However, recent studies using more refined techniques have confirmed that marsupials lack a distinct motor cortex (Beck et al., 1996). Moreover, axon tracing studies have shown that the ventral lateral and ventral posterior thalamic nuclei, which in placental mammals project to the primary motor and somatosensory cortices, respectively, project to overlapping regions in marsupials (see Figure 7.7A)(Killackey and Ebner, 1973; Haight and Neylon, 1979). These data are consistent with Lende's hypothesis, but there is another problem, namely that monotremes (the mammals most closely related to placentals and marsupials) do have separate motor and somatosensory cortices (Lende, 1969). This suggests either that the marsupial condition is not primitive or that monotremes have independently evolved separate motor cortices. Since monotreme brains are odd in many ways (see Rowe, 1990), and neither reptiles nor birds are known to have distinct motor cortices (see Medina and Reiner, 2000), the latter hypothesis is more likely. Therefore, Lende's idea remains consistent with the evidence: Separate motor and somatosensory cortices probably evolved out of a sensorimotor amalgam.

(A) Brush-tailed possum

(B) Laboratory rat

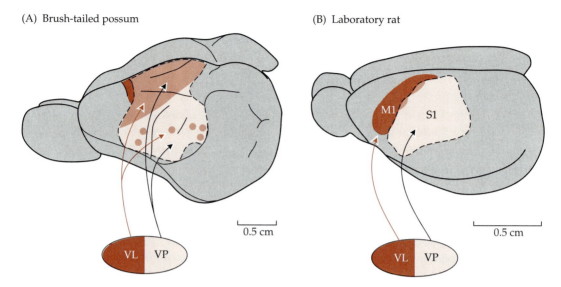

Figure 7.7 The Sensorimotor Amalgam Hypothesis The observation that distinct motor and somatosensory cortices exist in placental mammals, but not in marsupials, is consistent with Richard Lende's hypothesis that distinct motor and somatosensory cortices evolved from a "sensorimotor amalgam" (Lende, 1963). (A) In brush-tailed possums and other marsupials, motor (dark red) and somatosensory (light red) regions overlap extensively within the neocortex, shown here from a slightly dorsolateral perspective. This conclusion is based in part on the observation that the motor and somatosensory nuclei in the dorsal thalamus (VL and VP, respectively) project to overlapping cortical regions in marsupials. (B) In most placental mammals, including the laboratory rat, the primary motor and somatosensory cortices (M1 and S1) are distinct from one another, overlapping only in the distal limb representations. They receive nonoverlapping projections from VL and VP in the dorsal thalamus. Although these data are consistent with Lende's hypothesis, they need not imply that Ebbessonian parcellation was at work. (A after Haight and Neylon, 1978; B after Donoghue and Wise, 1982; Northcutt and Kaas, 1995.)

But does the phylogenetic segregation of motor and somatosensory cortices really represent an instance of Ebbessonian parcellation? Did it really involve only *losses* of connections from one or both of the daughter aggregates? Not really, because the motor cortex probably gained more direct connections to the motor neurons in the spinal cord as it evolved from the sensorimotor amalgam. This conclusion derives mainly from the observation that only placental mammals with a well-segregated motor cortex have well-developed corticospinal projections (Nudo and Masterton, 1990a; Wild and Williams, 2000). It is also in accordance with the results of cortical stimulation experiments, which have consistently revealed that much higher current intensities are needed to elicit movements from the sensorimotor amalgam of marsupials than from the motor cortex of placental mammals (Beck et al., 1996; Frost et al., 2000). Since indirect pathways are generally more difficult to excite electrically than direct ones, these physiological data support the notion that the motor cortex of placental mammals, but not the sensorimotor amalgam, has direct projections to the motor neurons of the spinal cord (and, possibly, the medulla). In other words, the motor cortex evolved not only by losing its connections from the somatosensory region of the dorsal thalamus (i.e., the ventral posterior nucleus), but also by gaining new connections to what we sometimes call the "lower motor neurons." Therefore, the phylogenetic segregation of the sensorimotor amalgam into discrete somatosensory and motor cortices is not a good example of parcellation as Ebbesson (1980) envisioned it.

Perhaps the best example of Ebbessonian parcellation is the evolution of topographically organized projections from the midbrain's red nucleus to the spinal cord. In an elegant series of double-labeling experiments (Figure 7.8), Margaret Huisman and her colleagues showed that individual neurons in the red nucleus of cats and monkeys project either to rostral (cervical) or to caudal (lumbar) levels of the spinal cord, but not to both (Huisman et al., 1982). They also showed that the rostrally projecting neurons lie in a different portion of the red nucleus than the caudally projecting ones (see Figure 7.8B), which implies that the "rubrospinal" projection is topographically organized. Remarkably, when Huisman performed precisely analogous experiments in rats, she obtained very different results (Huisman et al., 1981): Rostrally and caudally projecting rubrospinal neurons were not well segregated within the rat's red nucleus, and numerous neurons projected to both rostral and caudal levels of the spinal cord via axon collaterals (see Figure 7.8A). These data indicate that rubrospinal projections are not as topographically organized in rats as they are in monkeys or cats. They do not reveal which condition is primitive, but George Martin and his collaborators (Martin et al., 1981; Martin, 1983) showed that the rubrospinal projections in opossums are even less topographically organized than they are in rats (see Figure 7.8B). Given these data, and the observation that rubrospinal projections are also nontopographical in birds (Wild et al., 1979), we may conclude that nontopographical, extensively collateralizing rubrospinal projections are probably primitive for mammals. This, in turn, implies that topographically organized rubrospinal projections evolved at least once within mammals and probably did so by the selective elimination of some axon collaterals. In other words, the scenario is consistent with Ebbesson's parcellation hypothesis insofar as no "unusual" connections were formed. Of course, this scenario is not exactly what Ebbesson envisioned, since the number of distinct cell groups did not increase.

Finally, let us consider whether there is some developmental evidence for Ebbesson's hypothesis. As Ebbesson himself had pointed out, many neurons project to a more diverse array of targets early in development than they do later on (Innocenti, 1981; O'Leary, 1992). These developmentally transient or

(A) Double labeling of rubrospinal neurons with collaterals

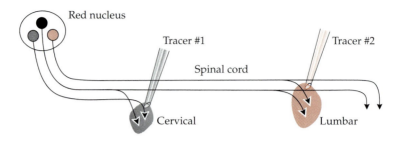

(B) Species differences in collateralization and somatotopy

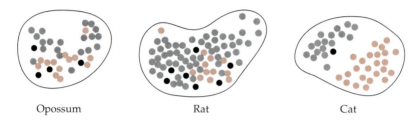

Opossum   Rat   Cat

Figure 7.8 Parcellation and Topography in the Rubrospinal System   In the rubrospinal system, projections that were not topographic ancestrally probably became topographic through the selective elimination of axon collaterals. Shown here are the experimental design and results of the work that supports this hypothesis. (A) Huisman et al. (1981, 1982), as well as Martin et al. (1981), injected two differently colored axon tracers at two different levels of the spinal cord (cervical and lumbar) and then looked for single- or double-labeled cells in the red nucleus. With this design, one can infer that double-labeled cells in the red nucleus must have collaterals that project to both levels of the spinal cord. (B) This experimental design yields different results in different species. In opossums and rats, rostrally and caudally projecting cells (gray and red dots, respectively) are intermingled within the red nucleus (outlined) and many double-labeled cells (black dots) are seen. In cats and monkeys, rostrally and caudally project-ing cells occupy different subregions of the red nucleus and only a few cells are double labeled. A parsimony analysis of these data indicates that nontopographical rubrospinal projections are probably primitive for mammals and that the topographical projections in cats and primates probably evolved by the selective elimination of some axon collat-erals. (B after Huisman et al., 1981; Martin et al., 1981; Huisman et al., 1982.)

"exuberant" connections generally consist of axon collaterals that are selec-tively lost during development. This developmental reduction of exuberant connections is analogous to Ebbessonian parcellation, since both processes involve the selective loss of connections. However, as we discussed in Chap-ter 2, development is not simply evolution on "replay." Ontogeny does not always, or even very frequently, recapitulate phylogeny, and this is as true for neuronal connections as it is for other kinds of characters. Consider, for instance, the projections from retina to lateral geniculate (LGN). As you may recall from Chapter 6, in adult primates and carnivores the two retinas project to well-segregated layers within the LGN (see Figure 6.7). Because this specificity develops gradually from an early embryonic condition of significant overlap between the axons from both eyes (Rakic, 1977), we can say that *ontogenetic* parcellation has occurred. However, if we look across species, we find that in most adult nonmammals the retinogeniculate projections are almost exclusively unilateral (e.g., Kenigfest et al., 1997), which means that the condition of having both eyes project heavily to the LGN is probably not primitive for adult mammals. Therefore, we must conclude that no (or very little) *phylogenetic* parcellation has occurred. Moreover, we now know that in developing cats the retinogeniculate projections become more specific mainly because additional collaterals grow to "appropriate" targets, not merely because "inappropriate" collaterals are lost (Figure 7.9) (Stretavan and Shatz, 1986). Thus, even if we believed that ontogeny must recapitulate phylogeny, the developmental data would be more consistent with connectional invasion than with Ebbesson's parcellation theory.

Overall, we must conclude that the available evidence provides only minimal support for Ebbesson's parcellation theory. Even the weak version of Ebbesson's hypothesis, according to which brains become more complex *predominantly* by selectively losing connections (Ebbesson, 1984), remains unconfirmed. Why, then, did I just review Ebbesson's ideas in such detail? Basically,

(A) Day 40   (B) Day 46   (C) Day 53   (D) Day 63 (adult)

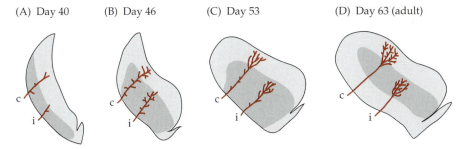

Figure 7.9 Ontogenetic Parcellation in the LGN   Ontogenetic parcellation of retinogeniculate projections in cats by collateral elimination and sprouting. (A–D) David Stretavan and Carla Shatz (1986) injected single retinal axons in cat embryos and examined their termination pattern within the lateral geniculate nucleus (LGN) at different stages of development. Initially, axons from ipsilateral and contralateral retinas (i and c, respectively) have overlapping distributions within the LGN. As development proceeds, their axon terminals overlap less and less until, in the adult condition, the projections from the two eyes do not overlap at all (D). This increased specificity results from the elimination of "inappropriate" axon collaterals, but it also involves a significant amount of new collateral sprouting in "appropriate" regions. (After Stretavan and Shatz, 1986.)

there are two reasons. The first is that Ebbesson's parcellation hypothesis received widespread attention when it was originally proposed (Ebbesson, 1980) and remains one of the best known "theories" of brain evolution. Since it is not particularly well supported by the evidence, we must set the record straight. The second reason for dealing at length with Ebbesson's hypothesis is that its weak version is probably correct *despite* the poor empirical support. Once we consider at an abstract level of analysis what happens to neuronal connectivity as neuron number increases, we realize that brains almost certainly became less densely interconnected as they increased in size (see Chapter 4) and that is equivalent to what Ebbesson referred to as the "selective loss of connections." We shall explore this argument later. For now, it suffices to point out that collecting the empirical evidence we need to test Ebbesson's hypothesis is technically difficult and that, therefore, relevant data sets are rare. From a strictly empirical point of view, the strong version of Ebbesson's hypothesis is dead, but the weak version remains alive.

Connectional Invasion and its Consequences

Connectional invasion may be defined as the evolution of projections to unusual targets or, more technically, as the phylogenetic appearance of projections to targets that did not ancestrally receive homologous inputs. Despite Ebbesson's claims to the contrary (discussed in the last section), empirical evidence for invasion is not difficult to find. For instance, we have known for more than twenty years that direct projections from the spinal cord to the thalamus occur in amniotes and sharks but not in hagfishes, lampreys, ray-finned fishes, or amphibians (Northcutt, 1984). Considering the phylogenetic relationships of these species (see Figure 2.13), it is most parsimonious to conclude that direct spinothalamic connections evolved independently in sharks and amniotes and did so by invasion. Similarly, the appearance of direct projections from the brain stem trigeminal nucleus to the telencephalon in birds (Dubbeldam et al., 1981; Arends and Zeigler, 1989) is almost certainly due to connectional invasion because in all other vertebrates, the trigeminal nucleus projects to the telencephalon only indirectly, via the dorsal thalamus. Another excellent example of invasion is provided by songbirds and parrots, which are unusual not only in being good vocal imitators, but also in having direct projections from the telencephalon to the brain stem vocal motor neurons (see Striedter, 1994; Farries, 2001). Many more examples could be adduced, not just from motor systems (as Ebbesson had claimed) but also from the sensory systems. Therefore, it is fair to say that connectional invasion has occurred repeatedly as brains evolved. But how and why? To answer that, we need to consider how neuronal connections develop.

Recent advances in molecular neurobiology have shown that developing axons respond to a multitude of signals that tell them where to go, when to stop growing, when and where to sprout collaterals, and which synapses to stabilize. Many of these mechanisms are highly conserved across both verte-brates and invertebrates (Chisholm and Tessier-Lavigne, 1999). This is exciting but also raises an important question: If the mechanisms of axon development are so conserved, then how do species differences in neuronal connectivity arise? This question remains largely unanswered. We know that some mole-cules with axon guidance functions (e.g., some cadherins) are expressed dif-ferently in different species, but most of that work has focused on species sim-ilarities (Redies and Takeichi, 1996; Redies et al., 2002). We also know that unusual connections sometimes appear when development is experimentally perturbed (e.g., Asanuma and Stanfield, 1990; Bhide and Frost, 1992) but the mechanistic details of how or why those connections form remain unknown. Thus, much remains to be discovered. However, we can also take a more holis-tic approach to understanding how and why unusual connections form: We can step back from the mechanistic details of axon guidance and ask whether there are any general rules or principles of brain development that help to explain how and why connectional invasion sometimes occurs.

This approach was pioneered by Terrence Deacon (1990) who proposed, as part of his "displacement hypothesis," that connectional invasion generally occurs when a brain region becomes disproportionately large in evolution (Figure 7.10). I shall refer to this idea as "Deacon's rule" or, more descriptively, as the notion that "large equals well-connected." It is based on two funda-mental principles of brain development, namely (1) that developing axons often compete with one another for access to target sites; and (2) that this com-petition is generally won by those axons that participate in "firing" the target cells (see Rakic, 1986; Purves, 1988). If both principles apply, Deacon reasoned, then a phylogenetic increase in a region's proportional size should provide axons from that region with a competitive advantage over axons from other regions (since the more axons a region can send to a target, the more likely it is to cause excitation in the target cells), and that should allow the hypertro-

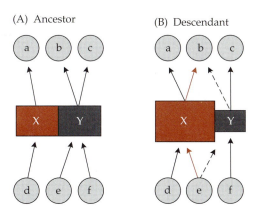

(A) Ancestor

(B) Descendant

Figure 7.10 Deacon's Rule of "Large Equals Well-Connected"   This schematic shows (A) a hypothetical ancestor and (B) a hypothetical descendant. In both brain regions X and Y project to one or more targets (a–c) and receive inputs from one or two sources (d–f), but the details of those connections vary according to Deacon's rule of large equals well-connected. This rule states that in evolution, when a brain region becomes disproportionately large (region X), it tends to invade novel targets (red arrow to region b) and receive some novel inputs (red arrow from region e). Conversely, when a brain region becomes dis-proportionately small (region Y), some of its inputs and outputs may be lost (dashed arrows).

phied brain region to invade targets that it did not innervate ancestrally. These new connections might even "displace" some old connections, particularly if the source of those old connections has decreased in proportional size. By the same reasoning, disproportionately large brain regions should attract some new inputs, and disproportionately small regions should lose some old inputs. The idea makes a great deal of sense from a theoretical point of view but has not yet been subjected to a comprehensive test. That is, there have been no broad comparative analyses to determine how frequently Deacon's rule applies. This book is not the place for such an analysis, but I will note that Deacon's hypothesis is logically sound and supported by some evidence—specifically, the observation that the mormyrid valvula and the primate neocortex, both of which are enormously hypertrophied (see Chapter 5), exhibit several "unusual" connections, as they should if the "large equals well-connected" rule applies. Let us look at these test cases in more detail.

We already encountered the cerebellar valvula of mormyrid electric fishes in Chapter 5, as a prime example of mosaic evolution. This valvula is so much larger in mormyrids than in other teleosts that it covers the entire brain (see Figure 3.4)(Nieuwenhuys and Nicholson, 1969). Its neuronal connections have not been studied exhaustively, but some data are available. These reveal that the mormyrid valvula has numerous connections that are typical of other valvulas, but also some that are unusual: only the mormyrid valvula has reciprocal connections with the midbrain torus semicircularis, receives somatosensory inputs from the brain stem, and projects to the trigeminal motor nucleus (Finger et al., 1981; Meek et al., 1986; Nieuwenhuys et al., 1998). Based on my reading of the literature, this is an uncommonly high number of unusual connections. Of course, some of the mormyrid valvula's unusual connections might be proven less unusual by subsequent research (see Wullimann and Rooney, 1990; Vonderschen et al., 2002). However, the *available* data do confirm what Deacon's rule predicts, namely, that the mormyrid valvula has an exceptionally large number of unusual connections.

Arguably the best test case for the idea that large equals well-connected is the neocortex. As we saw in Chapter 5, the mammalian neocortex enlarges disproportionately as overall brain size increases, and it is significantly larger in simians than in other mammals, even after we take overall brain size into account. Therefore, Deacon's rule predicts that the neocortex of large-brained primates should have several connections that are not seen in other mammals. Again, a definite analysis has not yet been performed, but the available data are consistent with the prediction. Some of these data are discussed in Chapter 9; here let us focus on the projections from the neocortex to the spinal cord. For many years now we have known that different species exhibit different corticospinal axon termination patterns (e.g., Verhaart, 1970; Heffner and Masterton, 1975). Particularly interesting is that, with increasing neocortex size, corticospinal axons penetrate deeper into the ventral horn (where the motor neurons are; see Figure 7.1B) and further down the spinal cord (Figure 7.11).

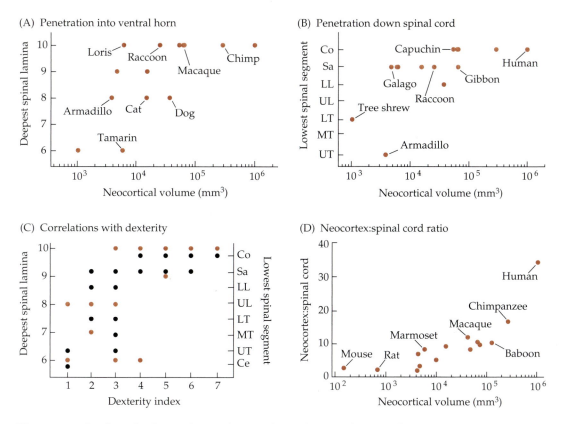

Figure 7.11 Corticospinal Invasion and Dexterity The spatial extent of neocortical projections to the spinal cord correlates, at least roughly, with both neocortical volume and manual dexterity. (A) The larger the neocortex, the more deeply corticospinal axons penetrate into the ventral horn, where the motor neurons lie (see Figure 7.1). (B) Neocortex size also correlates with how far corticospinal axons penetrate down the length of the spinal cord. (C) The typical level of manual dexterity within a species correlates with the extent to which corticospinal axons penetrate into the ventral horn (red data points) and down the spinal cord (black data points). (D) This invasion of the spinal cord by neocortical axons makes sense in terms of Deacon's rule because the ratio of neocortex:spinal cord increases dramatically with neocortical volume. Abbreviations: Ce = cervical; Co = coccygeal; LL = lower lumbar; LT = lower thoracic; MT = middle thoracic; Sa = sacral; UL = upper lumbar; UT = upper thoracic. (Data on spinal penetration and dexterity from Heffner and Masterton, 1975, spinal cord volumes from MacLarnon, 1996; neocortical volumes for most species from Stephan et al., 1981, or Mangold-Wirz, 1966; for others, neocortical volume was estimated from data on closely related species with similar overall brain weights.)

In relatively small-brained tree shrews, for instance, corticospinal axons penetrate only to lower thoracic levels and stay out of the ventral horn, while in chimpanzees and humans, the neocortex projects throughout the spinal cord. To see why this variation is consistent with Deacon's rule, you need two more pieces of information. First, you need to know that, as neocortex size increases, the spinal cord becomes disproportionately small (see Figure 7.11D). Second, you need to know that, as neocortex size increases, the number of corticospinal axons increases as well (Figure 7.12). Combining these two observations, we can spot a dilemma: An ever-increasing number of corticospinal axons must be "squeezed" into a spinal cord that fails to keep up as the size of the neocortex increases. According to Deacon's hypothesis, that should increase competition for spinal target sites and cause neocortical axons to invade additional regions of the spinal cord, just as the data indicate they do.

Now, you may be tempted to object to the above analysis on the grounds that it is based on the outdated notion of the neocortex increasing in size from "mouse to man" (see Chapter 2), but that is not what the rule of "large equals well-connected" implies. Instead, it stipulates a correlation between region size and neuronal connectivity that is explicitly *independent* of phylogeny. In the case at hand, this means that proportional neocortex size should correlate with spinal cord invasion no matter how often neocortex size increased independently among mammals and no matter which species we are examining. Whether this is strictly true remains to be seen, since a proper statistical analysis (Garland et al., 1992) has not yet been performed. It is, however, likely to be true because neocortex size appears to correlate with corticospinal neuron number both across different taxonomic orders and within them (see Figure 7.12). Another potential problem for Deacon's rule is that rats seem to be an exception to the rule, for they have a relatively low ratio of neocortex:spinal cord but reportedly possess direct projections from the neocortex to cervical motor neurons (Liang et al., 1991). This is not what one would predict if large equals well-connected. However, recent data obtained with the electron microscope have revealed that corticospinal axons in rats do not synapse on spinal motor neurons after all (Yang and Lemon, 2003). Therefore, we may conclude that rats are not exceptions to Deacon's rule. More comprehensive studies will be needed to determine just how broadly Deacon's rule of large equals well-connected applies but, for now, we may accept it as a promising hypothesis that is consistent with a considerable amount of evidence.

Before we leave the topic of corticospinal axons, let us briefly examine whether those axons displaced any other projections to the spinal cord (as Deacon's original displacement hypothesis predicts). The developmental data indicate that corticospinal axons do compete with one another for access to spinal cord neurons (Martin and Lee, 1999; Li and Martin, 2002). However, whether they also compete with other spinal projection systems remains unclear. Circumstantial evidence suggests that corticospinal axons compete with rubrospinal axons in humans, for the division of the red nucleus that projects to the spinal cord (see Figure 7.8) is present in human embryos but atrophied in adults (Schoen, 1964; Ulfig and Chan, 2001). Most other spinal projection systems apparently persist, however, even when the corticospinal tract is huge. This is nicely illustrated by recent neurophysiological studies, which showed that an intraspinal projection system commonly referred to as

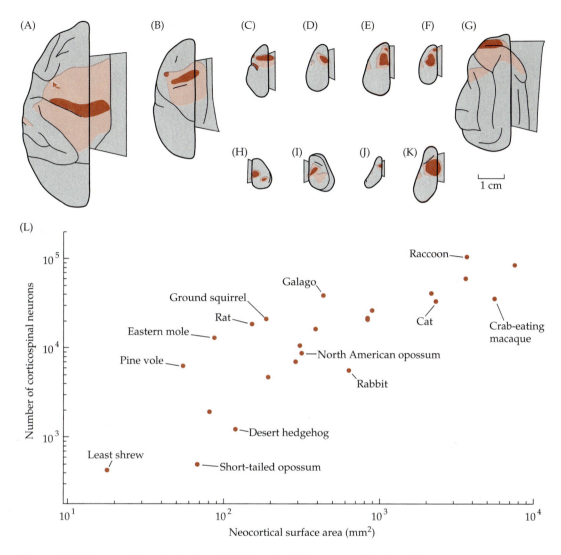

Figure 7.12 Comparative Analysis of Corticospinal Neurons The size of the mammalian corticospinal tract is directly proportional to the size of the neocortex. Ralph Nudo and Bruce Masterton (1990a,b) injected tracers into the cervical spinal cord of 22 different mammals and found that the area containing retrogradely labeled cells in the neocortex was directly proportional to the proportional size of the entire neocortex. This finding is here summarized in schematic dorsal views of the neocortex in (A) a rhesus macaque, (B) a squirrel monkey, (C) a bush baby, or galago, (D) a tree shrew, (E) a squirrel, (F) a rat, (G) a domestic cat, (H) a mole, (I) a hedgehog, (J) a short-tailed opossum, and (K) a Virginia opossum. The region illustrated in dark red contains a higher density of labeled cells than the region shown in light red. The detailed pattern of corticospinal neuron locations differs between species (e.g., compare primates and rodents). However, the total number of corticospinal cells correlates quite well with neocortical surface area (L). (After Nudo and Masterton, 1990a,b.)

Figure 7.13 Physiological Consequences of Corticospinal Invasion   Corticospinal ▶
invasion has major consequences for brain function and behavior, but need not imply
the anatomical "displacement" of other spinal projection systems. (A) Schematic illustration of how neocortical axons (red) in the corticospinal tract (CST) project more
directly and more heavily to lower cervical spinal motor neurons (MN [C6]) as absolute
brain size increases from cat to squirrel monkey to macaque to human (note that this is
not a phylogenetic series!). These corticospinal axons also terminate on excitatory and
inhibitory "propriospinal" neurons that are located in the upper cervical spinal cord
(EPN [C3] and IPN [C3], respectively). Although some authors have suggested that this
propriospinal system was "displaced" by the corticospinal system in macaques and
humans (Nakajima et al., 2000), more detailed studies indicate that this is not so.
Instead, the propriospinal system becomes more and more inhibited by the corticospinal
system as the latter increases in size (Alstermark et al., 1999; Pierrot-Deseilligny, 2002).
This suggests that the neocortex "subsumes" the propriospinal system in large-brained
primates. (B) Electrical recordings from the spinal cord during corticospinal-tract stimulation confirm that there are no direct projections from neocortex to lower cervical
motor neurons in cats and that those connections are stronger in the large-brained
macaques than in the relatively small-brained squirrel monkeys. Recordings from muscle fibers during cortical stimulation in humans suggest that in humans the cortico-
motoneuronal projections are even stronger than they are in macaques. (Dashed line
indicates that human data are based on electromyographic rather than direct motor neuron recordings.) (C) The strength of the cortical projections to the lower cervical motor
neurons (which innervate the forelimb musculature) correlates rather well with measures of manual dexterity (indicated by numbers). Overall, these data are consistent with
a revised neocorticalization hypothesis. (After Nakajima et al., 2000.)

the propriospinal system did not disappear as neocortical axons invaded the
spinal cord (Alstermark et al., 1999; Nakajima et al., 2000; Pierrot-Deseilligny,
2002). These physiological investigations also revealed that, as the corticospinal system expands, the propriospinal system becomes more and more
inhibited by it (Figure 7.13). This is an important finding because it suggests
that corticospinal axons may have "functionally displaced" the propriospinal
system, even though they did not displace it anatomically.

With that we have arrived at an important question: What are the functional consequences of connectional invasion? In the case of corticospinal
invasion, the commonly accepted answer is that increased invasion leads to
increased manual dexterity (Heffner and Masterton, 1975; Heffner and Masterton, 1983). This conclusion is debatable because only some aspects of corticospinal invasion correlate in a statistically significant manner with manual
dexterity (Iwaniuk et al., 1999), and because manual dexterity surely results
from a variety of factors, including hand morphology. Nonetheless, there is
at least some fairly general correlation between motor skill and the presence
of direct projections from telencephalon to spinal cord and brain stem motor
neurons. Among primates, for instance, the ability to execute diverse laryn-

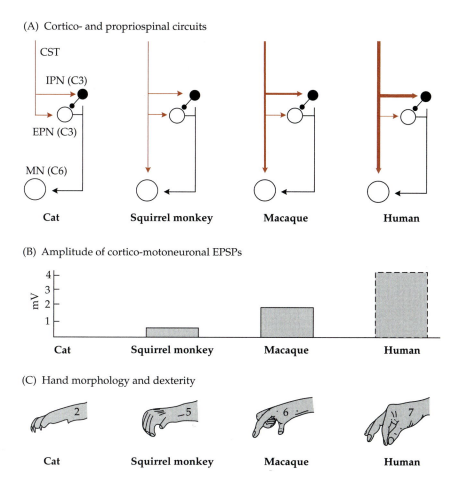

(A) Cortico- and propriospinal circuits

CST

IPN (C3)

EPN (C3)

MN (C6)

Cat   Squirrel monkey   Macaque   Human

(B) Amplitude of cortico-motoneuronal EPSPs

mV

Cat   Squirrel monkey   Macaque   Human

(C) Hand morphology and dexterity

Cat   Squirrel monkey   Macaque   Human

geal and tongue movements seems to correlate positively with the presence of direct neocortical projections to the motor neurons for those organs (see Chapter 9). Similarly, among birds the ability to produce and learn diverse vocalizations correlates rather well with the presence of direct projections from the telencephalon to brain stem vocal motor neurons (see Striedter, 1994). These data suggest that having direct projections from telencephalon to motor neurons is "good" for behavior in the sense that it tends to increase the behavior's diversity and flexibility. This is an interesting idea that is itself worth further exploration, but it does not address the physiological question of how connectional invasion affects brain function. In other words, the discovery of correlations between invasion and behavior does not reveal how the evolution of "unusual" connections alters the internal dynamics of the brain. That question, it turns out, has a long and fascinating history.

Back in 1886, David Ferrier noted that animals he considered "low on the phylogenetic scale" are not nearly as sensitive to large telencephalic lesions as "higher" animals, (e.g., monkeys and humans; Ferrier, 1886). Over the next 50 years, Ferrier's observation was transformed into the doctrine of "functional encephalization," according to which brain functions progressively shift from the lower brain regions into the neocortex as one proceeds along the phylogenetic series (see Marquis, 1934; Striedter, 2002). This principle, also known as "neocorticalization," enjoyed wide support in the 1930s and 40s, but then faded from the scene as comparative neurobiologists began to decry any theoretical notions that made reference to a *scala naturae* (see Chapter 2). This was of course appropriate, but the proverbial baby may have been tossed out with the bath water, for the principle of neocorticalization can be rephrased in terms of Deacon's rule: As the neocortex increased in proportional size and invaded more brain regions, no matter in which lineage or how many times independently, it assumed a more dominant role in overall brain function. This revised version of the neocorticalization hypothesis contains no reference to a *scala naturae*, involves no explicit "shifting of functions," and need not imply that noncortical regions were rendered obsolete by the expanding neocortex. It simply states that, as the neocortex became more widely interconnected with other brain regions, it became more capable of influencing the activity of those brain regions. As we saw in the discussion of the propriospinal system (see Figure 7.13), that influence on other brain regions may be excitatory or inhibitory. Either way, a larger neocortex would have a greater "vote" in the brain's complex "democracy" (Heiligenberg, 1991; Schall and Bichot, 1991; Finlay et al., 2001). If that is true, then we might expect neocortical lesions to be most deleterious in species with the proportionately largest neocortices, which is essentially what Ferrier reported more than 100 years ago.

How well has Ferrier's notion stood the test of time? The truth is we do not yet know, for no true test of Ferrier's idea (stripped of its *scala naturae* overtones) has ever been performed. Some studies have reported that neocortical lesions in species with proportionately small neocortices (e.g., tree shrews) cause more severe long-term deficits than Ferrier might have predicted (Ward and Masterton, 1970). It has also become apparent that neocortical lesions in species with proportionately large neocortices (e.g., humans or macaques) spare at least some simple behavioral functions (Weiskrantz, 1982). These observations indicate that not all behavioral functions simply "shifted" to the neocortex as it increased in size. Therefore, they falsify the version of the neocorticalization hypothesis that was popular in the 1930s and 40s. They do not, however, falsify the revised neocorticalization hypothesis that stresses regional "interdependence" rather than functional "shifting." In fact, my own informal survey of the lesion literature confirms that species with proportionately large neocortices are generally more impaired after neocortical lesions than species with smaller neocortices (e.g., Tower, 1940). It is particularly interesting that animals with large neocortices generally take longer to recover from cortical lesions than animals with proportionately smaller neocortices. For instance, auditory cortex lesions cause prolonged hearing deficits in macaques but only temporary deficits in rats (Heffner and Heffner, 1986; Talwar et al., 2001). This is precisely what one would expect if the auditory cortex assumed a more dominant role within the auditory system as its proportional size increased. Unfortunately, these kinds of observations remain scattered in the literature and have not been analyzed from a comparative perspective. Most lesion studies do not even report how long it takes for animals to recover from lesions. Therefore, the lesion data are currently inadequate for testing the revised neocorticalization hypothesis. More detailed studies, comparing not only long-term deficits but also the time courses of post-lesion recovery, are needed before we can pass judgment on Ferrier's fundamental idea that neocortical influence correlates with neocortex size. In that context, direct investigations of how the neocortex modulates activity in other brain regions would also be useful (e.g., Nakajima et al., 2000; Suga et al., 2002).

Overall, we can conclude that connectional invasion probably occurred repeatedly and had significant, if still indefinite, effects on brain physiology and animal behavior. We have also seen that many instances of connectional invasion are probably explicable in terms of "large equals well-connected." This is important because it negates the view that the phylogenetic comings and goings of connections in the brain are a haphazard affair that can, at best, be understood as a multitude of independent adaptations to specific organismal needs. Of course, invoking Deacon's rule does not imply that neuronal connections are rarely, if ever, adaptive (just as the existence of concerted evolution does not negate the possibility of adaptation; see Chapter 5). The fundamental point is simply that the brain is a "complex system with many interacting parts" that cannot all evolve independently of one another. As one region changes in proportional size, its developmental interactions with other regions are likely to change; that, in turn, is likely to cause changes in the adult circuitry, which produces changes in brain function and behavior. We still know relatively little about the causal basis and functional consequences of these "rules" of neural circuit evolution, but the available evidence is consistent with their existence. As we learn more about those rules, we will probably discover that much of the phylogenetic variation in neuronal circuitry is explicable in terms of *both* functional adaptation and some other rules or principles, such as parcellation and/or Deacon's rule.

General Principles of Network Design

Thus far, we have dealt mainly with data and ideas that emerged directly from the study of brains, which is the natural thing for neurobiologists to do.

However, we have also seen that good comparative data on neuronal connectivity are scarce. Confronted with this scarcity, we can benefit from broadening our perspective. Specifically, we can try to learn from other complex systems by asking whether there are any general rules or principles that govern variation in diverse systems with many interacting parts. With that goal in mind, let us revisit what I reviewed in Chapter 4 about how neural networks scale with absolute brain size. As you saw in Figure 4.16 (see also Figure 7.14A), the number of connections in a fully interconnected network increases exponentially with neuron number, which means that such networks quickly become axon-dominated as they increase in size (Deacon, 1990; Ringo, 1991). In Chapter 4, we also discussed the fact that real brains do not scale like that. Instead, the "average neuron" in real brains projects to roughly the same number of other neurons, no matter how large the brain (Figure 7.14B,C). This tendency to maintain absolute rather than proportional connectivity with increasing brain size helps to keep down the metabolic and physical costs of axonal "wiring." It also makes sense in terms of what we know about dendrites: If proportional connectivity were maintained as neuron number is increased, then the number of synapses received by an average neuron would increase exponentially. This would require a significant increase in average dendrite size, but increasing the size of dendrites is problematic because it dramatically alters a neuron's electrical properties (Bekkers and Stevens, 1970) and increases the metabolic demands on individual neu-

(A) Fully connected   (B) Sparse and random   (C) Sparse and minimal wiring

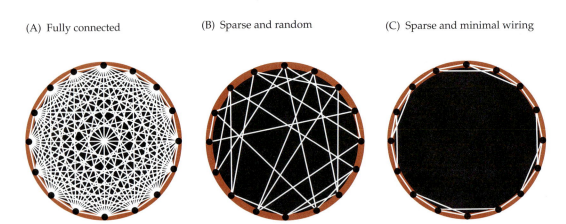

Figure 7.14 Principles of Network Scaling and Design   Schematic diagram comparing networks whose connectivity is (A) dense, (B) sparse and random, and (C) sparse and minimal. The figure illustrates that, if brains remained densely interconnected as they increased in size, then large brains would be dominated by axons (for more details, refer back to Figure 4.16). It also shows that randomly wired brains would have longer axons than minimally wired brains (by definition!), and that minimally wired brains tend to fragment into clusters that are densely interconnected internally but only sparsely connected to other clusters.

rons. Therefore, it is not surprising that real brains become more sparsely interconnected as they increase in size.

Now, let us think more deeply about the arrangement of those sparse connections in large brains. First, consider what would happen if those connections formed between randomly selected neurons. In that case, most axons would course between neurons that are relatively far apart within the brain (see Figure 7.14B). As I mentioned earlier, such long axons would take up a great deal of precious space and would be metabolically expensive. Moreover, a random wiring scheme would make it difficult for signals to be exchanged between neighboring neurons and would make local circuit interactions almost impossible—signals would tend to ricochet randomly through the network before arriving at a neighboring neuron. Because of that, randomly wired brains would not work well. Therefore, let us consider the alternative scenario, which consists of brains being wired in such a way that all connection lengths are minimized (see Figure 7.14C). In that case, axonal wiring costs would be minimal and signals could readily travel between neighboring cells. There might be a problem with exchanging signals between distant cells (discussed later in this chapter), but we know from the designers of computer chips that minimizing connection lengths is generally a good idea when you have a large number of interacting components. More important, several lines of evidence suggest that nervous systems do adhere to some kind of minimum-wire principle. In the roundworm *C. elegans*, for instance, the connections of all 302 neurons have been described and, given that connectivity, the physical arrangement of those neurons (into various ganglia) turns out to minimize the network's overall connection length (Cherniak, 1995). This does not imply that all *C. elegans* neurons only project to their immediate neighbors (as in Figure 7.14C), but it strongly suggests that the nervous system of *C. elegans* has evolved to minimize connection costs.

In vertebrates, some kind of "save-wire rule" seems also to apply (Ramón y Cajal, 1909). As I reviewed in Chapter 6, laminated brain regions probably evolved at least in part because they represent a minimum-wire solution to the problem of how to connect corresponding points in multiple topographical maps. Indeed, topographical maps themselves may have evolved because they effectively minimize the amount of wiring needed to interconnect neurons that are activated by similar stimuli (Durbin and Mitchison, 1990). Additional evidence for a save-wire rule comes from mammalian cortex. Malcolm Young and his colleagues (Young, 1992; Scannell et al., 1995) showed that, when you consider all of the connectional data published for cat and primate visual cortices, the majority of the connections are from "nearest-neighbor" and "next-door-but-one" areas. In addition, the "wire fraction" in various cortical regions consistently hovers around 60%, which is just what you would predict if wire length and, therefore, conduction delays were minimized (Chklovskii et al., 2002). Finally, the observed scaling relationship between cortical gray matter and white matter is most readily explicable if we assume that connection lengths are minimized (Zhang and Sejnowski, 2000). All of these data imply that cortical axons are significantly shorter than we would expect them to be if their connectivity were random. Again, it may not be the case that all neurons project *only* to nearby cells, but cortical networks and, most likely, all central nervous systems do apparently exhibit a significant bias toward short connections.

This observation is important because it probably explains why large brains tend to be divided into so many discrete "modules." To grasp this argument, consider the work of Jacobs and Jordan (1992), who designed an intriguing kind of artificial neural net. Most artificial neuronal nets begin their "life" as densely interconnected networks of neuron-like elements that are then modified by a learning rule that modifies the strengths of various connections depending on how the networks performed on some specific simulated task. Jacobs and Jordan, however, did something more. They asked their neural nets to perform two different tasks and to minimize connection lengths *simultaneously*. Under those conditions, the networks consistently fractionated into two subnets, each of which became specialized for one of the two tasks; that is, the networks became "modular." Subsequent work by Di Ferdinando et al. (2001) showed that this kind of modularity emerges "automatically" when artificial networks are evolved by simulated artificial selection for optimal performance on multiple tasks (see also Nolfi, 1997). Overall, these data indicate that selection for minimal wiring and effective multitasking "push" a network toward modularity. If that is true, then it is not surprising that modularity is so commonly seen in all kinds of nervous systems (Leise, 1990). Extending this line of reasoning, we would predict that, as neural networks increase in size, their individual modules would tend to fractionate into submodules. This is precisely what seems to have happened in real brains, for as absolute brain size increased in various lineages, the number of distinct brain regions increased as well (see Chapter 6). Thus, the tendency for brains to become more subdivided over evolutionary time probably results at least in part from natural selection for both efficient wiring *and* excellence at multiple skills (for related ideas, see Cowey, 1981; Barlow, 1986).

Balancing this drive toward minimum wiring and modularity is the fact that network modularity comes at a cost. Specifically, the problem is that a network's average "degrees of separation" (i.e., the average number of steps it takes to exchange information between randomly selected sites) increases as connection lengths are minimized. This presents a problem for real brains: if they become larger and more modular, then they become less efficient at exchanging information between distant brain regions. Since the exchange of information is crucial to brain function, excessive modularity is deleterious. So, how can brains increase in size, as they have repeatedly (see Chapter 4), without losing the ability to exchange information between their various component parts? The answer is that as brains enlarged, they did not minimize the length of all connections but, instead, retained some very long axons that serve as "shortcuts" between distant sites. That is, brains have adopted the kind of "small-world architecture" that is characteristic of many complex systems, including human social relations and the western United States power grid (Watts and Strogatz, 1998). This architecture is remarkable for its ability to accommodate both a high degree of modularity (or clustering) and a low degree of separation (meaning that few steps are needed to interconnect any two randomly selected network nodes) (Figure 7.15). To put it another way: Small-world networks are good at processing information both locally and globally (see also Newman and Watts, 1999; Latora and Marchiori, 2001).

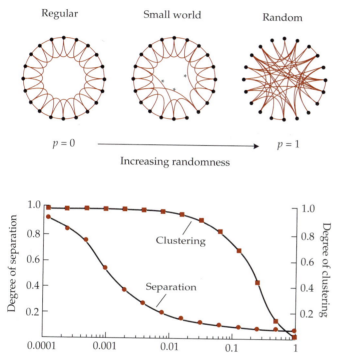

Figure 7.15 The Small World Architecture   Watts and Strogatz (1998) investigated large, sparsely interconnected networks that vary from being completely regular (each node projects only to its nearest neighbors) to being completely random. They noted that the regular, minimally wired networks exhibit a high degree of clustering but also many degrees of separation (many steps to go from one point to any other, randomly selected point). Random networks, on the other hand, exhibit little clustering and few degrees of separation. Most important, adding a few "shortcuts" (asterisks) to a regular network dramatically lowers the network's average degree of separation without bringing down its clustering index. Such "small-world networks" are efficient at both local and global information processing. Watts and Strogatz concluded that this "small-world phenomenon is not merely a curiosity of social networks nor an artifact of an idealized model—it is probably generic for many large, sparse networks found in nature" (Watts and Strogatz, 1998, p. 441). (After Watts and Strogatz, 1998.)

(A) Macaque visual cortex connectivity

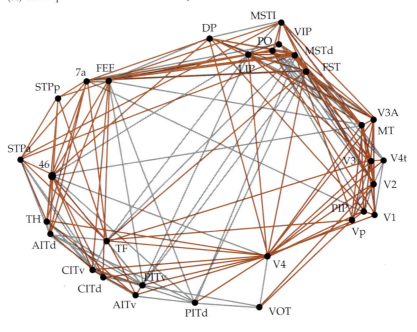

(B) Visual cortex is a "small world"

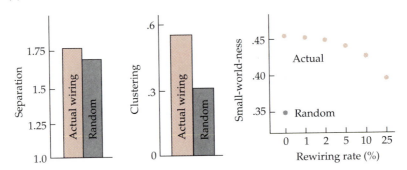

Given that processing efficiency, it makes sense that nervous systems have evolved into "small worlds" (Figure 7.16) (Young, 1992; Scannell et al., 1995; Sporns et al., 2000). Of course, this does not mean that brains can increase in size indefinitely (Ringo et al., 1994). Their average degree of separation still increases with increasing network size, but that increase is considerably slower than it would be if brains were not "small worlds."

◀ **Figure 7.16 Visual Cortex as a Small-World Network**   The visual cortical areas of macaques exhibit a small-world architecture. (A) When Malcolm Young (1992) asked a computer to find the two-dimensional topology that minimizes all connection lengths for the visual cortex of macaques, it produced this diagram (bidirectional connections are dark red, unidirectional connections are gray). Since this diagram's topology is very similar to the actual topology of the visual areas within the brain, we may conclude that the macaque visual cortex obeys a "save-wire" rule. The diagram also indicates that most areas project to their immediate neighbors or neighbors-but-one, but some areas also project to more distant targets. Such wiring is indicative of a small-world architecture. (B) Olaf Sporns and his collaborators (2000) confirmed that macaque visual cortex is a small world by showing that its average degree of separation (or path length) is only slightly higher than it would be if the network were wired randomly, while its degree of clustering is significantly higher than it would be in the random condition. Randomly rewiring the visual cortex circuitry (which can be done in a computer simulation) decreases the network's "small-world-ness," which Sporns quantified as "complexity." This finding implies that the macaque's visual cortex is optimized for both local and global information processing. The same is true of visual cortex in cats (Scannell et al., 1995; Sporns et al., 2000; Latora and Marchiori, 2001). (After Young, 1992; Sporns et al., 2000.)

Synthesis and Conclusions

Understanding how neural circuitry evolves is difficult because there are innumerable connections to consider and because changes in any one region may (but need not) lead to changes in other, interconnected regions. Brains simply do not develop or evolve "point by point and character by character" (Thompson, 1959) but as "complex systems with many interacting parts" (Alberch, 1984).

In this chapter we discussed a variety of hypotheses and "theories" about how neural circuitry evolves, but we have not arrived at a single, all-encompassing answer. Instead, we are left with several ideas that seemingly conflict with one another. For instance, the notion that trophic dependencies between interconnected neurons lead to "epigenetic population matching" seems to be contradicted by the conclusion that changes in the size of one brain region do *not* generally lead to corresponding increases of all components in the overall network. This conflict is more apparent than real, however. The problem is that the principle of epigenetic population matching, like all scientific rules or principles, applies only under a limited set of conditions. It applies only if the interconnected regions depend trophically on one another, if other sources of trophic support are insufficient, and if compensatory innervation is not a countervailing force. That is why the principle applies most reliably when we are considering motor neurons and their target musculature, or highly linear, rather than reticulate, circuits. It is not a "universal law" (see Chapter 1), but a rule or principle that generally cooperates with other rules and principles to govern the brain's ontogenetic and phylogenetic dynamics. We still know little about how common trophic dependencies are in the various brain regions and species, or the mechanisms that account for compensatory innervation, but we now have a causal framework for thinking about how evolutionary changes in the size of one brain region might—or might not—cascade through an entire brain. This is an important step forward.

Similarly, "parcellation" and "invasion" are not mutually exclusive hypotheses, as Ebbesson (1980) had originally argued. We now know that "unusual" connections do sometimes appear in evolution, but this does not imply that the principle of parcellation, of brains selectively losing some connections as cell groups proliferate, is totally false. In fact, our theoretical considerations about how networks generally scale actually imply that parcellation must be fairly common, even if it is difficult to demonstrate empirically. To see why this is true, recall that as brains increase in size they tend to maintain absolute rather than proportional connectivity (see Figure 7.14). That is, they tend to become less densely interconnected as brain size increases (see also Young et al., 1995). This, in turn, amounts to saying that, as brains enlarge, they tend to "lose" some of the connections we would expect them to have if they maintained proportional connectivity (the red arrows in Figure 7.5). Thus, our theoretical considerations help us realize that Ebbesson implicitly constructed his parcellation theory in opposition to a "straw man" scenario that requires brains to maintain proportional connectivity as they increase in size. Because that straw man scenario is prohibitive in terms of wiring and metabolic costs, we can conclude that Ebbessonian parcellation probably occurred quite frequently. In other words, we can conclude that the weak version of Ebbesson's hypothesis is probably correct: As brains became larger and cell groups proliferated, they may have evolved some "unusual" connections but, overall, their proportional degree of connectivity decreased.

Finally, let us look more closely at Deacon's rule of "large equals well-connected." It, too, is not a universal "law" but a general principle that applies under a specific set of conditions and may coexist with other rules or principles. As we discussed, Deacon's rule applies only when developing axons actively compete for access to target sites. At this point, it remains unclear how widespread such competition is, but it is probably safe to say that it is not a universal phenomenon. Therefore, Deacon's rule will have its exceptions, and in those cases, other rules are likely to apply. What those other rules might be remains unclear. However, it is already evident that invasion via Deacon's rule is not antithetical to Ebbessonian parcellation. To see why this is true, recall that under Deacon's rule the "invading" brain region increased in proportional size. Where should those "additional" neurons project? Unless their *ancestral* target region underwent a corresponding change in size (which is unlikely since Deacon's rule stipulates a change in *proportional* size), the additional neurons must project to "unusual" targets. Those unusual projections are needed just to maintain the brain's absolute connectivity and, therefore, need not increase proportional connectivity (see Figure 4.16); in fact, they actually may be associated with an overall decrease in proportional connectivity. This, in turn, implies that invasion via Deacon's rule is consistent with the weak version of Ebbesson's parcellation hypothesis. This insight has general significance because it illustrates that several different rules or principles, such as Deacon's rule and parcellation, may coexist quite peacefully within a larger theoretical framework. Although some philosophers of science accuse biologists of being "theoretical pluralists" who endlessly debate the relative merits of various conflicting and nonuniversal "laws" (Beatty, 1997), such critiques are unfounded. As the present discussion illustrates (and the entire book is meant to demonstrate), many biologists do strive to synthesize their diverse rules and principles into a larger causal framework that can be used to comprehend the "myriad complexity" of their systems (see Chapter 1). Those efforts at synthesis may still be incomplete, but they are encouraging.

Now that we have reviewed the panoply of rules and principles that govern various aspects of vertebrate brain evolution, from conservation of molecular mechanisms to changes in neuronal circuitry, it is time to consider how those principles can be applied to some specific brains. Therefore, let us turn our attention to mammalian and human brains (Chapters 8 and 9, respectively). Since complete accounts of how mammal and/or human brains evolved would burst the limits of this book, let us focus on the question of what makes mammal and human brains unique. Adopting this focus implies that we shall emphasize species differences rather than similarities. This bias is intentional because, in my opinion, species differences are often neglected in comparative analyses (see Preuss, 1995). Furthermore, I simply find species differences to be more interesting than similarities when it comes to contemplating our own neuroanatomy. What, if anything, makes our brains different from those of other primates, and what makes mammal brains "special"? In answering these questions, we will consider diverse facts about mammalian and human evolution, such as when those taxa first appeared and what kind of environments they probably lived in, but simply documenting facts is not enough. Instead, we must strive to tie those facts together into a coherent whole, and that is where the rules or principles come in. They are the glue that binds the facts and, without them, "facts are stupid things" (Scudder, 1874). As you will see, the full story of how mammalian and human brains evolved is not yet known, but its outlines are discernible.

8

What's Special about Mammal Brains?

We have come to know fishes as strictly palaeëncephalic animals. In reptiles and birds, a small neëncephalon coöperates. Finally, in mammals we meet a brain which has so large a neëncephalon that we may well expect a subordination of reflexes and instincts to associative and intelligent actions.

—L. Edinger, 1908a

We are mammals and so are most of the animals we know well: our dogs, cats, pigs, rabbits, and horses—indeed, most domesticated animals. Given this preponderance of mammals in our conscious lives, it is natural for us to ask: What distinguishes mammals from other animals? They are, of course, unique in having hair and mammary glands, but what about their brains, their behavior—are those "special" as well?

The most commonly given answer to that question is simple, if imprecise: Mammals are more "intelligent" than other animals and this increased intelligence is somehow linked to the fact that only mammals have a large and "proper" neocortex. This account of mammalian distinctiveness goes back at least to Ludwig Edinger (see Chapter 2) and was impressed upon several generations of subsequent psychologists and morphologists. Maier and Schneirla, for instance, wrote in an influential book on animal psychology that "the bird, though very highly developed in many ways, has a rudimentary cortex, and it is probable that it is largely incapable of reorganizing its experiences in order to adapt to a new situation." (Maier and Schneirla, 1935, p. 479). Similarly, the most widely read textbook on comparative anatomy, Alfred Romer's *The Vertebrate Body* (1955, Romer and Parsons, 1977), presents a clear synopsis of this view, which is worth quoting at length:

> *In birds we see a complex series of action patterns which may be called forth to meet a great variety of situations. They are all, however, essential-*

ly stereotyped; the actions are innate, instinctive. The bird, its brain dominated by its basal nuclei, is essentially a highly complex mechanism with little learning capacity which in mammals is associated with the development of the expanded cerebral cortex. The first faint traces of mammalian cortical development are to be seen in certain reptiles. In the hemispheres of these forms we find, between the paleopallium and archipallium, a small area of superficial gray matter of a new type, that of the neopallium. Even at its inception it is an association center, receiving, like the basal nuclei, fibers which relay to it sensory stimuli from the brain stem. The evolutionary history of the mammal brain is essentially a story of neopallial expansion and elaboration. The cerebral hemispheres have attained a bulk exceeding that of all other parts of the brain, particularly through the growth of the neopallium, and dominate functionally as well. This dominance is apparent in mammals of all types, but is particularly marked in a variety of progressive forms, most especially in man. (Romer and Parsons, 1977, p. 584; emphasis in the original)

This notion—that mammals are more intelligent than nonmammals solely because they have a well-developed neocortex—persists in popular culture, but its behavioral component is clearly obsolete. As long as we define "intelligence" broadly (e.g., as the ability to relate different unconnected pieces of information in new ways and to apply the results in an adaptive manner, see Chapter 4; see also Reader and Laland, 2002; Sol et al., 2002), we cannot deny that some nonmammals exhibit "mammal-like" intelligence. Birds in particular are quite intelligent. Some crows, for example, can make and use tools (Hunt, 1996), some parrots can use learned vocalizations to label material objects (Pepperberg, 1981), and some jays can remember thousands of locations where food is stored (Tomback, 1980). Of course, keen observers have always known that birds and other nonmammals can be remarkably clever (e.g., Kawamura, 1947), but this insight began to permeate the behavioral literature only after 1960 or thereabouts. Back in 1932, for example, O. L. Tinklepaugh was astonished to discover that "the learning of the turtle [in a multiunit T-maze] equaled the expected accomplishment of a rat in the same maze" (Tinklepaugh, 1932, p. 205), but this would hardly shock today's comparative psychologists.

Over the last 50 years or so, it has become apparent that some nonmammals perform just as well as mammals in various learning and "intelligence" tests, as long as the tests are designed with the animal's "special needs" in mind. Davidson (1966), for example, showed that alligators fail to learn a simple discrimination task if the reward is food, but readily master the same task, in the same apparatus, if they are offered the opportunity to escape from excessive heat. Such a finding might have surprised Tinklepaugh or Edinger, but it makes perfect sense once you realize that alligators (as ectothermic creatures with low metabolic rates) can go without food for long periods of time but must frequently move out of the sun to prevent heatstroke. In other words, comparative psychologists have realized that it is blatantly unfair to run reptiles or other nonmammals through intelligence tests that were designed by mammals for mammals.

This insight has, however, come at a cost, for once you acknowledge that intelligence tests must be designed in a species-appropriate manner, the task of comparing "intelligence" across species becomes more difficult. As Euan Macphail put it in his landmark book on animal intelligence:

> Suppose that a task has been devised which does obtain different rates of learning for various species of animals—may we assume that the species are now ranked in intelligence? Clearly, we cannot. Obvious alternative explanations are that the task merely distinguishes between their sensory capacities (as, for example, in the acquisition of a visual pattern discrimination), or between their motor skills (as, for example, in learning to fit one object into another), or reflects differences in motivation or incentive—some species may find the reward offered more attractive than others. It seems obvious that such "contextual variables" (Bitterman, 1965) would indeed generate species differences in many learning situations—how could we rule out the possibility that one of them was responsible for a given difference in any situation? (Macphail, 1982, p. 7)

One way to overcome this "contextual variable" quagmire is to test each species on a whole variety of tasks that were designed as species-appropriately as possible, and then to compare across species only the maximum levels of performance. Macphail himself performed this kind of meta-analysis and concluded that "there are no differences, either quantitative or qualitative, among the mechanisms of intelligence of non-human vertebrates" (Macphail, 1982, p. 330). Humans are special, Macphail argued, because they have language, but all other vertebrates are "equally intelligent"; or at least, Macphail claimed, we cannot reject this hypothesis on the basis of the available evidence. I personally think this conclusion is too radical, for it is probably no accident that within each vertebrate class the best performers consistently come from a few taxonomic groups that also exhibit relatively large and complex brains. Among birds, for instance, most reports of "remarkable intelligence" involve relatively large-brained parrots and songbirds (see Figure 4.8B). Emus, ostriches, chickens, or ducks may not be as "stupid" as some people believe, but they are probably less intelligent than parrots or songbirds (by most definitions of intelligence). Similarly, the most intelligent ray-finned fishes seem to be those with relatively large and complex brains (e.g., triggerfishes) and, among amphibians, frogs generally seem more intelligent than salamanders. Given this rather spotty distribution of "remarkable intelligence" and its apparent correlation with "large and complex brains," I conclude that the capacity for intelligent behavior probably increased several times independently among the vertebrates, in parallel with the increases in brain size and complexity that I discussed in Chapters 4 and 5. At least, this hypothesis ought to receive more attention than Macphail gave it.

I shall return to this argument at the chapter's end but, for now, let us avoid the mire of comparative intelligence analyses. After all, the subject of this book is the evolution of brains, not intelligence (see Lefebvre et al., 1997, 2002 for more detailed analyses of intelligent behavior in animals). For our present purposes, it suffices to point out that mammals are not generally more intelligent than all other vertebrates. Clearly, some nonmammals do just as well on various intelligence tests as many mammals do. This conclusion is important because it challenges the hypothesis of mammalian brain distinctiveness that Edinger, Romer, and others had proposed. If some nonmammals are just as intelligent as many mammals, do they also have a well-developed neocortex? Were the original reports of birds possessing a "rudimentary cortex" wrong? Or do birds and other nonmammals use other, noncortical structures to attain "mammal-like" levels of performance in complex problem-solving tasks? As you shall see shortly, the last of these scenarios turns out to be correct. That answer, however, also raises many new questions (e.g., what is special about avian brains?) and forces us to readdress the original question: What is special about mammalian brains? If Romer was correct in claiming that "the first faint traces of mammalian cortical development are to be seen in certain reptiles," then what changed between reptiles and mammals? What makes mammalian neocortex different from its reptilian precursor, and what are the functional consequences of those innovations? What advantage did mammals derive from evolving a "proper neocortex"? Was the story of mammalian brain evolution really just "a story of neopallial expansion and elaboration," as Romer had proposed?

This chapter is an attempt to answer some of those questions. I begin by reviewing what early mammal brains probably looked like and then discuss two major hypotheses that have been proposed to explain the phylogenetic origins of the neocortex. Although both hypotheses retain some adherents, I here focus on the notion that mammalian neocortex evolved from a dorsal pallial region that is relatively small (though hardly rudimentary) in reptiles and birds. If this hypothesis is correct, then we may conclude that mammalian neocortex evolved mainly through the addition of some new cell types, a realignment of its incoming axons (from tangential to radial), and the acquisition of at least some novel connections. Regions other than the neocortex also changed as mammals evolved, but many of these non-neocortical changes seem to be linked, both structurally and functionally, to the "invention" and subsequent expansion of the neocortex. Therefore, Romer's conclusion that "the evolutionary history of the mammal brain is essentially a story of neopallial expansion and elaboration" is at least partially correct. But let us begin the story at its beginning, with a consideration of "early mammals" and their brains.

Early Mammals and their Brains

Mammals first evolved more than 100 million years ago in the early Mesozoic, from mammal-like reptiles (also known as cynodonts). Because these early mammals and protomammals are now extinct, most of our knowledge about them is based on fossil finds. Fortunately, the fossil record of Mesozoic reptiles and mammals is quite rich (Rubidge and Sidor, 2001; Luo et al., 2002). In addition to those paleontological data, we can use our knowledge of extant mammals to infer what the earliest mammals might have looked like. As you might recall from Chapter 2 (see Figures 2.1, 2.11, and 2.13), there are three major groups of extant mammals, namely monotremes (platypus and echidna), marsupials (e.g., opossums and kangaroos), and placentals (e.g., humans, rabbits, and rats). By most accounts, placentals and marsupials are more closely related to each other than either is to monotremes. Therefore, any features that are found in monotremes and placentals or monotremes but not in reptiles or birds (living sauropsids), are likely to have been present in early mammals as well. Such cladistic analyses of data from extant species have major limitations (e.g., deciding how large the neocortex was in early mammals; discussed later in the chapter), but those limitations are mitigated if we let the fossils speak. As an illustration of this confluence of neontological and paleontological data (data from extant and extinct species), consider the following.

A cladistic analysis of extant vertebrates suggests that birds and mammals independently evolved the ability to hold their legs underneath the body rather than having them extend out laterally. This upright posture probably allowed them to move their legs parallel to the vertebral column, and this probably enhanced their ability to breathe and run at the same time—something that lizards, for example, cannot do (Carrier, 1987). The uncoupling of breathing from locomotion probably also allowed both birds and mammals to expand their use of oxygen to generate internal body heat and, thus, to become endotherms (Bennett, 1991). Collectively, these data suggest that early birds and mammals independently evolved into highly mobile creatures that could live in relatively cold climates and be active at night (when it is too cold for most ectothermic vertebrates to run around). The paleontological data strengthen this hypothesis. They indicate, for instance, that some protomammals could also move their hind limbs underneath the body and, therefore, probably walked more upright than early reptiles (Blob, 2001). The fossils further reveal that many protomammals had the same complexly folded intranasal bones that are used by extant mammals to warm inhaled air and reclaim water from what they exhale (Hillenius, 1994). In combination, all these data indicate that an upright gait, oxidative metabolism, and endothermy all evolved just prior to the origin of mammals. This is interesting because it suggests that the major increase in relative brain size that

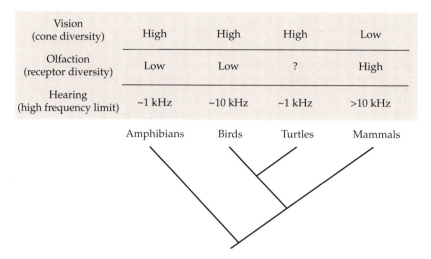

Vision (cone diversity)	High	High	High	Low
Olfaction (receptor diversity)	Low	Low	?	High
Hearing (high frequency limit)	~1 kHz	~10 kHz	~1 kHz	>10 kHz
	Amphibians	Birds	Turtles	Mammals

Figure 8.1 Differential Development of Tetrapod Distance Senses   Amphibians, reptiles, and birds have a greater diversity of cone photoreceptors than placental mammals do, which is why the former generally have better color vision. On the other hand, placental mammals do possess an unusually diverse set of olfactory receptor genes, which probably allows them to detect a greater variety of odorants. In the auditory realm, birds and mammals exceed most other vertebrates in their ability to detect high-frequency, airborne sounds. This "new and improved" hearing capacity probably evolved independently in those two lineages.

occurred with the origin of early mammals (discussed in Chapter 4 and later in this chapter) might have followed on the heels of an even earlier increase in metabolic rate. This interpretation would be consistent with the notion that overall brain size is limited or "constrained" by metabolic rate (see Chapter 4). However, before we get more deeply into early mammal brains, let us discuss what we can infer about the sensory capacities of those Mesozoic creatures (Figure 8.1).

We will begin with vision. According to the fossil record, early mammals were similar to protomammals in having fairly large, laterally placed orbits (eye sockets). Unfortunately, the fossils reveal little else about early mammal vision, but a comparative analysis of eyes in living tetrapods is informative. For instance, the colored oil droplets that are found in the conical photoreceptors (cones) of most vertebrates and which are known to bolster color vision are not found in placental mammals, and they are colorless in monotremes and marsupials (Ahnelt and Kolb, 2000). This prompted Gordon Walls (1942) and others to suggest that early placental mammals might have been color-blind. This conclusion has been overturned in recent years as comparative molecular and behavioral studies revealed that most placental mammals actually have two different kinds of cones that, collectively, allow for dichromatic color vision (see Jacobs, 1993; Hemmi, 1999). Simply put, most placental mammals can distinguish blue from yellow-green and bright from dark. It is still true, however, that in contrast to most nonmammals (and primates; see Chapter 9), most placental mammals cannot distinguish red from green; they are red–green color-blind (Jacobs, 1993). Thus, the ability of early placental mammals to see color was severely reduced compared to what it had been ancestrally. Obviously, we must wonder: why?

The best available explanation for why early placental mammals lost some of their color vision is that they became nocturnal. As you probably know from experience, color perception is virtually impossible at night anyway— mainly because at low light intensities, photoreceptors become too "noisy" for accurate comparisons between activity in different kinds of cones (see Land and Osorio, 2003). Indeed, most extant nocturnal species have retinas that are nearly devoid of cones and, instead, are packed with "rod" photoreceptors that are extremely sensitive to light in general but are not color-sensitive (e.g., Wikler and Rakic, 1991). These considerations suggest that, if early placental mammals were nocturnal, they would have needed rods more than cones and might well have lost some of the latter. The resultant decrement in color vision would have been offset by the improvement in night vision. One important implication of this "nocturnal bottle-neck" hypothesis (Walls, 1942) is that good color vision reevolved within some placental mammal lineages, notably primates (see Chapter 9). This idea is consistent with the observation that many mechanistic details of color vision in primates are quite different from what you find in nonmammals. Apparently, some mechanisms were irreversibly lost (for more on "irreversibility" in evolution, see Marshall et al., 1994). Further support for Walls's bottleneck hypothesis derives from the observation that placental mammals also lost most of the structures that nonmammals use for changing their eyes' focal length (Walls, 1942). This loss of focusing ability (which reevolved in primates) makes sense, because focusing an image on the retina is useful only if the retina can process high-resolution images, which is not the case in most nocturnal animals, who pool information from multiple photoreceptors to improve sensitivity to light. Finally, the inference that early mammals were endotherms (discussed earlier) supports the nocturnal bottleneck hypothesis because endothermy would have allowed early mammals to run around on chilly nights, avoiding competition for resources with their ectothermic reptilian cousins (see Ruben et al., 1998).

If early mammals were nocturnal, then they probably relied heavily on senses other than vision for perceiving distant stimuli. For example, they might have paid special attention to their sense of smell (as nocturnal birds are known to do; see Healy and Guilford, 1990). Indeed, several lines of evidence suggest that early mammals did increase their ability to detect volatile odorants. We may note, for instance, that early mammals, with their high metabolic rates (discussed earlier), probably breathed through their noses more frequently than their ancestors did. Because this would have increased the airflow across the olfactory epithelium, it probably made early mammals more sensitive to airborne odorants (Michael Leon, personal communication). Whether early mammals also increased the size of their olfactory epithelium is difficult to determine, but the fossil data indicate that the nasal cavities of early mammals and some protomammals were more complex than those of early reptiles. This increase in nasal complexity may have been linked to the evolution of endothermy (Hillenius, 1994), but it might also be related to the evolution of a better sense of smell—at this point, we simply do not know. The most direct evidence for increased olfactory abilities in early mammals comes from comparative molecular studies, which show that mice have more than 1,000 different olfactory receptor (OR) genes, while nonmammalian vertebrates consistently have less than 100 such genes (Freitag et al., 1998; Dryer, 2000). Since monotremes and marsupials apparently resemble nonmammals in having relatively few OR genes (Glusman et al., 2000), we can infer that OR gene diversity probably increased dramatically with the origin of (or within) placental mammals. Intriguingly, the increase in OR gene diversity seems to be limited to that branch of the OR gene family tree that is specialized for the detection of airborne odorants (Freitag et al., 1998; Mezler et al., 2001), just as one would expect given the aforementioned increases in mammalian breathing rate. Collectively, these data strongly suggest that early mammals, particularly early placental mammals, expanded their ability to use olfaction as a distance sense.

Early mammals also improved their ability to detect airborne sounds. Comparing extant species, we find that mammals are much better than most reptiles or amphibians at hearing sounds above 1 kHz. This upward extension of the mammalian hearing range is due to several remarkable innovations, including the appearance of ear drums, long coiled cochleas, and "active" hair cells that amplify the intracochlear vibrations (Webster et al., 1992; Dallos and Evans, 1995). Remarkably, ear drums and active hair cells also evolved in some other tetrapod lineages (Manley, 2000), but these specializations are lacking in both anamniotes and turtles (which are probably representative of early reptiles in this context). Parsimony therefore suggests that the auditory system became more sensitive as mammals evolved. This hypothesis can be refined by looking at the fossil record, which shows that the chain of tiny middle ear bones (which transfers vibrations from the ear drum to the cochlea) existed in the earliest mammals but not prior to them. Protomammals certainly had homologues of the middle ear bones, but those homologues were massive and were used to join the lower jaw to the skull (Allin and Hopson, 1992); they could not have been used to amplify airborne vibrations. This, in turn, implies that protomammals were probably quite bad at detecting airborne sounds and, instead, relied on hearing vibrations through bone conduction (which occurs when you hear your teeth grinding) (Kermack and Mussett, 1983). It also implies that protomammals could not hear much of the outside world while they were chewing (a potentially serious problem for the many herbivores among them; see Sues and Reisz, 1998). Only later, as true mammals evolved, did hearing and chewing become structurally and functionally uncoupled from one another, allowing jaw and middle ear bones to become specialized for chewing and hearing, respectively. Indeed, the jaws of early mammals differ dramatically from those of protomammals (see Kermack and Kermack, 1984). But let us stay focused on the sensory world of early mammals and look more carefully at the evolution of mammalian middle ears.

As I mentioned earlier, the middle ear bones of early mammals were separated from the lower jaw. This frequently cited "fact" is actually an inference, since the mammalian middle ear bones are generally too small to be part of the fossil record. The inference is reasonable, however, because all protomammals but no true mammals have a small groove or trough on the caudomedial aspect of the lower jaw that most likely housed part of the most lateral middle ear bone. Since early mammals lack this "postdentary trough," their middle ear bones were probably separated from the lower jaw. If this line of reasoning is correct, then the skull of a small insectivorous proto-mammal called *Hadrocodium* becomes particularly interesting, for it also lacks the postdentary trough (Figure 8.2). Since *Hadrocodium* appears to be the closest known relative of all true mammals (Luo et al., 2001), this lack of a postden-

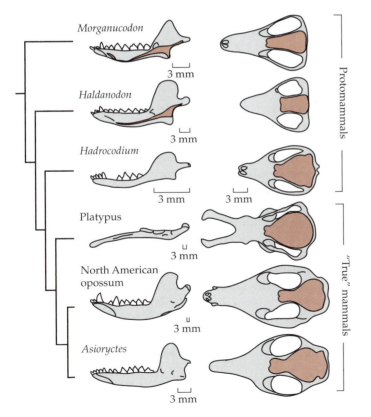

Figure 8.2 Mammalian and Protomammalian Skulls In early protomammals, the braincase (or endocranium) is narrower than it is in either the late protomammal *Hadrocodium* or in true mammals, and a "postdentary trough," which probably housed part of the middle ear, is evident on the medial aspect of the lower jaw. In this figure, the lower jaws are shown at left, with the postdentary trough shaded in red. Dorsal views of the skull and endocrania (red) are depicted at right. Anterior is always to the left. *Morganucodon, Halanodon,* and *Hadrocodium* are Mesozoic protomammals; *Asioryctes* is one of the earliest known placental mammals. Platypuses and opossums are monotremes and marsupials, respectively. (After Luo et al., 2001.)

tary trough implies that the middle ear bones became detached from the lower jaw prior to the evolution of true mammals. This is an important modification of the traditional story, which holds that detached middle ear bones are a uniquely mammalian feature. Nonetheless, middle-ear detachment was almost certainly a relatively recent innovation for the earliest mammals. This, in turn, implies that the ability to hear high-frequency, airborne sounds was also relatively new in early mammals. Overall, the available data suggest that early mammals could hear faint, high-pitched sounds better than early reptiles or amphibians, but not as well as the placental mammals of today (Frost and Masterton, 1994; Meng and Wyss, 1995).

Turning to the brain, we find that the appearance of detached middle ear bones correlates quite well with a widening of the braincase, or endocranium (see Figure 8.2). The *Hadrocodium* specimen, for instance, has not only detached middle ear bones (as true mammals do), but also a much larger endocranium than we find in other protomammals. This observation must be interpreted with care because, at an estimated body weight of just 2 g, *Hadrocodium* is unusually small and, as we saw in Chapter 4, small animals always have disproportionately large brains. However, once those allometric effects are accounted for, the braincase of *Hadrocodium* remains significantly larger than one would expect for a protomammal of that body size; it falls squarely onto the "true mammal" allometric line (Luo et al., 2001). This finding suggests that the detachment of the middle ear bones from the jaw is causally related to increases in relative brain size. We might suppose, for instance, that the prolongation of brain growth that is needed to produce phylogenetically larger brains actively pushes (or pulls) the developing middle ear bones away from the lower jaw (Rowe, 1996, 1997). This hypothesis is intriguing but most likely incorrect, since some early mammals have both small brains *and* detached middle ear bones (Wang et al., 2001). As an alternative hypothesis, I propose that brains might have increased in size in order to exploit the newly evolved potential for hearing high frequencies. This hypothesis is consistent with the observation that the mammalian auditory cortex has no obvious homologue in nonmammals and may, therefore, have been "added" as mammals evolved (see the discussion later in this chapter). At this point, however, there is no direct evidence to support or refute this hypothesis, and we cannot exclude the possibility that the temporal coincidence, or near coincidence, of middle-ear detachment and increased brain size was just that—coincidence. In any case, what is fairly clear is that early mammals already had brains that were larger than those of similarly sized early reptiles (Luo et al., 2001).

Now let us think about the shape and structure of those early mammal brains. For starters, let us try to determine whether they were smooth (lissencephalic) or complexly folded (gyrencephalic). This question is difficult to answer if we look at extant species only, because in each of the three major mammalian lineages we find some lissencephalic and some gyrencephalic species (Figure 8.3). Within monotremes, for example, the platypus has a rel-

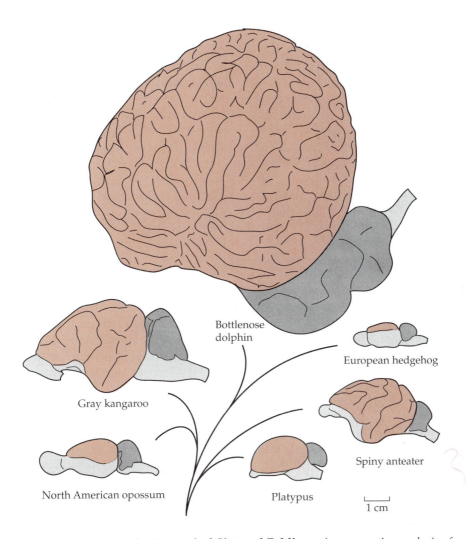

Bottlenose dolphin

European hedgehog

Gray kangaroo

Spiny anteater

North American opossum

Platypus

1 cm

Figure 8.3 Variation in Neocortical Size and Folding   A comparative analysis of gross brain shape reveals that neocortical folding correlates better with absolute brain size than with phylogenetic allegiance. Within each of the three major mammalian orders—monotremes (platypuses and spiny anteaters, or echidnas), marsupials (e.g., kangaroos and opossums), and placental mammals (e.g., dolphins and hedgehogs)—we find both large- and small-brained species. The large-brained species all tend to have proportionately large and highly folded neocortices (red). In all small-brained species, the neocortex is proportionately small and smooth. All brains are drawn to the same scale. (After images on the Comparative Mammalian Brain Collections website [http://brainmuseum.org], from the University of Wisconsin–Madison Brain Collection.)

atively smooth telencephalon but the spiny anteater's telencephalon exhibits numerous gyri. Therefore, a simple cladistic analysis cannot tell us whether early mammals had lissencephalic or gyrencephalic brains; both scenarios are equally parsimonious. We can, however, use the fossil record to extend our analysis. Specifically, we can utilize the finding that early mammals were all small (e.g., *Asioryctes*; see Figure 8.2). They may not have been as small as *Hadrocodium,* but they were certainly smaller than a North American opossum or a platypus (see Figure 8.2). Since brains generally scale predictably with body size (see Chapter 4), we can infer that the brains of those early mammals must have been quite small as well, weighing in at considerably less than 5 g (probably closer to 0.5 g than to 5 g). That, in turn, allows us to infer that early mammal brains must have been lissencephalic because, among extant mammals, only brains that weigh at least 5 g are gyrencephalic (von Bonin, 1941; Zilles et al., 1989). Thus, by combining what we know about the bodies and brains of both extant and extinct mammals, we can infer that early mammal brains must have been relatively smooth.

Similarly, we can use information from both extant and extinct species to estimate the proportional size of individual brain regions in early mammal brains. From the shape of fossil endocasts we can determine that the olfactory bulbs were quite large in early mammals and protomammals. Most likely, they comprised about 10% of the brain, as they do in extant hedgehogs and opossums (see Figure 8.3)(Stephan et al., 1981). This is consistent with the idea that early mammals had a well-developed sense of smell. Other regions are more difficult to delineate in fossil endocasts but, again, we can use what we know about brain scaling in extant mammals to derive additional insights. Specifically, we can use the observation that most mammalian brain regions scale predictably with absolute brain size (see Chapter 5) to infer that early mammals probably had brains that were similar, in terms of regional proportions, to those of small extant mammals. We can, for instance, use the observation that early mammals had small brains to infer that their neocortex was most likely small relative to the remaining brain, as it is in extant hedgehogs, tenrecs, and opossums (Figure 8.4). Of course, this inference is only approximate because the scaling rule on which it is based is imperfect (see Chapter 5). However, the neocortex in an early mammal brain almost certainly occupied less than 20% of the total brain volume, as it does in modern insectivores, and not 40% or more, as it does in extant primates (Clark et al., 2001; see Chapter 9). The same logic allows us to derive rough estimates for the size of other brain regions that are known to scale predictably with absolute brain size (see Figure 5.4D) (Finlay and Darlington, 1995). Overall, we can conclude that a cross section through the forebrain of an early mammal would probably look similar to one or more of the sections shown in Figure 8.4. That is, early mammal brains were probably similar, at a gross level of analysis, to the brains of extant hedgehogs, tenrecs, and/or opossums.

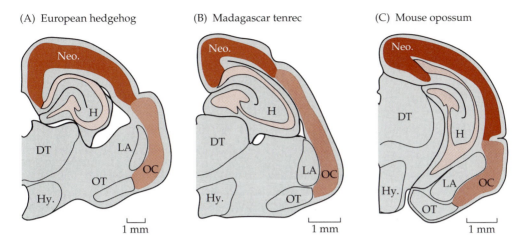

(A) European hedgehog   (B) Madagascar tenrec   (C) Mouse opossum

Figure 8.4 The Forebrains of Small Insectivores   In small extant insectivores, such as (A) hedgehogs, (B) tenrecs, and (C) mouse opossums, the neocortex (dark red) is not much larger than the olfactory cortex (medium red) or the hippocampal formation (light red). This is an important finding because the brains of early mammals were probably quite similar, in terms of absolute size, to those of small extant insectivores. Since proportional neocortex size varies predictably with absolute brain size (see also Figure 5.4D), we can infer that early mammals probably also had proportionately small neocortices. This is good to know because neocortical volume, in turn, correlates with other attributes, such as the number of neocortical areas (see Figure 6.16). Nonetheless, it is important to remember that these small insectivores are far from all alike. For example, European hedgehogs and mouse opossums are nocturnal or crepuscular, whereas Madagascar tenrecs are active in daylight. These differences are not unexpected, since extant insectivores belong to several different taxonomic groups (e.g., tenrecs belong to the Afrotheria; see Figure 2.11). Abbreviations: DT = dorsal thalamus; H = hippocampus; Hy. = hypothalamus; LA = lateral amygdala; Neo. = neocortex; OC = olfactory cortex; OT = olfactory tubercle. (After images on the Comparative Mammalian Brain Collections website [http://brainmuseum.org], from the University of Wisconsin–Madison Brain Collection.)

This conclusion sounds suspiciously similar to the old discredited notion that some extant species can serve as stand-ins for truly ancestral species (see Chapter 2), but that is not what I mean. My point is simply that we can combine our factual knowledge about early mammalian body size with our understanding of how brains and brain regions scale to derive a cogent estimate of what the brains of early mammals probably looked like. This analysis reveals that early mammal brains probably looked similar to the brains of extant "basal insectivores" (Stephan and Andy, 1970; see also Kaas, 2002). Of course, this does not mean that early mammal brains looked exactly like hedgehog brains, for instance, because a cladistic analysis tells us that early mammals almost certainly lacked a corpus callosum (see Chapter 2). Further, it is clearly misleading to talk about "the" basal insectivore, since the hedgehog, tenrec, and opossum families are only distant relatives of one another (see Figure 2.11; note that tenrecs are within the Afrotheria) and each family comprises dozens of species. Indeed, the "basal insectivores" are an eclectic group. However, most of them are small, nocturnal, and predominantly insectivorous (just like early mammals). Therefore, it is not unreasonable to suppose that the brains of small insectivores resemble early mammal brains, at least in terms of gross morphology. Early mammals' brains were certainly more similar to the brain of a mouse opossum (see Figure 8.4) than to the brain of a duck-billed platypus (see Figure 8.3), which sports a tiny olfactory bulb, a proportionately large neocortex, and numerous specializations that are related to its electrosensory ducklike beak (Rowe, 1990). Thus, we have arrived at a gross but justifiable estimate of early mammal brain morphology.

The Phylogenetic History of Neocortex

Having obtained a general idea of what early mammal brains looked like, we can try to determine how those brains differed from early reptile brains; that is, we can try to find out what made mammalian brains "special." In order to simplify that task, let us begin by looking specifically at neocortex, which, as we discussed earlier, is widely reputed to be "unique" to mammals. Do non-mammals have anything that is homologous to mammalian neocortex? What kind of changes in development and adult structure might prompt us to call the neocortex "new"? As you will see, these questions are not yet resolved to everyone's satisfaction, but some answers have emerged. Let us begin with a brief description of the structure whose history we're trying to explain.

The mammalian neocortex typically develops in the dorsal portion of the embryonic forebrain, which comparative embryologists generally refer to as the dorsal pallium (see Chapter 3; Figures 3.8 and 3.14). It is bounded ventrolaterally by the olfactory (or piriform) cortex and ventromedially by the hippocampal formation (see Figure 8.4). In contrast to these neighboring structures, which tend to be trilaminar, the neocortex generally exhibits six major laminae. Although the precise number of neocortical layers and sublayers differs somewhat from species to species and from area to area, the six-layered pattern is probably primitive for all mammals (Brodmann, 1909).

The most common type of neuron found in the neocortex is the pyramidal cell, which typically has a long "apical" dendrite that courses radially from the cell body toward the neocortical surface (Figure 8.5A), several "basal" dendrites that ramify (branch) deep within the neocortex, and an axon that projects to both local and relatively distant sites. These pyramidal cells (which are also found in the olfactory cortex and the hippocampus) come in various shapes and sizes but are clearly distinct from neocortical interneurons, which tend to have smaller dendritic trees and project more locally, but are otherwise diverse (Nieuwenhuys, 1994b). Axons coming from the dorsal thalamus typically enter the neocortex from below (i.e., from its ventricular surface), ascend radially through the various cortical layers, and then terminate in the middle and upper laminae (see Figure 8.5A). The detailed pattern of how these afferent axons terminate within the neocortex varies between species as well as between cortical areas (e.g., Valverde et al., 1986), but that need not concern us here. In fact, rather than delving deeper into neocortical details (see Valverde, 1986 and Nieuwenhuys, 1994b for excellent reviews), we will proceed directly to comparisons with nonmammalian vertebrates.

Amphibians, the only anamniotes among the tetrapods (see Figure 3.1), have a relatively simple telencephalon with no obvious neocortex. On the other hand, amphibians do have a pallium that is divisible into several longitudinal zones (see Figure 8.5B). The precise number and location of these zones remains disputed but medial, dorsal, and lateral pallia are generally recognized (Neary, 1990; Northcutt, 1995). The medial pallium is thought to be homologous to the mammalian hippocampus (discussed in more detail later in this chapter), and the lateral pallium, with its heavy input from the olfactory bulb, is probably homologous to the mammalian olfactory cortex. That leaves the dorsal pallium as the most likely homologue of mammalian neocortex. Like the neocortex, it is located dorsally within the pallium; it also receives some inputs from the dorsal thalamus, which is a widely known neocortical "hallmark." Because of these similarities, most authors agree that the amphibian dorsal pallium is homologous to the mammalian neocortex. One should note, however, that the amphibian dorsal pallium is at best bilaminar rather than hexalaminar, and its "pyramidal neurons" generally lack basal dendrites (see Figure 8.5B) (Nieuwenhuys, 1994b). The amphibian dorsal pallium also differs radically from the mammalian neocortex in terms of connectivity. Its thalamic inputs are predominantly multimodal (i.e., they generally combine information from several sensory modalities), it does not project out of the telencephalon, and it has strong reciprocal connections with the olfactory bulb (Neary, 1990). Most important, the amphibian dorsal thalamus actually projects more heavily to the medial and lateral pallia than to the dorsal pallium (see Figure 8.5B). This is not what you see in mammals. Therefore, the hypothesis of homology between the amphibian dorsal pallium and the mammalian neocortex rests not on the discovery of numerous detailed similarities between the two structures, but on the observation that both occupy similar relative positions within the telencephalon. It is a classic case of homology between dissimilar structures (see Chapter 6). Since amphibians are not the closest relatives of mammals, that lack of similarity is understandable. So, what about reptiles? Do they have a neocortical homologue that more closely resembles mammalian neocortex?

(A) Mammalian neocortex

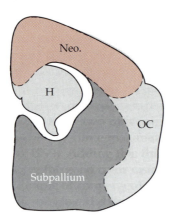

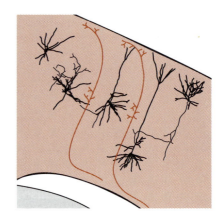

(B) Amphibian dorsal pallium

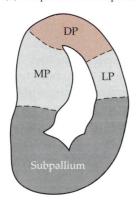

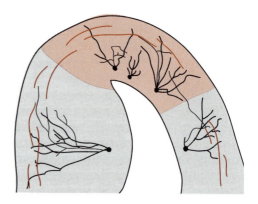

(C) Reptilian dorsal cortex

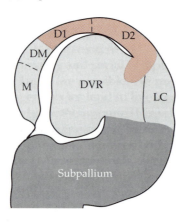

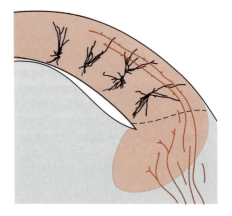

◀ **Figure 8.5 Neocortex, Dorsal Pallium, and Dorsal Cortex**   The mammalian neocortex (Neo.), the amphibian dorsal pallium (DP), and the reptilian dorsal cortex (D1 and D2) all occupy similar positions within the telencephalon and contain some similar neurons; they are probably homologous to one another. Shown here are schematic cross sections through the right telencephalon of (A) a hedgehog, (B) a salamander, and (C) a turtle. Low-magnification views are illustrated on the left, close-ups on the right. Some pyramidal (or "pyramidal-like") neurons are shown, as are some axons ascending from the dorsal thalamus (red lines). An important derived feature of mammalian neocortex is that its thalamic inputs course radially through a neocortical column. Other abbreviations: H = hippocampus; DM = dorsomedial cortex; DVR = dorsal ventricular ridge; LC = lateral cortex; LP = lateral pallium; M = medial cortex; MP = medial pallium; OC = olfactory cortex;. (A after Valverde, 1986; Valverde et al., 1986; B after Herrick, 1948; C after Connors and Kriegstein, 1986; Heller and Ulinski, 1987; Mulligan and Ulinski, 1990.)

The simple answer is yes, but we need to look at some details. Reptilian forebrains are diverse but generally larger and more complex than those of amphibians. Among the simplest reptile brains are those of turtles. This is not particularly surprising, since turtles first evolved more than 200 million years ago and were fairly conservative in their subsequent evolution. Cutting a transverse section through a turtle telencephalon, we are immediately struck by the presence of a large dorsal ventricular ridge (DVR) that protrudes into the ventricle from its lateral aspect (see Figures 8.5C, 8.6B,C). This DVR exists in all reptiles, though it varies considerably in size (Northcutt, 1978; Ulinski, 1983) and is difficult to homologize to anything in amphibian or mammalian brains; we shall discuss it in due course. Capping the turtle's DVR is a thin sheet of neural tissue that contains three laminae, of which the middle one contains most of the neuronal cell bodies (Ulinski, 1990). This trilaminar cortex is divisible into four or more longitudinal zones. The two most dorsal zones, called D1 and D2 of the dorsal cortex (Desan, 1988), receive major inputs from the dorsal thalamus. The D1 region receives inputs from limbic regions of the dorsal thalamus (which are neither sensory nor obviously motor in function), and the D2 region receives its major input from the reptilian dorsal lateral geniculate nucleus (LGN; see Chapter 6). As you can see in Figure 8.5C, these LGN axons enter D2 from its lateral edge and then course tangentially through D2 (Mulligan and Ulinski, 1990). Most of the neurons contacted by the LGN axons are quite similar to mammalian pyramidal neurons, but turtle cortex also contains some neurons that resemble neocortical interneurons (Connors and Kriegstein, 1986).

Thus, a turtle's dorsal cortex is structurally intermediate between an amphibian dorsal pallium and a mammalian neocortex. Its neurons are more similar to mammalian pyramidal neurons and more clearly arranged into laminae. It has only three laminae, rather than the six we see in neocortex, but its dendrites and axons course more rectangularly (i.e., radially and tangentially; see Chapter 6) than they do in the amphibian dorsal pallium. The dorsal cortex of turtles also receives significantly stronger inputs from the dorsal

(A) Turtle DVR in cross section

(B) Turtle DVR in horizontal section

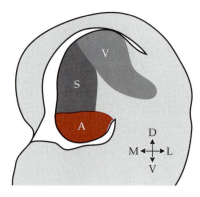

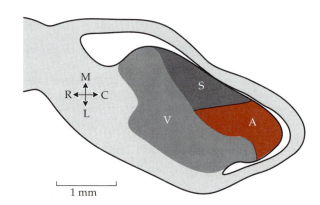

(C) Iguana DVR in a rostrocaudal series of cross sections

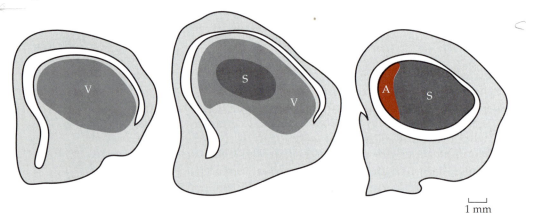

Figure 8.6 Sensory Zones within the Reptilian DVR   The reptilian dorsal ventricular ridge (DVR) contains several large sensory zones that receive their principal inputs from the dorsal thalamus. Shown here are schematic cross sections and a horizontal section, their orientations indicated by the arrows. (A,B) In turtles the visual zone (V) is located dorsally within the DVR, the auditory zone (A) lies ventrocaudally, and the somatosensory zone (S) occupies an intermediate position. (C) In iguanas, the same sensory zones are evident in cross sections (rostral to caudal equals left to right). As you can see, the auditory zone in this species lies not so much ventral as caudomedial within the DVR. Nonetheless, the topological arrangement of the three sensory zones within the DVR is roughly similar between iguanas and turtles, suggesting that it is a conserved aspect of the reptilian DVR. (A after Balaban and Ulinski, 1981; B is my interpretation of data published by Balaban and Ulinski, 1981; C after Manger et al., 2002.)

thalamus than the dorsal pallium does in amphibians, lacks reciprocal connections with the olfactory bulb, and projects to several extratelencephalic targets. In these respects, dorsal cortex is quite similar to neocortex. On the other hand, the dorsal cortex of turtles does not project directly to the spinal cord (see Chapter 7), lacks reciprocal connections with the contralateral dorsal cortex (but see Bruce and Butler, 1984), and shows no sign of auditory inputs from the dorsal thalamus. Since these connections are typical of mammalian neocortex, the dorsal cortex of turtles seems like an ideal "missing link" between the amphibian dorsal pallium and the mammalian neocortex, which is why many early comparative neuroanatomists referred to it as a "primordial neopallium" (Ariëns Kappers et al., 1936). There is, however, one major fly in this ointment, namely the aforementioned, wonderfully enigmatic DVR. It is a major component of the telencephalon in all reptiles and birds (i.e., sauropsids; see Chapter 3), and, like the dorsal cortex, it also resembles mammalian neocortex.

Soon after experimental axon-tracing techniques were invented (see Chapter 2), it became apparent that the sauropsid DVR is a major target of ascending sensory pathways (Karten, 1969). In turtles, the rostrolateral DVR receives visual projections from a dorsal thalamic nucleus that receives its principal input from the optic tectum (Balaban and Ulinski, 1981). Medial to this visually receptive zone is a somatosensory region, and caudoventral to that lies the major target of auditory projections from the dorsal thalamus (see Figure 8.6A,B). Similar pathways exist in birds (the avian DVR is discussed in more detail later in the chapter) and in various lizards (see Guirado et al., 2000), though the size, absolute position, and cytoarchitectural appearance of those sensory zones vary significantly between species (see Figure 8.6C). Therefore, it is fair to say that the DVR is *the* major target of ascending sensory pathways in the sauropsid forebrain; it is certainly larger than the dorsal cortex (i.e., D1, D2, and their homologues), and it is the only pallial region in reptiles that is known to process auditory information. Processing within the DVR is complicated, but in general, information is sent from primary regions to various secondary and tertiary regions that, ultimately, project to the caudalmost DVR, from which axons emanate to various extratelencephalic targets, including the hypothalamus, the midbrain, and the medulla (Ulinski, 1983; Lanuza et al., 1998; Husband and Shimizu, 1999). In aggregate, these connectional data reveal a striking similarity between the DVR and mammalian neocortex. This leaves us with a fascinating dilemma: Can DVR and dorsal cortex *both* be homologous to neocortex?

Before we try to answer this question let us go back to amphibians, where a related puzzle lurks: What, if anything, is the amphibian homologue of the sauropsid DVR? As we saw in Figure 8.5B, amphibians have no intraventricular ridge that looks like a DVR. Topologically, we would expect an amphibian DVR to lie ventrolateral to the dorsal pallium, but in that location we find only the lateral pallium, which does not bulge into the ventricle and receives its principal inputs from the olfactory bulb (see Figure 2.10). Essentially, the problem is that reptiles have four major divisions in their pallium (medial, dorsal, lateral, and DVR), while amphibians have only three. Since the pallia of most other anamniotes also contain only three major pallial divisions, we can infer that a phylogenetic proliferation of pallial areas occurred. But was this proliferation due to segregation or addition (see Chapter 6)?

One argument in favor of the segregation hypothesis is that the amphibian lateral pallium receives not only olfactory projections but also some (meager) inputs from the dorsal thalamus (Kicliter and Northcutt, 1975). That makes the amphibian lateral pallium at least somewhat similar to *both* the reptilian lateral cortex and the DVR (Northcutt and Kicliter, 1980; Neary, 1990), which is exactly what you would expect to find if lateral cortex and DVR evolved by phylogenetic segregation from an ancestral amalgam of both. According to this hypothesis, the lateral cortex would have lost its old thalamic inputs, while the DVR lost its direct connections with the olfactory bulb. The principal problem with this otherwise attractive hypothesis is that the amphibian lateral pallium is far more similar to the reptilian lateral cortex than to the DVR. By almost any measure, the DVR is the odd-man out in this comparison. Therefore, given the formalism I proposed in Chapter 6, the reptilian DVR could well be considered "novel by addition." This hypothesis implies that DVR development should be due to a novel constellation of developmental mechanisms, but the data needed to test that prediction are not yet available.

Now, as we try to compare the telencephalons of reptiles and mammals, we encounter an analogous problem: Mammals have no structure that is obviously homologous to the reptilian DVR. Prior to the 1960s, many neurobiologists had thought that the DVR was homologous to the mammalian striatum (which is why many components of the DVR were named with the suffix "-striatum"), but that hypothesis became untenable when it was discovered that reptiles (and birds) have a perfectly good striatum *ventral to* the DVR (see Figure 2.9). This left the DVR without an obvious mammalian homologue because the only structure that lies dorsal to the mammalian striatum is the neocortex, which does not look at all like the DVR, at least at first glance. This is important because, if mammals really have no DVR homologue, then it would be most parsimonious to conclude that the DVR evolved after reptiles and mammals diverged from one another. That is, we would have to conclude that the DVR is unique to "late" rather than "early" reptiles. In support of this hypothesis, we might cite that most early reptiles had rather narrow endocrania (see Figure 8.2), but this is weak evidence because even narrow endocrania may contain at least a small DVR. In fact, most evolutionary neurobiologists have not endorsed this "late reptilian DVR" or "outgroup" hypothesis (Northcutt and Kaas, 1995). Instead, they have proposed that early reptiles, including the last common ancestor of mammals and reptiles, had a small DVR that became obscure in extant mammals (see Striedter,

1997; Aboitiz et al., 2003). Where do we find this cryptic DVR? Well, on that point, disagreements remain rife. Basically, there are two competing hypotheses (Figure 8.7A,B). One says that the DVR in mammals became part of the neocortex; the other argues that the mammalian DVR is the claustroamygdalar complex. As you will see, both hypotheses have some evidence in their favor, but I favor the second. Let us examine both hypotheses in turn.

The first hypothesis, which I call the "neocortical DVR hypothesis" (see Figure 8.7A), was originally proposed by Harvey Karten (1969) and subsequently modified by Ann Butler (1994) and Anton Reiner (2000). It states that the sauropsid DVR is homologous to the most lateral portion of the mammalian neocortex (which consists mainly of auditory cortex; see Figure 6.13) and that the reptilian dorsal cortex is homologous to the more medial neocortex (which includes mainly the primary visual and somatosensory cortices) (see Medina and Reiner, 2000). The strongest evidence in favor of this hypothesis is that the DVR, like the lateral neocortex, receives auditory inputs from the dorsal thalamus, while the dorsal cortex does not (Karten, 1968; Pritz, 1974; Pritz and Stritzel, 1992). Additional support derives from the observations that the DVR, like the neocortex, is the principal integration center for all modalities other than olfaction and that it plays a major role in motor control. According to Karten and other supporters of the neocortical DVR hypothesis, these similarities exist because they were inherited from a common ancestor (most likely from an early reptile with a small DVR; see Figure 8.7A). This implies that, in adults, the positional differences between the DVR and the lateral neocortex, with the former lying deep to olfactory cortex and the latter dorsal to it, must be due to phylogenetic divergence. Furthermore, the neocortical DVR hypothesis implies that the cytoarchitectural differences between DVR and lateral neocortex, with only the latter being prominently laminar, evolved after reptiles and mammals had diverged from one another. Thus, at the core of the neocortical DVR hypothesis lies the notion that neuronal function and connectivity are more conserved in evolution than adult position and cytoarchitecture.

Diametrically opposed is the "claustroamygdalar DVR hypothesis," which was developed by Nils Holmgren (1925) and then modified by Bruce and Neary (1995), Striedter (1997), and Puelles et al. (2000). According to this hypothesis, the reptilian DVR is homologous to the mammalian claustroamygdalar complex (CA), which includes both the claustrum and the pallial, or "basolateral," amygdala. The principal evidence in favor of the claustroamygdalar DVR hypothesis is that the embryonic locus of the CA is similar to that of the DVR. That is, both structures develop just dorsal to the pallial–subpallial boundary and deep to the olfactory cortex (which is called the "lateral cortex" in most reptiles; see Figure 8.7B). This was apparent to Holmgren, who looked at Nissl-stained sections through embryonic brains, and it has been confirmed by various modern techniques, including stains for radial

(A) The neocortical DVR hypothesis

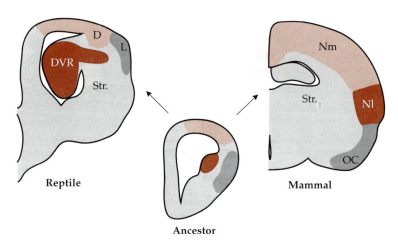

(B) The claustroamygdalar DVR hypothesis

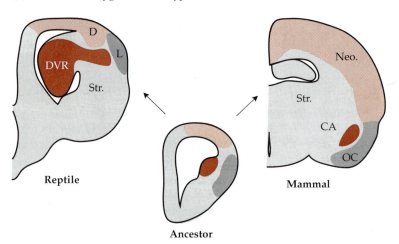

glia, developmental fate mapping, and analyses of developmental gene expression (Figure 8.8) (Misson et al., 1991; Striedter and Beydler, 1997; Smith-Fernandez et al., 1998; Striedter et al., 1998; Puelles et al., 2000). According to the claustroamygdalar DVR hypothesis, this similarity in embryonic origin is retained from a common ancestor. In contrast, the differences in adult position and morphology are interpreted as the result of a phylogenetic divergence in relatively late development. Specifically, the hypothesis implies that the mammalian CA is "squeezed" between the enlarging striatum ventrally

◀ **Figure 8.7 Two Hypotheses of Pallial Homologies in Amniotes** Several different hypotheses about how to homologize the pallial regions of amniotes have been proposed, but most of them cluster around one of two very different "prototypical" hypotheses. (A) The first of these is the neocortical DVR hypothesis, which basically states that the reptilian DVR is homologous to a lateral portion of the mammalian neocortex (Nl), which includes the auditory cortex. This hypothesis was originally proposed by Harvey Karten (1969) but comes in several different "flavors." (B) The second prototype hypothesis is the claustroamygdalar DVR hypothesis, which postulates that the reptilian DVR is homologous to the mammalian basolateral amygdala and to the ventral claustrum, which is also known as the endopiriform nucleus. This second hypothesis dates back to Nils Holmgren (1925) but has also been modified over the years. An important difference between these two hypotheses is that, according to the neocortical DVR hypothesis, mammalian neocortex is homologous to two distinct pallial regions in reptiles (DVR and dorsal cortex), whereas according to the claustroamygdalar DVR hypothesis, the neocortex is homologous to just a single region in reptiles (dorsal cortex). Abbreviations: CA = claustroamygdalar complex; D = dorsal cortex; L = lateral cortex (the most likely reptilian homologue of mammalian olfactory cortex); Neo. = neocortex; Nl = lateral neocortex (including auditory, extrastriate visual, and secondary somatosensory cortices); Nm = medial neocortex (including primary visual and somatosensory cortices); OC = olfactory cortex; Str. = striatum. (After Reiner, 2000.)

and the neocortex dorsally until it finds itself far from the ventricle, whereas the reptilian DVR enlarges in situ until it becomes the largest region of the telencephalon (see Figure 8.7B). The claustroamygdalar DVR hypothesis also implies that the reptilian DVR acquired some new neuronal connections that ultimately allowed it to become functionally similar to the mammalian neocortex. Thus, a central tenet of the claustroamygdalar DVR hypothesis is that early embryonic development is more conserved than either adult position or neuronal connectivity.

As you might imagine, these two hypotheses about how to homologize the DVR between reptiles and mammals are the subject of considerable debate (see commentary in Aboitiz et al., 2003). How do we decide between them? Both hypotheses embody efforts to homologize marginally similar structures and, as we noted in Chapter 6, such efforts are bound to be controversial because different investigators are likely to have different ideas about what kind of similarities are the most reliable "criteria" of putative homology. Those who favor connectional similarities are likely to endorse the neocortical DVR hypothesis because it is, after all, based almost exclusively on connectional data. On the other hand, those who consider similarities in embryonic origin to be more conserved than connectivity are likely to champion the claustroamygdalar DVR hypothesis. As you probably gathered from the preceding chapters, I have found the comparative neurobiological literature to be replete with species differences in neuronal connectivity, and full of similarities in embryonic patterning (see also Striedter, 1992; Striedter, 1994;

(A) Radial glia in embryonic mice and chicks

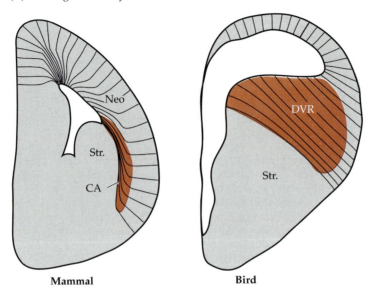

Mammal　　　　　　　**Bird**

(B) Regulatory gene expression in mice and chicks

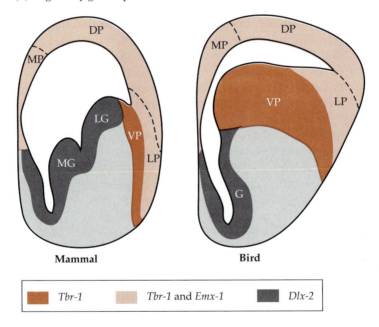

Mammal　　　　　　　**Bird**

■ *Tbr-1*	*Tbr-1* and *Emx-1*	■ *Dlx-2*

◀ **Figure 8.8 Embryology and the Claustroamygdalar DVR Hypothesis**   Recent embryological studies have tended to support the hypothesis of homology between the DVR and the claustroamygdalar complex. (A) Radial glial cells (black lines) are found in the forebrain of both embryonic mammals (left) and birds (right). Their spatial distribution and orientation is quite different in mammals and birds, but we can mentally morph one pattern into the other without "cutting or pasting." This implies that the developing claustroamygdalar complex (CA) is topologically equivalent to the DVR. Since radial glia are known to indicate the major routes of developing neuron migration (see Figure 8.9), this finding indicates that CA and DVR are likely to develop from homologous precursor regions. (B) The transcription factors *Tbr-1*, *Dlx-2*, and *Emx-1* are expressed in the developing forebrain of both mice (left) and chicks (right). In both species, *Dlx-2* marks the subpallium, *Tbr-1* the pallium; *Emx-1* is interesting because it is expressed throughout all of the pallium, except in its most ventrolateral portion (Smith-Fernandez et al., 1998), which Puelles et al. (2000) have called the ventral pallium (VP). Since this ventral pallium apparently develops into part of the claustroamygdalar complex in mammals and into part of the DVR in birds, these gene expression data support the claustroamygdalar DVR hypothesis. Other abbreviations: DP = dorsal pallium; G = ganglionic eminence; LG = lateral ganglionic eminence; LP = lateral pallium; MG = medial ganglionic eminence; MP = medial pallium; Neo. = neocortex; Str. = striatum. (A after Misson et al., 1991; Striedter and Beydler, 1997; B after Puelles, 2001.)

Striedter and Beydler, 1997). Therefore, I am generally inclined to support the claustroamygdalar DVR hypothesis. In addition, I would argue that the large equals well-connected rule actually predicts some of the DVR's unusual connections, because the DVR of modern reptiles is almost certainly hypertrophied relative to the early reptilian DVR and the mammalian CA. Unfortunately, both of these arguments are less than decisive. Perhaps the DVR is one of those exceptions to the rule in which connections are more conservative than embryology and structural hypertrophy occurred without changes in connectivity. That possibility exists! Therefore, let us pursue an additional and time-honored strategy. Let us ask whether either hypothesis makes predictions that were not used to construct the hypothesis initially but are now available for inspection. If the predictions are borne out, the hypothesis would be strengthened. If not, then the original hypothesis ought to be modified or abandoned.

One such prediction, first made by Harvey Karten back in 1969, is that lateral neocortical neurons should differ in embryonic origin from medial neocortical neurons. Specifically, Karten predicted that an embryonic DVR expands in situ in reptiles to form the adult DVR but migrates dorsolaterally in mammals, inserting itself between olfactory cortex and the medial neocortex (Figure 8.9A). This prediction remained untested for years but now has been falsified. Detailed embryological studies in mammals have revealed that, although some neuronal precursors do migrate dorsolaterally into the neocortex, the source of those migrating neurons is not the embryonic DVR

(A) Hypothetical migrations

(B) Actual migrations

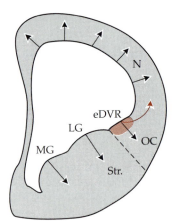

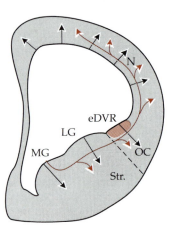

Figure 8.9 Neuronal Migration in Developing Forebrains Vertebrate neurons are typically "born" adjacent to the ventricle and then migrate to their adult location. The bulk of that migration is along the radial dimension (black arrows), from the ventricular zone directly toward the brain surface. The ventrolateral portion of the telencephalon thickens early in development to form two intraventricular ridges, which are called the lateral and medial ganglionic eminences (LG and MG, respectively). Just dorsolateral to them lies the lateral pallial–subpallial boundary (dashed line) and dorsal to that lies, at least in birds, the embryonic DVR (eDVR). (A) In order to square these embryonic data with the neocortical DVR hypothesis (see Figure 8.7A), Harvey Karten (1969) proposed that in mammals the cells that correspond to the avian eDVR (shaded red) migrate dorsolaterally until they lie dorsal to the olfactory cortex (OC). (B) Subsequent studies showed that many neurons do indeed migrate dorsolaterally in mammals (as well as in birds!), but most of those cells come from the medial ganglionic eminence (red arrows), not from the eDVR. Furthermore, those neurons migrate not only to the lateral neocortex, but also to a variety of different areas, where they ultimately differentiate into inhibitory interneurons. These findings are not consistent with the neocortical DVR hypothesis. Other abbreviations: N = neocortex; Str. = striatum. (A after Nauta and Karten, 1970; B after Marín and Rubenstein, 2001.)

but a more ventral *subpallial* structure, the medial ganglionic eminence (Figure 8.9B) (Lavdas et al., 1999). The dorsolaterally migrating cells also become inhibitory interneurons rather than neocortical projection neurons, as Karten had predicted. Therefore, we must either abandon the neocortical DVR hypothesis or amend it in some way. One feasible amendment is to postulate that the embryonic DVR in mammals develops not deep to olfactory cortex, as it does in reptiles, but dorsal to the olfactory cortex. If that were the case, its cells would not have to migrate far to reach their adult lateral neocortical position, and the neocortical DVR hypothesis would be reconciled with the data on neuronal migration. This amendment comes at a cost, however, for it implies that the relative position of the embryonic DVR changed radically during tetrapod phylogeny. Such changes in embryonic topology are conceivable (as I reviewed in Chapter 3) but rare. In fact, I know of no well-established instances where homologous brain regions develop from embryonic precursors that are not topologically equivalent (or at least similar) to one another. It is far more common for homologous brain regions to develop from topologically similar precursors but then migrate to topologically disparate positions in the adult forms.

The claustroamygdalar DVR hypothesis has a somewhat better (though also imperfect) record when it comes to testing its predictions. This more positive record stems mainly from the fact that all of the molecular evidence in favor of the claustroamygdalar DVR hypothesis (see Figure 8.8) was gathered after the hypothesis was first proposed (Holmgren, 1925; see Striedter, 1997). Indeed, molecular studies supporting the claustroamygdalar DVR hypothesis continue to be published at a rapid-fire rate (e.g., Brox et al., 2004). In addition, a comparative study of ascending auditory pathways in mammals (Frost and Masterton, 1992) has revealed that the main auditory region of the dorsal thalamus (the medial geniculate nucleus) projects more prominently to the basolateral amygdala in small-brained marsupials and insectivores than it does in rats and most other placental mammals (Figure 8.10). Considering that marsupials are the sister group of placental mammals and that early mammals were probably small-brained (discussed earlier), these results strongly suggest that the basolateral amygdala was a more important auditory region in early mammals than it is today, which is precisely what one would expect if the auditory dorsal thalamus "invaded" the lateral neocortex early on in mammalian phylogeny and then shifted its ascending projections increasingly away from the subcortical pallium (i.e., the claustroamygdalar complex) into the neocortex. As a further test of this hypothesis, one might want to examine the ascending auditory pathway in monotremes but, unfortunately, those data are incomplete. Monotremes clearly have an auditory cortex, curiously located caudal to the visual cortex rather than lateral to it (Rowe, 1990), but we do not yet know whether the medial geniculate nucleus in monotremes has major projections to the basolateral amygdala, as it does in small-brained marsupials and insectivores (Frost and Masterton, 1992). The recent demonstration that monotremes have no obvious claustrum (Butler et al., 2002) is intriguing but mute with respect to auditory projections to the basolateral amygdala. Therefore, most of what we know about the ascending auditory pathways in mammals is compatible with Holmgren's (1925) old hypothesis that at least some part of the mammalian claustroamygdalar complex is homologous to the reptilian DVR.

These considerations have, in aggregate, convinced me that the claustroamygdalar DVR hypothesis is preferable to its rival, the neocortical DVR hypothesis. Others may disagree, and the debate is likely to continue as new

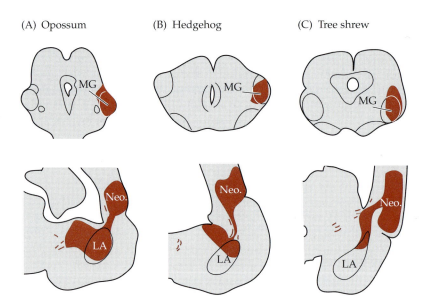

(A) Opossum   (B) Hedgehog   (C) Tree shrew

Figure 8.10 Variation in the Auditory Pathways to the Telencephalon   Frost and Masterton (1992) injected axonal tracers into the main dorsal thalamic auditory nucleus, the medial geniculate (MG), in various mammals. Illustrated here are representative cross sections showing tracer injection sites (top row) and the major telencephalic projections (bottom row); all structures labeled by the tracers are shown in red. (A) The researchers found that in short-tailed opossums, roughly 50% of MG cells project to the subcortical telencephalon, including the lateral amygdala (LA). (B) In hedgehogs, the percentage of MG neurons with subcortical projections is reduced to 25%. (C) In tree shrews, less than 5% of the medial geniculate neurons project to subcortical targets. Frost and Masterton (1992) interpreted these data to mean that the telencephalic neurons that receive MG inputs migrated from a subpallial location into the pallium as mammals evolved. However, it is simpler to suppose that the MG projections "shifted" from subcortical targets into the neocortex (Neo.), leaving their old target neurons behind. If the latter is true, then the data lend support to the claustroamygdalar DVR hypothesis, which posits the invasion of the neocortex by the auditory dorsal thalamus as one of its central elements. (After Frost and Masterton, 1992.)

data emerge and fresh minds reinterpret what is already known. To illustrate the latter point, I should point out that my analysis of the ascending auditory pathways in mammals differs dramatically from that of Frost and Masterton (1992), who used their data to support Karten's (1969) notion that auditory neurons migrated from a subcortical location (i.e., from the basolateral amygdala and/or the DVR) into the neocortex as mammals evolved. That interpretation cannot be squared with what we know about development, but not everyone sees it that way (see discussion in Aboitiz et al., 2003). In addition,

I suspect that some biologists find it simply inconceivable that functional similarities as striking as those between DVR and neocortex could be due to anything but common ancestry. To rebut this argument I would point out that several equally astonishing examples of convergent evolution are already widely accepted. There is little doubt, for instance, that birds and mammals independently evolved upright body postures, endothermy, and large brains (discussed earlier and in Chapter 4). Similarly, it is rapidly becoming clear that birds, lizards, and mammals independently evolved more efficient auditory systems with proper "ear drums," enlarged cochleas, and "active" hair cells (see Manley, 2000). If those convergences are "real," why should the brain be exempt from similar convergences? Indeed, would we not expect that functional improvements in the peripheral auditory system should lead to major expansions of the central auditory system, such as the formation of "new" auditory projections? Once we admit that possibility, we are primed to think that convergent evolution could be rampant rather than rare. Several evolutionists have recently arrived at that conclusion for non-neural structures (Morris, 2004), but I think it may apply to brains as well. Therefore, I conclude that it is certainly *conceivable* that the reptilian DVR is *functionally similar* to the mammalian neocortex but *homologous* to the claustroamygdalar complex. Let us explore some implications of that thought.

The most important corollary of the claustroamygdalar DVR hypothesis is that the mammalian neocortex probably evolved from a structure that looked quite similar to the dorsal cortex of a turtle. Since a turtle's dorsal cortex features only one prominent layer of cells (sandwiched between two layers of dendrites and axons), the emergence of a six-layered neocortex almost certainly involved a phylogenetic increase in the degree of lamination. Indeed, Anton Reiner (1993) showed that the turtle's dorsal cortex lacks several of the histochemically defined neuron types that are found in the upper layers of the mammalian neocortex. Those neocortical cell types were probably "added" as mammals evolved, just as the auditory input from the dorsal thalamus was added in phylogeny.

The other major innovation of mammalian neocortex is that dorsal thalamic axons began to course radially, rather than tangentially, through the neocortex (Diamond and Ebner, 1990; Supèr and Uylings, 2001). Going back to Figure 8.5C, you can see that in a turtle's dorsal cortex, the incoming thalamic axons course parallel to the brain's surface, contacting numerous pyramidal neurons along the way, whereas in the mammalian neocortex, the incoming axons ascend radially through the cortex. Consequently, individual thalamic axons actually project to fewer pyramidal neurons in mammals than they do in turtles. We know little about what caused thalamocortical axons to change their trajectory through the cortex (see Molnár, 1998), but the functional consequences of that change were surely considerable: It probably facilitated the emergence of neurons with small receptive fields (i.e., neurons that respond to a narrow range of stimuli), of radially organized "columns" of cells with similar response profiles, and of fine-grained topographical maps. All of these attributes are typical of neocortex and other highly laminated structures (e.g., the goldfish vagal lobe; see Chapter 6) but not evident in the dorsal cortex of turtles. Thus, if the mammalian neocortex evolved from something like a turtle's dorsal cortex, then what makes it "special" is that it evolved from a state of fairly simple lamination to become one of the most highly laminar brain structures ever seen (Raizada and Grossberg, 2003).

What can mammals do with that more highly laminated neocortex and its columnar connectivity? Did the "new cortex" enable early mammals to process sensory information more precisely and to learn more from experience? Perhaps (see Allman, 1990), but we cannot be sure because (as we have discussed) comparing behavioral capacities across species is notoriously difficult and because the neocortex is not the only brain region that differs between reptiles and mammals. In contemplating these issues, however, it is crucial that we refrain from directly comparing turtles (or other reptiles with relatively simple dorsal cortices) to primates or other mammals with large neocortices. As we already discussed , the earliest mammals probably had small neocortices, similar in size to those of today's small insectivores (see Figure 8.4). Therefore, the critical question is how the behavioral capacities of turtles differ from those of small insectivores, and to what extent those differences can be attributed to mammals having a neocortex. That question is likewise difficult, but I suspect that the functional advantage afforded by the neocortex in this comparison was probably minor, relative to what it would be if we compared turtles to large-brained primates.

This brings up an interesting thought: Could it be that one important attribute of mammalian neocortex is that it is eminently scalable? Could it be that what makes mammalian neocortex "special" is, at least in part, that it can be increased in size without impairing functionality? Evidence supporting this hypothesis was provided by Murre and Sturdy (1995), who showed that distributing "theoretical neurons" across the surface of a cube (or a sphere), with their connections running through the cube, is far more efficient in terms of "wiring costs" than having the same number of neurons scattered randomly throughout the cube, especially as neuron number increases. Applying this insight to real brains, we can conclude that as brain size increases, the wiring pattern we see in the neocortex, where most long connections run through underlying white matter, is more efficient than the wiring we see in the dorsal cortex, where axons are more interspersed among neuronal cell bodies. That would mean neocortex is more expandable than dorsal cortex. This increased expandability may not explain why neocortex *originally* evolved, but it may well explain why mammals with large neocortices evolved more frequently than reptiles with large dorsal cortices.

Before we leave the topic of neocortical phylogeny, let us briefly consider what has happened to the DVR and dorsal cortex homologues in birds (the "flying reptiles"). As we saw in Chapter 4, bird brains are significantly larger than reptilian brains at the same body size (see Figure 4.8B). Much of that increase in overall brain size is due to an increase in the relative size of the DVR, which grows so large in birds that it virtually obliterates the telencephalic ventricles and is scarcely recognizable as an intraventricular ridge (see Striedter and Beydler, 1997). In terms of connections, the avian DVR is similar to that of earthbound reptiles but also exhibits some rather unusual connections, which one would expect given the rule of "large equals well-connected" (see Chapter 7). Only in birds, for instance, do we find auditory and somatosensory pathways that course directly from the isthmic region to the DVR, bypassing midbrain and thalamus (Arends and Zeigler, 1986; Wild and Farabaugh, 1996; Wild et al., 2001). The avian DVR also contains more subdivisions than the DVR of other reptiles. Pigeons, for instance, have at least six distinct visual areas within their DVR, whereas turtles and lizards possess only one to three (Ulinski, 1983; Husband and Shimizu, 1999; Manger et al., 2002). Many of these DVR subdivisions are arranged like the layers of an onion, or geological strata (Figure 8.11A), and connected by relatively short axons that course at right angles to those laminae. This laminar organization is not as obvious in the avian DVR as it is in neocortex, but it is quite prominent in the DVR's principal sensory zones (e.g., Veenman and Gottschaldt, 1986). Since there is relatively little evidence for that kind of lamination in turtles, lizards, or snakes, a partially laminated DVR probably evolved specifically in birds. Most likely, it evolved as an elegant solution to the problem of how to interconnect an increasing number of neurons with a minimum of wiring (see Chapter 6).

The avian dorsal cortex is generally called the hyperpallium, or "Wulst" (see Reiner et al., 2004). Lying like a crest atop the telencephalon, this Wulst does not look much like a turtle's dorsal cortex, for it is much thicker, devoid of pyramidal neurons, and not obviously trilaminar. Yet in terms of embryonic origin, adult topology, and general connectivity, Wulst and dorsal cortex are so similar that their homology is not in doubt. Assuming that dorsal cortex is homologous to neocortex (discussed earlier), the Wulst is a bird's most likely homologue of mammalian neocortex. Given that homology, it is perhaps surprising that the Wulst is relatively small in the majority of birds (see Figure 8.11A). However, as I mentioned already, birds have a huge DVR that probably fulfills some functions similar to those of mammalian neocortex. Moreover, the Wulst is hardly "rudimentary" in birds. Owls in particular have an enormous Wulst (see Figure 8.11B). Since the Wulst is mainly visual in function, with a small somatosensory component lurking rostrally, its hypertrophy in owls probably relates to owls' excellent visual acuity, depth perception, and talon–eye coordination. The most fascinating aspect of an owl's Wulst is that it is clearly divisible into four laminae that are interconnected by many short axons (see Figure 8.11B) and that axons from the dorsal thalamus come into the owl's Wulst not parallel to the laminae, as they do in a turtle's dorsal cortex (see Figure 8.5C), but orthogonal to them (see Fig-

(A) A pigeon's DVR

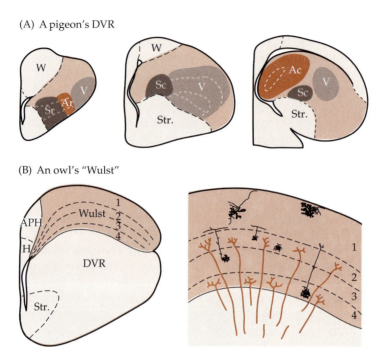

(B) An owl's "Wulst"

Figure 8.11 Expansion and Elaboration of the Avian Telencephalon Most bird brains are larger than reptilian brains (see Figure 4.8) and much of that enlargement involves the telencephalon. (A) The avian DVR is so large that it is scarcely recognizable as an intraventricular ridge for most of its extent. Within the avian DVR, we find homologues of the visual, somatosensory, and auditory zones that exist in turtles and lizards (see Figure 8.6), but we also find some novel somatosensory and auditory regions in the rostral DVR (Sr and Ar, respectively), which receive their sensory inputs directly from the isthmic region (see Chapter 3). Note that many of the subdivisions of the avian DVR (indicated by white dashed lines) are stacked atop one another or arranged like the layers in an onion. That is, much of the avian DVR has a distinctly laminated appearance. (B) In most birds the neocortex homologue, generally called the "Wulst" (W), is relatively small, but in owls it is hypertrophied. The owl's Wulst is also clearly laminar, even though those laminae are not homologous to the laminae of neocortex (Medina and Reiner, 2000). Most striking is that dorsal thalamic axons enter and course through the Wulst at right angles to those laminae, an arrangement that is reminiscent of mammalian neocortex but very different from that seen in dorsal cortex (see Figure 8.5). Overall, we may conclude that the telencephalons of mammals and birds have undergone a remarkable amount of convergent evolution. Other abbreviations: Ar = rostral auditory zone; Ac = caudal auditory zone; APH = parahippocampal area; H = hippocampus; Sr = rostral somatosensory zone; Sc = caudal somatosensory zone; Str. = striatum; V = visual zone. (A after Wild et al., 1985; Wild, 1987; Arends and Zeigler, 1989; Husband and Shimizu, 1999; B after Karten et al., 1973; Pettigrew, 1979.)

ure 8.11B)(Karten et al., 1973). In other words, the wiring diagram of an owl's Wulst is stunningly similar to that of the mammalian neocortex. As one might expect, this anatomical similarity translates into physiological similarity, as many Wulst neurons respond to visual stimuli just as neocortical neurons are known to do (Pettigrew and Konishi, 1976; Pettigrew, 1979; Wagner and Frost, 1994; Liu and Pettigrew, 2003). Since these similarities are not shared by dorsal cortex in most other reptiles (nor even in most birds), we can safely conclude that many of the striking similarities between an owl's Wulst and the mammalian visual cortex are due not to common ancestry but to independent evolution (Striedter, 1997; Medina and Reiner, 2000). It is yet another example of brains "inventing" similar solutions to similar biological problems, and it confirms my earlier assertion that convergent evolution may be common, if not rampant, in brains.

Beyond the Neocortex

In the preceding discussion, we saw that mammalian neocortex really is quite different from reptilian dorsal cortex even though the two are likely homologues, making it appropriate to call the neocortex "new" (see Chapter 6). I also think it fair to say that the neocortex was *the* key innovation of mammalian brains, especially in light of how that neocortex has expanded in large mammals such as *Homo sapiens* (see Chapter 9). But what about the rest of the brain? How does it differ between mammals and nonmammals? One answer is that, compared to the differences between dorsal cortex and neocortex, all other differences between mammalian and nonmammalian brains are relatively insignificant. This point is implicit in the disproportionate amount of attention comparative neurobiologists have lavished upon the evolution of the neocortex relative to the rest of the brain. However, mammalian and reptilian brains do differ even at subcortical or, more accurately, non-neocortical levels. Many of those non-neocortical differences are poorly understood and quantitative rather than qualitative, but this need not make them insignificant. To illustrate the point, let us focus on two non-neocortical regions that comparative neurobiologists have studied in considerable detail, namely, the forebrain's basal ganglia and hippocampal formation. How do they differ between mammals and nonmammals and, more generally, how much do they contribute to making mammalian brains "special"?

The forebrain components of the basal ganglia are well conserved across the vertebrates, particularly across tetrapods (Parent, 1986; Northcutt, 1995; Reiner et al., 1998). The striatum, the largest single structure in the basal ganglia, is present in all tetrapods and always receives at least some dopaminergic projections from the posterior diencephalon and/or the midbrain tegmentum (see Chapter 3). Similarly, all tetrapods seem to have a "nucleus

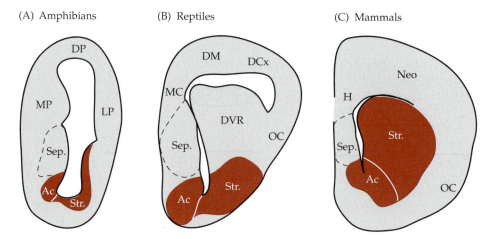

Figure 8.12 The Position of the Forebrain's Basal Ganglia in Tetrapods The striatum (Str.) and nucleus accumbens (Ac; also called the ventral striatum) occupy topologically equivalent positions in the telencephalon of (A) amphibians, (B) reptiles, and (C) mammals. The nucleus accumbens is generally located in the medial subpallium, just ventral to the septum (Sep.). The striatum, in contrast, generally lies lateral to the nucleus accumbens and occupies most of the ventrolateral subpallium. Other abbreviations: DCx = dorsal cortex; DM = dorsomedial cortex; DP = dorsal pallium; DVR = dorsal ventricular ridge; H = hippocampus; LP = lateral pallium; MC = medial cortex; MP = medial pallium; Neo. = neocortex; OC = olfactory cortex. (After Reiner et al., 1998; Marín et al., 1998.)

accumbens (Figure 8.12) and a globus pallidus, or pallidum (Marín et al., 1998). Even the subthalamic nucleus is present in both mammals and birds (Medina et al., 1997; Jiao et al., 2000). Given how conserved these nuclei and their principal connections are across the tetrapods, it seems fair to say that a fundamental scheme" of basal ganglia organization evolved with, or prior to, the origin of tetrapods and was retained thereafter (Marín et al., 1998). On the other hand, the size of the various pathways through the basal ganglia does vary considerably across species (Figure 8.13). Thus, in amphibians, the striatum receives its sensory input mainly from the dorsal thalamus and sends its output mainly to the pallidum, which then projects both directly and indirectly to the optic tectum (see Figure 8.13A). In reptiles, the striatal output circuitry is similar to what it is in amphibians, but the sensory inputs to the striatum come mainly from the DVR rather than directly from the dorsal thalamus (see Figure 8.13B) (see Guirado et al., 2000). This probably implies that the striatum receives more highly processed sensory information in reptiles than in amphibians. In mammals, the situation is different yet again (see Figure 8.13C), for in them, most striatal sensory inputs come from the neocortex (Parent and Hazrati, 1995a,b). Moreover, mammals direct their striatal output

(A) Amphibians

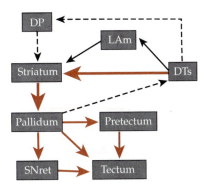

(B) Reptiles

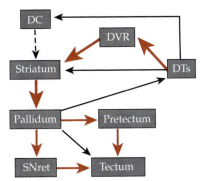

(C) Mammals

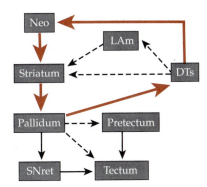

Figure 8.13 Connectivity of the Basal Ganglia in Tetrapods In this schematic diagram, connection strengths are represented by line width and arrowhead size, with thin, dashed arrows indicating the weakest connections. Based on these connection strengths, it is possible to estimate the predominant lines of information flow through the basal ganglia (red arrows). Such an analysis suggests that the striatum's major input comes (A) from the dorsal thalamus in amphibians, (B) from the DVR in reptiles and birds, or (C) from the neocortex in mammals. The striatal outflow also differs between taxonomic groups, being directed mainly toward the tectum in reptiles and amphibians, and toward the dorsal thalamus in mammals. Omitted from this figure are the dopaminergic inputs to the striatum, which are highly conserved across the vertebrates (see Chapter 3). Abbreviations: DC = dorsal cortex; DP = dorsal pallium; LAM = lateral amygdala; DTs = sensory dorsal thalamus; SNret = Substantia nigra, pars reticularis; Neo = Neocortex; Pretec = Pretectum; DVR = Dorsal ventricular ridge.

mainly to the dorsal thalamus, and from there back to the neocortex, rather than toward the optic tectum. Thus, we may conclude that, although most of the cell groups and connections in the basal ganglia can be homologized across all tetrapods, the manner in which information flows through the basal ganglia differs considerably between amphibians, reptiles, and mammals.

The aforementioned variation in basal ganglia circuits implies that the basal ganglia did not evolve as one unit. This is intriguing because a simplistic interpretation of the epigenetic population-matching hypothesis (see Chapter 7) might have led us to predict that all components of the basal ganglia network should have increased or decreased in size coherently. This is not what happened in phylogeny. Instead, some regions increased while others decreased; some connections waxed while others waned. As we discussed in Chapter 7, this kind of differential circuit elaboration prevails in most of the brain's more complex, nonlinear circuits. It is also interesting that the flow of information through the basal ganglia consistently targets the brain's most prominent, hypertrophied structures, namely the neocortex in mammals and the optic tectum in reptiles. This observation is consistent with the view that a region's proportional size correlates with the amount of influence it wields within the brain (see Chapter 5). As a structure enlarges in size, other regions apparently direct more and more of their output toward the growing behemoth. These shifts in information flow may result from the combined effects of numerous quantitative changes in connection strength, as the basal ganglia data suggest, or they may involve outright gains and losses of neuronal connections, as one would expect from Deacon's rule. Either way, the behemoth assumes more control. Summing up, it seems quite fair to say that, while there may be a fundamental *anatomical* scheme of basal ganglia organization, the *functional* schema of how the various basal ganglia components interact with one another and with the remaining brain is likely to be more varied. Does that conclusion also apply to the hippocampal formation?

The mammalian hippocampal formation (HF), which includes the dentate gyrus, the CA fields, the subiculum, and the entorhinal cortex, has long fascinated neuroscientists. Its circuitry and physiology are well worked out in primates and rats, but whether nonmammals also have an HF was unclear for many years. That uncertainty arose, at least in part, from the observation that the HF's adult position varies considerably across mammals. In humans it lies ventrolaterally within the brain, in rats caudally, and in opossums dorsomedially. Given that diversity, where should one look for an HF in nonmammals? This question was resolved when Nils Holmgren (1922, 1925) and others realized that the HF in mammals always develops from the medial edge of the telencephalic pallium, regardless of where it ends up in the adult brain. Armed with that insight, and knowing that homologous structures typically develop from topologically equivalent precursors (see Chapter 3), those "classical" comparative biologists began to look for nonmammalian homologues of the HF among the derivatives of the embryonic medial pallium (see also Smith, 1910). They proposed, for instance, that the "medial pallium" of adult amphibians and the "medial cortex" of turtles (see Figure 8.5) are both homologous to parts of the mammalian HF. On the whole, those hypotheses have been borne out by subsequent research. Axon tracing studies, for example, showed that all putative HF homologues have strong projections to the ipsilateral septum, which is itself highly conserved (Krayniak and Siegel, 1978; Neary, 1990). More recent studies also showed that lesioning the HF homologues in reptiles, birds, and teleosts causes deficits in spatial memory that are similar to those observed after HF lesions in rodents (Colombo and Broadbent, 2000; Salas et al., 2003). Therefore, most comparative biologists agree that all vertebrates, or at least all tetrapods, have some kind of homologue of the mammalian HF. Since HF connectivity is also fairly well conserved, one might say, as we did for the basal ganglia, that a fundamental scheme of HF organization evolved early on in vertebrate phylogeny and was retained within the tetrapods.

Again, however, this does not imply that the hippocampal formation is entirely constant across tetrapods. In fact, the hippocampal formation exhibits a considerable amount of phylogenetic variation that is likely to be functionally significant. On the input side, we find that in amphibians the HF receives its major nonolfactory sensory inputs directly from the dorsal thalamus (Figure 8.14). In reptiles, on the other hand, the HF receives only minor inputs from the sensory nuclei of the dorsal thalamus, and in mammals, those projections are lacking entirely (Neary, 1990; Ulinski, 1990). Subjecting these data to a parsimony analysis, we can conclude that the route by which non-olfactory sensory information reaches the HF has become more roundabout as mammals evolved. Just as for the basal ganglia, the flow of ascending sensory information was re-routed through the neocortex. The functional implications of that re-routing have not yet been explored, but they are probably significant.

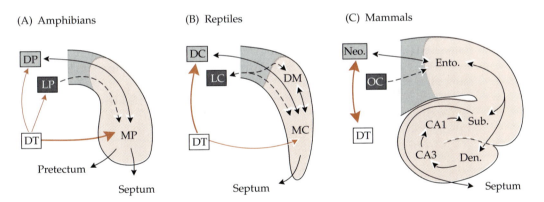

(A) Amphibians (B) Reptiles (C) Mammals

Figure 8.14 The Hippocampal Formation in Tetrapods   The hippocampal formation (light red shading) goes by different names in different tetrapods, but it always develops from the medial pallium and generally has a major projection to the septum. (A) In amphibians, the hippocampal formation is generally known as the medial pallium (MP). Its major sensory inputs come directly from the dorsal thalamus (DT). (B) In most reptiles, the hippocampal formation consists of medial and dorsomedial cortices (MC and DM, respectively). The medial cortex receives some direct projections from the dorsal thalamus, but most of its sensory input comes via the dorsal cortex (DC). (C) In mammals, the hippocampal formation consists of entorhinal cortex (Ento.), subiculum (Sub.), CA fields, and dentate gyrus (Den.). It receives no direct inputs from the sensory nuclei of the dorsal thalamus. Other abbreviations: DP = dorsal pallium; LC = lateral cortex; LP = lateral pallium; Neo. = neocortex; OC = olfactory cortex.

Another complication to the "conserved schema" view of hippocampal evolution is that the HF's intrinsic organization varies considerably across species. In mammals, both the neocortex and the olfactory cortex project only to the entorhinal cortex, making it an "input zone" through which all sensory information must pass on its way to the dentate gyrus and the CA fields (see Figure 8.14C). Such a distinction between input and processing zones is not apparent in the amphibian HF (Neary, 1990). In reptiles, the medial cortex qualifies as an "input zone" insofar as its neurons receive convergent inputs from the dorsal and lateral cortices, but the medial cortex also provides the major outputs from the reptilian HF, making it similar to the mammalian subiculum (Ulinski, 1990; Hoogland and Vermeulen-Vanderzee, 1995). In an effort to make sense of these data, we might propose that the reptilian medial cortex is homologous to both entorhinal cortex and subiculum, leaving the dorsomedial cortex as the most likely homologue of the dentate gyrus and the CA fields. This hypothesis is not consistent with the developmental data, however, since the medial cortex develops medial to the dorsomedial cortex, just as the dentate develops medial to the CA fields. Therefore, we are left with a by-now familiar dilemma: Connections and embryonic origins cannot both have been conserved—one or both must have diverged between mammals and reptiles. Again, I think that embryonic origins are less liable to change in evolution than neuronal connections, which is why I believe that the medial cortex is probably homologous to the mammalian dentate gyrus (see also Pérez-Clausell, 1988; Iglesia and Lopez-Garcia, 1997). If that is true, then the reptilian dorsomedial cortex is the most likely homologue of the mammalian CA fields. According to this hypothesis, entorhinal cortex and subiculum have no homologues in the reptilian HF. The hypothesis also implies that the evolution of the mammalian dentate gyrus involved the loss of direct sensory inputs from the dorsal thalamus and the loss of direct projections out of the HF. These changes would have transformed the dentate gyrus from a major sensorimotor integration center into part of a "processing zone" that communicates mainly with the neocortex.

One problem for the preceding hypothesis is that it is difficult to reconcile with recent reports that birds seem to have all of the major HF subdivisions found in mammals (Montagnese et al., 1996; Székely, 1999; Hough et al., 2002; Kahn et al., 2003). According to some authors, the subdivisions of the avian HF are even wired together in the same manner as their mammalian counterparts (Figure 8.15). These similarities suggest that all major HF subdivisions are homologous between birds and mammals, but we need to be careful here: The similarities could also be due to convergence. Indeed, the fact that turtles and other earthbound reptiles lack most of the avian HF subdivisions and circuits makes the "convergence hypothesis" most parsimonious. The alternative "conservation hypothesis" is less parsimonious because it implies that turtles, lizards, and several other reptile groups independently simplified their hippocampal formations. Therefore, I conclude that whatever

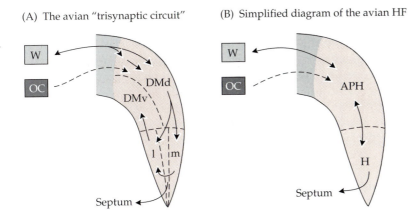

(A) The avian "trisynaptic circuit"   (B) Simplified diagram of the avian HF

Figure 8.15 The Hippocampal Formation of Birds The avian hippocampal formation (light red shading) is named and subdivided differently by different authors. (A) Some authors (e.g., Kahn et al., 2003) divide it into at least four subdivisions that are interconnected with one another in a curvilinear circuit that resembles the "trisynaptic" circuit through the mammalian hippocampal formation (Ento. → Den. → CA3 → CA1; see Figure 8.14C). (B) Other authors (e.g., Atoji et al., 2002) divide it into just two principal areas, which may themselves be further subdivided. The first naming and segregation scheme highlights similarities between birds and mammals; the second makes it easier to see the similarities between birds and earthbound reptiles (see Figure 8.14B). Once we consider the reptilian data, it becomes most parsimonious to conclude that the "trisynaptic circuit" in birds evolved independently from its mammalian analogue. Abbreviations: APH = parahippocampal area; DMd = dorsal portion of the dorsomedial area; DMv = ventral portion of the dorsomedial area; H = hippocampus proper; l = ventrolateral hippocampal layer; m = ventromedial hippocampal layer; OC = olfactory, or piriform, cortex; W = Wulst. (A and B after Hough et al., 2002; Kahn et al., 2003.)

features the hippocampal formations of birds and mammals share with one another—but not with the HF of other reptiles—probably evolved convergently. That raises another question, however: How did the avian HF attain its current level of complexity? That question has not yet been resolved. However, if we assume that the avian HF evolved from something like the turtle's HF, then we can infer that the avian HF probably increased in complexity by upping its cell density, dividing into superficial and deep components, and selectively losing some of the connections between those subdivisions (see Figure 8.15). These changes, which would amount to Ebbessonian parcellation (see Chapter 7), are very different from those that gave us the mammalian HF, but the end results are similar in terms of structural complexity and neuronal circuit design. Thus, the convergence hypothesis allows us to reconcile the avian data with what is known about HF anatomy in other reptiles and mammals.

Overall, this discussion illustrates that structures other than the neocortex also vary across tetrapods. The neocortex may be the most dramatically "new" structure in mammalian brains, but other regions also exhibit some noteworthy variation. It is quite common, for instance, for homologous brain regions to differ considerably in size, particularly when we focus on proportional brain region size (see Chapter 6). Those differences in size are often accompanied by differences in internal complexity, as reflected in subdivision number and intrinsic circuitry. They also tend to be associated with quantitative and/or qualitative differences in connectivity. As we have seen so many times throughout this book, these differences may seem either prominent or relatively insignificant, depending on your point of view. Since we still know little about the functional import of those species differences in region size, complexity, and connectivity, the issue remains indeterminate. However, I consider it likely that, even when you have a largely conserved "fundamental scheme" of brain regions and circuitry, many minor changes in brain region size, complexity, and connectivity may conspire to yield major changes in how information flows through the system and, consequently, in how the whole brain "works." A perfect illustration of that point is that both striatum and HF have apparently "shifted" their major source of input from thalamus to neocortex as mammals evolved, even though most of the components in this circuitry are highly conserved. This is interesting because it suggests that some of the non-neocortical variation is functionally related to the changes that occurred within the dorsal pallium (i.e., dorsal cortex and/or neocortex). However, until we have more complete comparative neurophysiological data, we cannot say much more than that. Indeed, the entire subject of non-neocortical variation across tetrapods remains poorly explored, perhaps because it is widely considered to be relatively insignificant.

Conclusions

In this chapter's epigraph, Ludwig Edinger (1908) suggested that mammals are more intelligent than nonmammals because only mammals have an exceptionally large neëncephalon or, as we say today, neocortex. This idea was widely accepted and does contain some grains of truth. Edinger was certainly right to point out that the vertebrate brain's most variable region is the dorsal telencephalon, where the neocortex resides. He was also correct in pointing out that nonmammalian homologues of mammalian neocortex are generally quite small, at least if we compare them to the proportionately large neocortices of large-brained mammals like us. Thus, it probably is fair to state that "the evolutionary history of the mammal brain is essentially a story of neopallial expansion and elaboration" (Romer and Parsons, 1977, p. 584). However, Edinger and his followers (see Chapter 2) were not infallible, and some of their errors have been perpetuated so insistently that they can still be heard today. Therefore, let us try to set the record straight.

One of Edinger's most egregious mistakes was to claim that nonmammalian forebrains were dominated by the basal nuclei. Instead, as Nils Holmgren (1922, 1925), Harvey Karten (1969) and many others have shown, virtually all vertebrates have a well-developed pallium that lies dorsal to the basal nuclei of the subpallium (e.g., the striatum; see Figures 2.9, 2.16 and 8.12). It is also incorrect to claim that a homologue of mammalian neocortex first appeared in the last common ancestor of reptiles, birds, and mammals (amniotes). Instead, it now appears quite clear that neocortex homologues are present in amphibians (see Figure 8.5) and many other vertebrates (Northcutt, 1981). Mammalian neocortex was not added to a preexisting complement of "old" brain regions; instead, it was radically transformed from a precursor that probably resembled the dorsal cortex of turtles. That transformation included the addition of new cell types and connections, as well as an important realignment of the incoming axons (see Figure 8.5), but the neocortex has its homologues in nonmammalian vertebrates. Finally, one should point out that this neocortex homologue in birds is hardly "rudimentary" (Maier and Schneirla, 1935, p. 479). In many birds, especially in owls (see Figure 8.11), the Wulst is large and clearly laminar. This complexity is intriguing because it probably evolved quite independently of the mammalian neocortical complexity.

With that in mind we can return to the question of nonmammalian intelligence. As I mentioned in this chapter's introduction, it is difficult to compare "intelligence" across species, but it is likewise difficult to deny that some nonmammals are at least as intelligent as many mammals are (if we define "intelligence" broadly). Euan Macphail (1982) concluded from this observation that, perhaps, all nonhuman vertebrates are equally intelligent, but now it is time to reconsider an alternative hypothesis. Could it not be that highly intelligent species evolved repeatedly, in several different lineages? We just saw that birds diverged from other reptiles in a whole host of ways, including the possession of more complex DVRs and hippocampal formations. We also saw that some birds, such as owls, evolved brains that rival many mammalian brains in their complexity. Combining these data with the more general observation that brain complexity is at least roughly correlated with behavioral complexity (see Chapter 4), we may infer that birds probably became more "intelligent" as they evolved from their early reptilian cousins. Moreover, some groups of birds (e.g., the parrots, crows, and owls) probably evolved degrees of behavioral complexity and/or "intelligence" that rival what most mammals can muster. Of course, no bird is as intelligent as human beings are, at least on average, but that is not the point. My fundamental argument is simply that we should not compare mammals to nonmammals generically because not all nonmammals are alike. Many birds are very smart, but they evolved that high intelligence independently of what happened in mammals. It is particularly interesting that birds apparently attained that high intelligence mainly by elaborating not their neocortex homologue (i.e., the Wulst) but the DVR. Thus, the evolutionary history of mammalian brains is just one of many different "stories" that may be told. It just happens to be a story that is of special significance to us—who are, after all, a very "special" kind of mammal.

Finally, let us review how well our "principles of brain evolution" apply to the emergence of specifically mammalian brains. Obviously, mammal brains are not just scaled-up reptile brains. The emergence of mammalian neocortex involved several highly specific changes in brain anatomy that cannot be explained by reference to some "scaling rule," and the proportionate reduction of the mammalian lateral pallium (i.e., the DVR) cannot, I think, be due to any kind of "concerted evolution" (see Chapter 5). Simply put, if mammal brains were scaled-up reptile brains, we would expect them to look far more similar to avian brains. Nor does it make sense to claim that the tiny brains of early mammals were "miniaturized" reptile brains. Clearly, the evolution of specifically mammalian brains was not due to any simple change in overall brain size. On the other hand, the preceding discussion of how early mammal brains evolved did make reference to several other "principles." For instance, the appearance of "new" auditory projections to the mammalian dorsal pallium (i.e., the lateral neocortex) is consistent with the rule of "large equals well-connected." Still, the principles of brain evolution that we discussed throughout this book are better suited to explaining the variation among mammalian brains than the variation between early mammalian and reptilian brains. This becomes particularly obvious as we turn our attention to the evolution of specifically human brains, for then absolute brain size becomes a major explanatory variable (see Chapter 9). As I mentioned in Chapter 1, and will take up again in Chapter 10, this makes sense: If evolutionary principles or rules are sometimes excepted, and if those rule/changes are propagated in phylogeny, then we would expect the utility of our principles to decrease with the phylogenetic distance of the species we are comparing. Comparing apples to oranges is more difficult (though hardly impossible) than comparing different apple breeds. As I said, we will come back to this. For now, let us ask how human brains evolved.

9 What's Special about Human Brains?

What a piece of work is a man, how noble in reason, how infinite in faculty?
in form and moving how express and admirable? in action, how like an
angel? in apprehension, how like a god? the beauty of the world; the
paragon of animals; and yet to me what is this quintessence of dust?

—W. Shakespeare, c. 1600

*I*nnumerable artists, scientists, philosophers, and even kids have pondered who we humans are. Are we "the paragon of animals" or, as Nietzsche put it, "clever beasts" (see Smith, 1987)? Is there an unbridgeable chasm between the animals and *Homo sapiens*, or are we just one of many brutes? Almost half of all Americans believe that God created humans, fully formed, within the last 10,000 years (according to the Gallup Organization, the percentage has hovered around 45% since 1982), but the scientific consensus holds that *Homo sapiens* evolved from other animals. Still, even modern biologists remain uncertain about what makes us humans so "uniquely unique" among the animals (Alexander, 1990).

Legions of uniquely human features have, at one time or another, been proposed, but many have turned out to be less unique than was originally supposed. For instance, the notion that only humans can remember specific episodes of their life (Tulving, 1985) was called into question by the discovery that some food-storing birds remember not only what they stored where but also when they stored it (Clayton and Dickinson, 1998, Clayton et al., 2003). Similarly, the image of "man as toolmaker" was tarnished by the finding that some crows and chimpanzees also use and manufacture tools (McGrew, 1992; Hunt, 1996; Weir et al., 2002). Faced with such humanlike abilities in nonhuman animals, one is tempted to conclude that there is no "qualitative discontinuity between people and animals" (Cartmill, 1990). That hypothesis is likewise dubious, however, for no other animal writes books, lectures on history, or tells stories around a fireplace. Man may not be a

297

paragon in all respects but, when it comes to knowing about the world and shaping its fate, no other animal can hold a candle to *Homo sapiens*. Therefore, the notion that humans are "special" will not go away. The crucial question for scientists is whether the question of human uniqueness can be studied without getting lost in fuzzy practices, such as redefining "human diagnostic traits" whenever they prove not to be diagnostic (Cartmill, 1990).

Neurobiologists tend to approach the question of human uniqueness by asking what, if anything, is special about human brains; this line of inquiry goes back more than 100 years. As you may recall from Chapter 2, Owen (1857) had proposed that only human brains have a posterior cerebral lobe, a hippocampus minor, and a posterior horn in the lateral ventricle (see Figure 2.3). Huxley (1861, 1863) quickly crushed this hypothesis by pointing out that chimpanzees and other apes do possess those supposedly uniquely human traits. In fact, Huxley ultimately claimed that there are no major qualitative differences between human and nonhuman primate brains at all. According to Huxley, even the quantitative differences between chimpanzee and human brains are of only minor significance because, after all, brain size is notoriously variable even within *Homo sapiens*. These conclusions carried the day but put Huxley in an interesting bind: If human brains hardly differ from those of other apes, then why are human minds so unusually complex? Huxley's answer was that language is what sets *Homo sapiens* apart and that the evolution of language required no major quantitative or qualitative changes in the brain. In other words, it must have been a minor, as yet undetected, change in brain anatomy that gave humans language and, thereby, transformed the human mind. Since Huxley could not specify the details of that change in brain anatomy, he essentially left unexplained the "mental chasm" (Cosans, 1994) that Owen tried to bridge by pointing to uniquely human features in the brain. Viewed from this perspective, Owen was hardly the conservative, ideologically driven dinosaur that some authors have described (Desmond, 1994); instead, he was a complex individual who tried earnestly to understand what makes human minds and brains special. Still, it is undeniable that Owen's three "uniquely human" features were not so. Can we do better nowadays?

In contrast to Owen and Huxley, who knew almost nothing about brain histology and connectivity (see Chapter 2), we have amazingly detailed knowledge of several nonhuman brains. For good ethical reasons, however, most of the methods that are routinely used to study neuronal connections and activity in nonhumans are, by and large, not feasible in humans. We do have a rapidly expanding set of functional imaging data on human brains, but those are difficult to compare with the more detailed anatomical and physiological data available for nonhumans. Thus, the data sets for human and nonhuman brains are nonoverlapping in many structural and functional respects, making it difficult to determine what is special about human brains. This problem will abate as better techniques for tracing neuronal connections in human brains are developed (Conturo et al., 1999; Brandt et al., 2001) and as brain imaging methods are increasingly applied to nonhuman brains. For the moment, however, comparative analyses of human and nonhuman brains are plagued by many missing data points. Most serious is that, if we want to find uniquely human neural traits, then we cannot compare humans to mice, rats, cats, or even macaques. Instead, we must compare our brains to those of our closest living relatives, the chimpanzees. As you might surmise, however, experimental data on chimpanzee brains are extremely rare (partly because chimpanzees are now officially an endangered species). Nor do we know much about the brains of any other apes, such as gorillas or gibbons. Therefore, our knowledge of how human brains differ from those of other apes is still quite rudimentary. This data scarcity explains why the literature on human brain evolution is so rife with speculation and uncertainty. Nonetheless, we can discern at least a tentative outline of how human brains evolved.

This chapter begins with a detailed look at primates and their brains because, unless we know what primate brains look like, we cannot know what features of the human brain are uniquely human rather than generically primate. After that, I turn to humans and their closest relatives, the fossil "hominins." Although many aspects of hominin phylogeny are controversial, the available literature allows us to determine that absolute brain size increased dramatically at several points (or periods). Next, I discuss whether human brains were modified in aspects other than size. Did they evolve some novel parts? Are some brain regions exceptionally large in humans, while other regions are meager? Do human brains differ from those of chimpanzees in connectivity? All of these questions are difficult to answer at the present time. However, it is safe to say that evolution has "reorganized" at least some aspects of human brain function and anatomy. Since at least some of those organizational changes are causally linked to changes in brain size, an old debate about whether human brains changed mainly in size or in organization (see Holloway, 1974) turns out to have been moot: They have changed in both. Just like the brains of other vertebrates, human brains are special in a multitude of different but causally entangled ways.

Primate Behavior and Overall Brain Size

The oldest known primate fossils are roughly 50 million years of age but, given the large gaps in that fossil record, the first primates may well have evolved more than 65 million years ago, when dinosaurs still roamed the Earth (Tavaré et al., 2002). Today's primates comprise roughly 235 species that are classified into a somewhat bewildering array of taxonomic groups (Purvis, 1995). The two main primate groups are strepsirhines and haplorhines. Strepsirhines are commonly known as prosimians (Figure 9.1), while

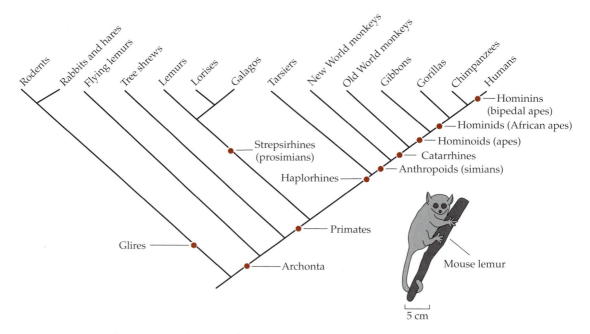

Figure 9.1 Phylogeny of Primates and their Closest Relatives   The phylogenetic relationships of primates have been studied extensively and are now fairly well established (see Purvis, 1995). Since many primate taxon names (e.g., strepsirhines, anthropoids, and hominoids) are neither commonly known nor especially euphonious, I substitute for them more common names (e.g., prosimians, simians, and apes). This maneuver is potentially confusing because many authors include tarsiers among prosimians and some exclude humans from simians, but it reduces the amount of jargon in the text. The mouse lemur, which probably resembles the earliest primates in terms of body size and build, is depicted in the inset.

haplorhines consist of tarsiers and simians (or anthropoids). Simians, in turn, comprise New World monkeys, Old World monkeys, and apes (or hominoids). Among the apes, gorillas, chimpanzees, and humans comprise the hominids (or African apes), which most likely shared a common ancestor that was not also shared by gibbons (the so-called lesser apes). Within the hominids, the closest living relatives of *Homo sapiens* are chimpanzees, whose DNA sequence differs from ours by only about 1.2% (Chen and Li, 2001). This degree of DNA divergence is sometimes portrayed as being insignificant, but it amounts to roughly 18 million base-pair differences (Carroll, 2003). Many geneticists are working very hard to tell us how those differences in DNA sequence relate to differences in brain structure and function, but at this point we know next to nothing about how the changes in our DNA made us different from the chimpanzees (see Enard et al., 2002; Cáceres et al., 2004; Stedman et al., 2004). Given that knowledge vacuum, I here focus not on differences in genes or other molecules, but on differences in brain anatomy and, to some extent, behavior.

What were early primates like? Judging from the fossil record and the distribution of features among living primates, the earliest primates were probably no larger than a squirrel, arboreal, and active at night or in twilight (Heesey and Ross, 2001). They probably had large, forward-facing eyes and long, flexible digits that were tipped with sensitive pads rather than claws (Hamrick, 2001). Given these features, as well as a fairly generalized dentition, we can infer that early primates probably ate mainly leaves, fruit, and insects they found hiding among the fine branches of the forest canopy (Sussman, 1991; Cartmill, 1992). They probably foraged alone but may have gathered periodically into small "coed" groups (Müller and Thalmann, 2000). In other words, the earliest primates were probably quite similar to some of the small nocturnal prosimians that are alive today (e.g., the grey mouse lemur; see Figure 9.1). From these early primates the simian lineage emerged approximately 30–40 million years ago. Most likely, the earliest simians were also both arboreal and relatively small (<1300 g; Ross, 1996). In contrast to early primates, however, the early simians were more diurnal. Early simians also decreased the size of their nasal cavities, moved their eyes closer together (Figure 9.2), and evolved more complex social systems (Ross, 1996). As you shall see below, all of these changes can be functionally related to the adoption of a more diurnal niche.

Within simians, the early catarrhines (i.e., Old World monkeys and apes) became even more diurnal than their ancestors. They also began foraging in

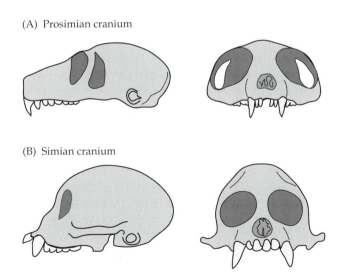

(A) Prosimian cranium

(B) Simian cranium

Figure 9.2 A Comparison of Prosimian and Simian Skulls
Compared to prosimians, simians generally have shorter snouts, more frontal eyes, and more vaulted crania. These skeletal differences correlate with simians having a poorer sense of smell, better binocular vision, and larger brains (relative to body size). (After images on the Animal Diversity website, http://animaldiversity.ummz.umich.edu.)

complex social groups, reduced their olfactory receptor repertoire (Rouquier et al., 2000), and reinvented trichromatic color vision. As you may recall from Chapter 8, early mammals had reduced their capacity for color vision when they became nocturnal. At the molecular level, this decrease in color vision was due to the loss of photopigments that are sensitive to long wavelengths (i.e., red light). Only the pigments sensitive to short and medium wavelengths were retained, giving early mammals dichromatic color vision (and making them red–green color-blind). Early catarrhines, however, duplicated one of those two photopigment genes and modified it so that the pigment it encodes became maximally sensitive to reddish hues (Jacobs, 1993; Dulai et al., 1999). Adding this third photopigment allowed early catarrhines to ree-volve the good color vision of their nonmammalian ancestors. Another good example of convergent evolution! But why did simians reevolve good color vision? How did they benefit from "seeing red" again? Most likely, the abil-ity to distinguish red from green allowed early catarrhines to find ripe fruit, which tends to be reddish rather than green, against a background of green leaves (Osorio and Vorobyev, 1996; Regan et al., 2001). It may also have helped them spy young leaves, which tend to be red in Africa, where the early catarrhines evolved (Dominy and Lucas, 2001). Either way, trichromatic color vision would have been useful to the early catarrhines because both ripe fruit and young leaves are more nutritious and digestible than most other plant foods.

Thus, as we go from early primates to early simians to early catarrhines, we see a trend of increasing diurnality, reduced olfaction, improved daylight vision, and more complex social behavior. These phyletic trends are probably interrelated, for color vision is not feasible at night (Land and Osorio, 2003), improved vision lessens the need for sensitive olfaction, and seeing one's peers probably facilitates both gregarious foraging and the emergence of more complex social systems. In other words, these trends apparently belong to a whole suite of trends that helped to shape the early catarrhines.

This is interesting because some other primate trends may also be part of that "catarrhine suite." Consider, for example, body size. Early simians may not have been much larger than early prosimians, but catarrhines as a group tend to be larger than other primates, and the highly social apes are larger still. Is this trend of increasing body size related to the trends of increasing diurnality and social complexity? A direct link is difficult to see, but indirect linkages are possible. We might note, for instance, that among primates the larger species tend to live longer, develop more slowly, and care more inten-sively for their young (Harvey et al., 1987). These size-related trends could be related to increased sociality, because learning how to navigate a complex social system requires both guidance and time. In other words, we would expect larger animals to be more capable of learning "social skills." Now, what about the brain? Learning about social relationships clearly requires brains and, as we reviewed in Chapter 4, large-brained vertebrates tend to be social. This line of reasoning suggests that, as primates became more social, more diurnal, larger, and more visual, their absolute and relative brain size should have increased as well. Indeed, the data do suggest that brain size increased repeatedly among primates; we will now examine this in some detail.

At any given body size, prosimian brains tend to be larger than those of rodents, rabbits, or flying lemurs, but no larger than the brains of tree shrews, which are generally considered the primates' closest living relatives (see Figures 9.1 and 9.3). Given these data, we may conclude that relative brain size probably increased (roughly twofold) just before tree shrews and true primates diverged from one another. Since early primates were arboreal and small, this increase in relative brain size probably reflects an evolutionary decrease in body size (Deacon, 1997). As we saw in Chapter 4 (see Figures 4.7 and 4.12B), such decreases in body size may be quite common among animals that live in trees. In any case, it is fairly clear that the brains of early primates were small in terms of absolute brain size. Given our best estimates of early primate body size and the general scaling relationship between primate brain and body size (Figure 9.3), we can conclude that early primate brains most likely weighed only 2–3 g. That is, their brains were probably no larger than those of modern laboratory rats. This is somewhat surprising because we tend to think of all primates as having hefty brains, but it follows logically from the discovery that early primate bodies were so small (i.e., smaller than the body of a laboratory rat). Only later, as body size increased, did absolute brain size increase in several primate groups, especially within the simians. Although it would be false to claim that all simian brains are larger than all prosimian brains, or that the brains of all hominoids are larger than all Old World monkey brains, a cladistic analysis does indicate that absolute brain size probably increased repeatedly along the lineage that leads from haplorhines to simians to catarrhines to hominoids to hominids and, finally, to hominins (see Figure 9.1).

One intriguing aspect of all primate brains is that they have unusually large neocortices. Since the proportional size of the neocortex generally tends to increase with increasing absolute brain size (see Chapter 5), we would expect the neocortex to be disproportionately large in the large-brained late primates. However, even the earliest primates are likely to have had neocortices that were larger than you would expect given their absolute brain size and the general mammalian scaling rule. This becomes apparent when you compare the brain of a large insectivore (the hedgehog *Erinaceus europaeus*) to the brain of a small prosimian (*Galago demidovii*). Both brains weigh roughly 3.4 g, but their neocortical fractions are 16% and 46%, respectively, and their neocortex:medulla ratios are 1.6 and 9.3 (Stephan et al., 1981). Of course, *Galago* did not evolve from *Erinaceus*, but the comparison does illustrate how disproportionately large the neocortex became as primate brains evolved (see Figure 5.7). We can argue about whether this departure from the scaling rule

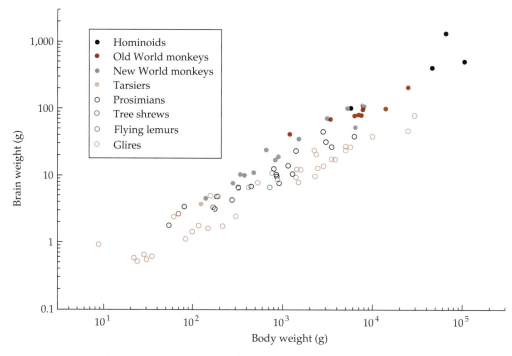

Figure 9.3 Primate Brain and Body Weights At any given body size, prosimians tend to have larger brains than glires (i.e., rabbits and rodents; see Figure 9.1). The brains of simians (including New World monkeys, Old World monkeys, and hominids) are larger still. The largest primate brains are those of humans, while the largest primate bodies typically belong to gorillas. (After Stephan et al., 1981; Mangold-Wirz, 1966; Pirlot and Kamiya, 1982.)

constitutes true rule-breaking or just noise within the rule (i.e., mosaic or concerted evolution; see Chapter 5). Either way, the neocortical enlargement undertaken by the early primates was considerable. It was then magnified as absolute brain and body size increased within the simians. In a 405-g chimpanzee brain, for instance, the neocortex comprises an astounding 76% of the brain and is 50 times larger than the medulla (45% and 30× if you consider only neocortical gray matter)(Stephan et al., 1981; Frahm et al., 1982). Overall, we may conclude that both absolute brain size and the proportional size of the neocortex increased enormously within primates, particularly within our own phylogenetic branch.

What was the functional significance of increasing brain and neocortex size? According to some authors (e.g., Dunbar, 1998) primates with large neocortices are socially more complex than their less cortically endowed cousins. Indeed, as we reviewed in Chapter 4, social group size correlates quite well with both proportional and absolute neocortex size among primates (see Figure 4.14). Intuitively, this makes sense. Highly social animals need to collaborate with one another and to outmaneuver their competitors (Humphrey, 1976), and the neocortex surely plays a major role in those behaviors. It is difficult, however, to pursue this notion further, because it is inherently implausible to suppose that the entire neocortex performs a function as specific as "social intelligence." Moreover, brain and neocortex size in primates also correlate with several other parameters, such as frugivory, longevity, home-range size, and metabolic rate (Sacher, 1973; Clutton-Brock and Harvey, 1980; Armstrong, 1985; Allman et al., 1993). As one observer quipped, "too many factors seem to be associated with brain size in primates" (van Dongen, 1998, p. 2127). For each factor we can construct a plausible explanation about why it might, or even should, be correlated with overall brain or neocortex size, but none of this helps very much. We are still left with a plethora of factors and parameters. One promising strategy for simplifying that complexity is to shift our focus away from overall brain or neocortex size and toward smaller regions and systems that are more readily linked to specific functions and behaviors (Harvey and Krebs, 1990). Therefore, we will now discuss how primate brains have changed in structural detail or organization.

Evolutionary Changes in Primate Brain Organization

Comparing the brains of primates to those of their closest relatives, we find many structural similarities but also some important differences. Some of those structural or organizational differences are evident only when we compare nonprimates to humans or to other large primates, but other differences are more general; they involve features that are found in all (or almost all) primates. Since those primate-typical features probably evolved early in primate phylogeny, we might expect them to be functionally related to the behavioral specializations of the early primates we discussed earlier. As you shall see, that expectation is, in general, fulfilled.

Most primate-typical features involve the visual system. As we already discussed, the evolution of trichromatic color vision probably occurred only when catarrhines arrived on the scene, but the evolution of more frontally placed eyes probably dates back to the very origin of all primates. Most intriguingly, that change in eye position is reflected in how the retina projects to the midbrain's superior colliculus (Figure 9.4). In most nonprimates, that projection is almost completely crossed. In all primates, however, each retina projects in roughly equal proportions to both superior colliculi (Kaas and Huerta, 1988), giving that structure unprecedented access to information coming from both eyes. Early primates probably benefited from that change in circuitry because combining information from both eyes allows for better depth perception, which is crucial when you climb around in trees, hunting

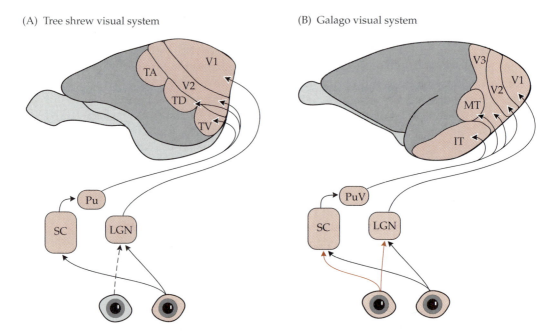

(A) Tree shrew visual system

(B) Galago visual system

Figure 9.4 Ascending Visual Pathways in Primates and Tree Shrews   The ascending visual pathways are more bilateral in primates than in most other mammals. (A) In tree shrews, which are close relatives of the primates (see Figure 9.1), the superior colliculus (SC) receives virtually all of its retinal input from the contralateral eye, and the lateral geniculate nucleus (LGN) receives more inputs from the contralateral retina than from the ipsilateral retina (dashed arrow). (B) In contrast, in galagos and other primates, the SC and LGN receive roughly equal inputs from both retinas (see also Figure 6.7). This increased bilaterality correlates with the more frontal placement of the eyes in primates. The figure also shows that, while many visual cortical areas are similar in tree shrews and galagos, only the latter have a large inferior temporal cortex (IT). The thalamic pulvinar nucleus in primates contains both dorsal and ventral divisions; only the ventral division (PuV) appears to be homologous to the pulvinar of nonprimates (Pu; see Figure 9.15). For the sake of clarity, many details are omitted from this diagram. Abbreviations: MT = middle temporal area; TA = anterior temporal area; TD = dorsal temporal area; TV = ventral temporal area; V1 = primary visual cortex; V2 = secondary visual cortex; V3 = tertiary visual cortex. (After Harting et al., 1972; Kaas and Huerta, 1988; Kaas and Lyon, 2001.)

for insects, fruits, and leaves. In other words, it is quite reasonable to suppose that frontally placed eyes and bilateral retinocollicular pathways were adaptations to the early primate's "fine-branch niche." Consistent with that hypothesis is that fruit-eating megabats (see Figure 4.13) also have large, forward-facing eyes and bilateral retinocollicular pathways (Pettigrew, 1986; Rosa and Schmid, 1994). Although these features were once thought to indicate that megabats are closely related to primates, the "flying primate" hypothesis has not stood the test of time (Lapointe et al., 1999). Instead, most current data indicate that all bats are only distantly related to primates (see Figure 2.11). Therefore, the most parsimonious conclusion is that megabats and primates independently (i.e., convergently) evolved frontal eyes and bilateral retinocollicular pathways. That makes sense once we consider that the niche of the fruit-eating megabats is fairly similar to the fine-branch niche once occupied by the earliest primates.

The cerebral motor system of primates is likewise specialized in several respects. Recent studies indicate, for instance, that all primates have nine or more premotor areas, whereas nonprimates tend to have just two to four (see Wu et al., 2000). Particularly interesting is that primates are unique in having a ventral premotor area that is specialized for arm and mouth movements (Preuss et al., 1996). This area may well be homologous to part of Broca's area in humans (discussed later in this chapter), but it appears to have no homologue in nonprimates. Its descending connections have not been studied in detail, but it seems to have unusually direct projections to the spinal cord (see Figure 7.12A,B) (Nudo and Masterton, 1990). Given that direct corticospinal projections are generally indicative of increased dexterity (see Chapter 7), we may infer that the evolution of this ventral premotor area probably helped primates to perform more dextrous hand and arm movements (Rizzolatti and Arbib, 1998). Additional improvements in hand and arm control were probably achieved by evolutionary changes in the primate somatosensory system. Specifically, the evolution of touch-sensitive fingertips and toes probably increased the amount of somatosensory information that was available for feedback during hand and foot movements. Furthermore, the somatosensory cortex expanded in primates, adding several areas that have no obvious homologues in non-primates (Kaas, 1983, 1988). Since these somatosensory cortical areas are intimately interconnected with the motor cortex, as well as with the spinal cord (Nudo and Masterton, 1990), they are likely to play a major role in movement control. Overall, we may conclude that diverse evolutionary changes in the visual, motor, and somatosensory systems all interacted to give early primates exceptionally good hand–eye coordination, which must have come in handy in the fine-branch niche (pun intended, forgive me).

The most controversial special attribute of primate brains is the prefrontal cortex, which lies rostral to the premotor cortex and plays a major role in making decisions (Krawczyk, 2002). Nonprimate mammals do have a prefrontal cortex, but it apparently consists of only two major regions, rather than three as in primates (Figure 9.5). The two conserved prefrontal regions are the orbital prefrontal region, whose neurons respond preferentially to external stimuli that are likely to be rewarding or otherwise significant (Tremblay and Schultz, 1999; Schoenbaum and Setlow, 2001), and the anterior cingulate cortex, which mainly processes information about the body's internal

(A) Macaque

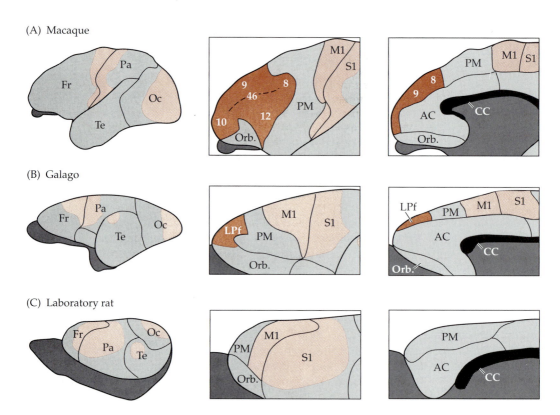

(B) Galago

(C) Laboratory rat

Figure 9.5 Comparative Anatomy of the Frontal Lobe   Shown here are lateral views of the entire brain (left column) and close-ups of the frontal lobe from both lateral and medial perspectives (middle and right columns, respectively). (A) In macaques and other simians, the frontal lobe (Fr) is large relative to other lobes and contains a large lateral prefrontal subdivision (red area with white numbers). (B) Galagos also have a lateral prefrontal cortex (LPf), but it is relatively small. Instead, the prefrontal cortex of galagos consists mainly of the orbital prefrontal cortex (Orb.) and the medially located anterior cingulate cortex (AC). (C) Rats and other rodents possess the orbital and anterior cingulate divisions of the prefrontal cortex, but whether they have a homologue of primate lateral prefrontal cortex remains controversial. According to Todd Preuss (1995a), they do not. If that is true, then the most rostral and dorsal portion of the neocortex in a rat is homologous to the premotor cortex (PM) in primates. (The numbers 8, 9, 10, 12, and 46 refer to individual cortical areas within the lateral prefrontal cortex.) Other abbreviations: CC = corpus callosum; GF = granular prefrontal cortex; M1 = primary motor cortex; Oc = occipital lobe; Pa = parietal lobe; S1 = primary somatosensory cortex; Te = temporal lobe. (A and B after Preuss and Goldman-Rakic, 1991a; C after Preuss, 1995a; Zilles and Wree, 1995.)

state (Nauta, 1971; Luu and Posner, 2003). Collectively, these two regions contribute to what we might call the "emotional" aspects of decision making (Damasio, 1994; Dias et al., 1996; Allman et al., 2001). The third prefrontal region, which is generally known as the lateral, or granular, prefrontal cortex, is apparently unique to primates (Preuss, 1995a) and is concerned mainly with the "rational" aspects of decision making. Its neurons respond less rapidly than orbitofrontal neurons to rewarding stimuli and are more selective for the physical attributes of stimuli, such as their spatial location (Wallis and Miller, 2003). Without those lateral prefrontal neurons, primates become less able to retrieve and manipulate information about objects in the outside world (Owen et al., 1999). In the context of decision making, this probably means that the lateral prefrontal cortex helps primates to consider alternative interpretations of external objects and to construct alternative scenarios of how to interact with them. To the extent that those external objects can be members of the same species, the appearance of a lateral prefrontal cortex may have been related to the evolution of more complex social lives, but it might also have facilitated various nonsocial interactions, such as extracting seeds from hard-shelled fruit (Parker and Gibson, 1977). Either way, it would be fair to say that the lateral prefrontal cortex probably helped early primates to exhibit more complex and flexible behavior.

This notion of the lateral prefrontal cortex being a primate innovation originated with Brodmann (see Chapter 2) and has recently been championed by Todd Preuss (1995a). Central to their argument is that the small-celled granular layer that characterizes the lateral prefrontal cortex in primates is lacking in most other mammals. This finding does indeed suggest that the lateral prefrontal cortex is unique to primates. Alternatively, however, one might claim that primates simply added a granular layer to part of their prefrontal cortex (Rose and Woolsey, 1948). If that were true, then the primate lateral prefrontal cortex would have evolved by "phylogenetic conversion from" rather than "addition to" (see Chapter 6) the agranular prefrontal cortex of nonprimates. Given how difficult it is in general to distinguish phylogenetic conversion from addition, it is not surprising that these two alternative hypotheses of lateral prefrontal cortex evolution remain at loggerheads. I favor the addition hypothesis mainly because homologizing the orbital prefrontal and anterior cingulate cortices between primates and nonprimates is much simpler than homologizing the lateral prefrontal cortex, which is what you would expect if the latter had been added as primates evolved (see Chapter 6). Also consistent with the addition hypothesis is that the lateral prefrontal cortex is heavily interconnected with several other regions that are likewise difficult to homologize between primates and nonprimates (i.e., posterior parietal cortex and dorsal pulvinar). The fact that this whole network is homologically dubious indicates that the evolution of the lateral prefrontal cortex involved considerably more than the mere addition of a granular layer.

However, further studies will be needed to shore up the hypothesis that non-primates lack a lateral prefrontal cortex homologue.

In addition to the aforementioned specializations, primate brains harbor several other unusual features (see Preuss, in press), but what I have mentioned so far probably suffices to establish primate brains as being specialized in numerous behaviorally relevant respects. If we want to understand how primate brains evolved, pointing to those special features seems more fruitful than attempts to correlate absolute or relative brain size with various behaviors. Indeed, since early primate brains were only slightly larger than the brains of their nonprimate relatives, most of the changes I reviewed above must have occurred independently of phylogenetic changes in absolute brain size. It is difficult to see, for instance, how the evolution of bilateral retinocollicular pathways could have been caused by increases in absolute brain size. I stress this now because a very different situation obtains within primates, particularly within the lineage that led to *Homo sapiens*. There, we find several changes in brain structure that are most likely linked to increases in absolute brain size (discussed in more detail later in this chapter). Before we pursue this thought, however, let us back up and consider hominin phylogeny in the same manner in which we just discussed primate phylogeny. That is, let us begin with the non-neuronal aspects of hominin phylogeny, then consider questions of overall brain size, and only in the final section explore how human brains have changed in structural detail.

Hominin Behavior and Overall Brain Size

The fossil record for humans and their closest relatives (i.e., bipedal apes, or hominins; see Figure 9.1) is more extensive than the early primate record, but it is also more hotly debated (see Conroy, 2002; White, 2003). My discussion here avoids most of those debates (for lack of space, not lack of interest) and simplifies some issues that do not relate directly to how human brains evolved; it is by no means a full account of hominin phylogeny in general. I begin with a brief description of the major hominin taxa, their bodies, and their principal behaviors. That is followed by a discussion of how hominin brains have changed in size, and how those changes might relate to changes in behavior. The issue of how hominin brains have been "reorganized" is considered in the next section.

According to most estimates, early hominins diverged from early chimpanzees roughly 6 million years ago in Africa. The most important aspect of that initial divergence is that the early hominins, who are generally referred to as early australopithecines, were bipedal. They had already lost the opposable big toe we see in other apes, and their legs and hips were far more suitable for walking on the ground than on tree limbs. Adopting this bipedal gait probably allowed the early australopithecines to move around more easily in

(A) *Australopithecus afarensis*   (B) *Homo erectus*

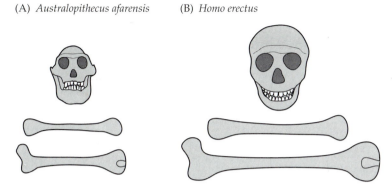

Figure 9.6 Skulls, Arms, and Legs of Fossil Hominins Compared to early australopithecines, specimens of *Homo erectus* are brainy and large. (A) Multiple adult specimens were used to reconstruct this skull of *Australopithecus afarensis* (White and Kimbel, 1988). Below it are the humerus and femur of "Lucy," the most famous of all australopithecines. (B) This *H. erectus* skull is based on a specimen called WT 15000 (most likely a young male). The humerus and femur below it come from ER 1808, an adult female that probably lived roughly 1.6 mya. Since these skulls and bones are all drawn to roughly the same scale, we may infer that *H. erectus* was much taller that *A. afarensis*, had longer legs than arms, and possessed a more vaulted cranium. Relative to the remaining skull, the jaw of *H. erectus* was also less robust than that of the australopithecines. (After Wolpoff, 1999.)

the open woodlands that were beginning to replace the dense forests of pre-hominin East Africa. Other data strongly suggest that early australopithecines ate mainly grasses, seeds, and roots, which they chewed with strong jaws and teeth (Figure 9.6). These australopithecines were probably gatherers, rather than hunters or scavengers, and likely shared their food with "friends and family." Their use of tools was probably quite limited, but approximately 2.5 million years ago (mya) some of the australopithecines began to manufacture simple tools from stone (Panger et al., 2002). By about 2 mya, some taller, more robust australopithecines appeared. They persisted until at least 1.4 mya, after which they went extinct.

The earliest representatives of the genus *Homo* appeared roughly 2 mya and were quite diverse (see Hawks et al., 2000; Conroy, 2002). Some, like *Homo rudolfensis* and *H. habilis*, were quite similar to the early, gracile australopithecines. Others, mainly *H. erectus*, were at least a foot taller than the australopithecines that shared their living space (see Figure 9.6). They also had more vertical faces, smaller molars, and larger brains. According to the fossil record, *H. erectus* made a variety of stone tools, such as cutting flakes and hand axes, which were probably used to cut meat and extract marrow from bones. In addition, *H. erectus* probably discovered how to control fire

(Rowlett, 2000) and make containers for water and/or food (see Wolpoff, 1999). Equipped with these new capabilities, *H. erectus* began to migrate out of Africa around 1.6 mya and into Asia and Indonesia. What ultimately happened to *H. erectus* is unclear. Some authors claim that the various populations of *H. erectus* gradually evolved into *H. sapiens* (Wolpoff, 1999); others argue that *H. sapiens* evolved in Africa and then displaced *H. erectus* as it, too, migrated out of Africa (Templeton, 2002). Either way, we can say that roughly 250,000 years ago, *H. erectus* disappeared and *H. sapiens* appeared.

Early *H. sapiens* lived where *H. erectus* had lived (i.e., Africa, Asia, and Indonesia) but one of its forms, namely the Neandertals, spread into northern Europe. At an average male height of 5′ 6″ (167 cm) the Neandertals were no taller than *H. erectus*, but their bodies were heavier and more compact, as one might expect from individuals who are adapted to a cold climate (recall Allen's rule; see Figure 1.4). Neandertals had smaller jaws than *H. erectus* and more sophisticated tools. They apparently "imported" raw materials from distant sites and invented standardized techniques for making spears and other special-purpose tools (Simek, 1992). Armed with these implements, they must have been formidable hunters of large game. Roughly 50,000 years ago, Neandertals began to decorate their bodies with pigments and to inter their dead (Smirnov, 1989). According to many authors, these activities imply some self-awareness and the beginnings of symbolic thought. Given that symbol use is the essential component of human speech (Deacon, 1997), these data indicate that human speech probably evolved 50,000–100,000 years ago (Henshilwood et al., 2004). However, the extent to which Neandertals were capable of full-blown human speech remains unclear (Arensburg, 1994; Tobias, 1996; Fitch, 2000). Be that as it may, Neandertals were clearly not the mindless brutes of lore. In fact, several authors (e.g., Hublin et al., 1996) have suggested that, for many years, Neandertals competed quite effectively with the anatomically modern form of *H. sapiens*, which dispersed out of Africa roughly 100,000–150,000 years ago (Templeton, 2002; Caramelli et al., 2003). So, what happened to Neandertals? Most likely they were gradually replaced by those anatomically more modern *H. sapiens*, but the possibility of interbreeding between Neandertals and modern humans remains hotly debated (see Tattersall and Schwartz, 1999; Kramer et al., 2001). Either way, it seems that the Neandertals coexisted with modern *H. sapiens* until roughly 30,000 years ago (Shea, 2003).

The anatomically modern forms of *H. sapiens* generally weighed less than Neandertals or other early *H. sapiens* (Ruff et al., 1997), and they had smaller jaws, teeth, noses, and brow ridges. Around 40,000 years ago, they learned to make tools from bones and shells as well as stone. They also began to conduct elaborate burials, paint images on rock, and adorn their bodies with a wide variety of artifacts, including seashells, animal teeth, and, perhaps, clothes (Kittler et al., 2003). Many of these cultural traits had precedents in early *H. sapiens*, but in aggregate they were quite new (Klein, 2000; McBrearty and

Brooks, 2000). Over the next few thousand years, human population size grew steadily. It increased even further after humans learned to cultivate the land and began to domesticate various animals, roughly 9,000 years ago (Vigne et al., 2004). The next important "revolutions" in human evolution were the invention of writing (which probably occurred within the last 4,000 years and independently in various regions of the world), the spread of *H. sapiens* into the New World, the industrial revolution, and, in 1947, the invention of the transistor, which brought us the computer age. Of course, other revolutions could likewise be cited, but the listed ones suffice to show that the pace of change in human behavior increased dramatically over the last 40,000 years. Language likely played the major role in this acceleration, but it probably was not its only cause. Increases in population density, migration, social complexity, and imitative prowess (Meltzoff, 1996) probably contributed as well to the increased rate of behavioral change in late *H. sapiens*.

This account of human evolution is obviously simplified, but it provides an adequate context for discussing hominin brain evolution. Now we can ask, for instance, whether hominin brains increased gradually or suddenly in size, and whether those changes in size correlate with changes in behavior. We can also ask whether hominin brain size changed independently of body size. However, before we try to answer these and several other questions, it is appropriate to contemplate some caveats. First, we must note that all measurements of absolute brain size in fossil hominins are derived from estimates of endocranial volume (see Chapter 1). These estimates are generally based on incomplete data and debatable assumptions, which explains why authors sometimes disagree on a fossil specimen's "true" endocranial volume (see Tobias, 1973; De Miguel and Henneberg, 2001). Second, all assertions about relative brain size in fossil specimens must be taken with a double grain of salt, because fossil body sizes are also estimated from incomplete data (e.g., from tooth, femur, and/or orbit size) (see Kappelman, 1996; Smith, 2002). Third, the fact that male and female hominins generally differ considerably in brain and body size complicates comparative analyses whenever a fossil's gender is uncertain or unknown. Finally, we have to realize that it is often difficult to ascertain the developmental age of fossil specimens; dental records help, but rates of tooth development themselves have changed as hominins evolved (Dean et al., 2001). These handicaps seriously undermine our attempts to comprehend how human brains evolved. Nonetheless, the hominin fossil record has now been scrutinized so thoroughly that we can discern at least the outline of how hominin brains have changed in size.

Let us begin with the australopithecines. Plotting endocranial volume against body weight for various australopithecines, we see that the data form a reasonably straight line in log–log space (Figure 9.7). This line has approximately the same slope as the allometric line for nonhominin apes (~0.4) but it is vertically offset by about 30% (Hofman, 1983). In other words, the relative brain size of australopithecines is roughly 30% greater than that of nonho-

(A) Hominoid relative brain size

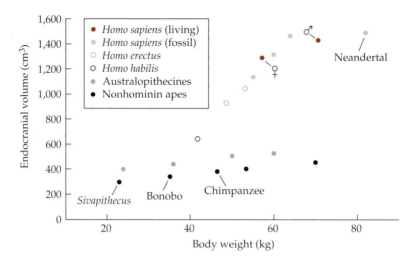

(B) Hominin absolute brain size

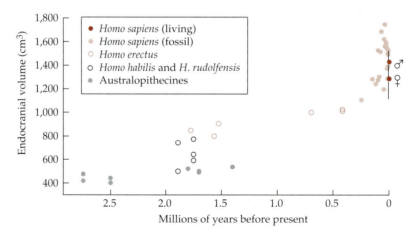

Figure 9.7 Endocranial Volumes of Fossil and Extant Hominoids Both relative and absolute brain size (as measured by endocranial volume) have changed during the course of hominoid phylogeny. (A) Compared to nonhominin apes, including the extinct *Sivapithecus* (also known as *Ramapithecus*) and the pygmy chimpanzee, or bonobo, australopithecines had larger brains at any given body size. Members of the genus *Homo* evolved even larger brains, relative to body size. Among living humans, males tend to have larger bodies and brains than women. (B) Plotting endocranial volume against the estimated age of fossil specimens, we find that brain size increased in fits and starts (see also Figure 9.8). (After Hofman, 1983; Kappelman, 1996.)

minin apes. Because the earliest australopithecines probably weighed about 30–40 kg, which is roughly the weight of today's bonobos, or "pygmy chimps," and considerably more than the body weight of early apes like *Sivapithecus* (see Figure 9.7A), we can infer that the australopithecine increase in relative brain size was not due to "phylogenetic dwarfism" (see Chapter 4). Instead, the data indicate that, when australopithecines first evolved, their average endocranial volumes expanded independently of body size, from less than 350 cm^3 to more than 400 cm^3 (see Brunet et al., 2002; Wolpoff et al., 2002). Over the next 4–5 million years, the endocranial volume of australopithecines increased even further (to around 500 cm^3 for the most robust australopithecines), but that increase in absolute brain size was accompanied by a significant increase in body size. Thus, it did not constitute an additional increase in relative brain size.

The most dramatic increase in hominin brain size occurred with the evolution of the genus *Homo* (Hawks et al., 2000). Comparing early specimens of *H. erectus* to early australopithecines, we see that the brains of *H. erectus* were roughly twice as large (see Figure 9.7). Although this increase in absolute brain size was accompanied by a major increase in body size (see Figure 9.6), it was not merely a result of increasing body size, for *H. erectus* brains were also considerably larger than the brains of the later, more robust, and heavier australopithecines. In fact, all *H. erectus* data points lie far above the allometric line for all australopithecines (see Figure 9.7A). This means that early *H. erectus* underwent a significant increase in absolute and relative brain size. After the initial appearance of *H. erectus*, relative brain size changed only slightly for about 1.5 million years. During that time, absolute endocranial volume did increase from about 850 cm^3 to roughly 1000 cm^3, but most of that increase was accompanied by an increase in body mass. Therefore, this increase in absolute brain size was distinct from the earlier increase in absolute *and* relative brain size. These data are consistent with the theory of "punctuated equilibrium" (Eldredge and Gould, 1972; Gould and Eldredge, 1993), which states essentially that over the long run evolutionary change occurs in bursts, separated by long periods of relative stasis (Hofman, 1983). This observation must be viewed with caution, however, since the fossil record of early hominins is too scarce, and the dating of its specimens too imprecise, to know exactly when and how *H. erectus* first evolved. Still, I deem it fair to say that brain and body size in early hominins increased in fits and starts, rather than gradually (Figure 9.8).

The appearance of *H. sapiens* was associated with additional increases in brain and body size (see Figure 9.7). Looking closely at the change in overall brain size, we can see that this was not the kind of "quantum jump" we saw between australopithecines and *H. erectus*, but a more gradual, curvilinear increase in absolute brain size (see Figure 9.7B). Essentially, the data indicate that the rate of change in absolute brain size was relatively low in early *H. sapiens* but then continually increased with time, setting up an exponential

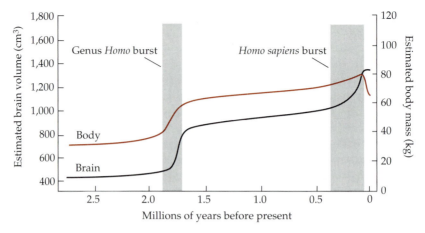

Figure 9.8 Nonlinear Evolution of Hominin Brain and Body Size   This highly simplified graph illustrates that hominin brain and body size (black and red lines, respectively) increased nonlinearly, showing two major growth spurts, or "bursts," separated by long periods of relative stasis. During the first burst, which coincided roughly with the origin of the genus *Homo*, absolute brain and body size both increased dramatically; relative brain size increased, but modestly. During the second burst, absolute brain size increased exponentially while absolute body size increased only slightly, yielding a considerable increase in relative brain size. This second burst differs from the first in that it seems to be less "sudden," starting slowly and accelerating gradually (Lee and Wolpoff, 2003). After that, absolute brain size plateaued, but relative brain size increased as human body size decreased (it may now be increasing again, but that is another story).

rise in overall brain size (Lee and Wolpoff, 2003). This is fascinating because it suggests that human brain size in this period changed according to some runaway process (discussed later). Also interesting is that many early human brains (e.g., those of Neandertals) were larger than most recent human brains. This suggests that absolute brain size might have decreased during recent human evolution, but the matter is not so simple. We must consider, for instance, that average endocranial volumes differ between present-day men and women and have large standard deviations (Kappelman, 1996). Since such variability is seen also among Neandertals (Ruff et al., 1997), the absolute brain weights of recent humans and Neandertals overlap considerably. Comparing many specimens, we see that Neandertal brains were larger, on average, than modern human brains, but that difference is small (~10%) and associated with a major difference in body size. Moreover, as I mentioned earlier, recent humans probably evolved not from Neandertals but from other early humans roughly 100,000–150,000 years ago. Since hominin endocranial volumes from that time period (mean = 1,354 cm^3; see Ruff et al., 1997) are roughly equal to those of recent humans, I conclude that absolute brain size probably did not decrease appreciably during the last 100,000 years of human evolution; instead, it essentially plateaued. During that same period, however, human body size (as estimated from postcranial remains) decreased by 11%–12%, yielding an increase in relative brain size. The significance of that increase remains debatable.

As I mentioned earlier, these data show that hominin brain size has not increased at a single, steady rate but in several bursts (see Figure 9.8). The first burst, which occurred with the origin of the australopithecines, was relatively small and did not involve a major change in body size. The second burst, near the origin of the genus *Homo*, was an enormous quantum jump that included a sizable increase in body size (see Figure 9.8). The third and final burst began slowly during the heyday of *H. erectus* and then accelerated in the early days of *H. sapiens*. It entailed a major expansion of the cranial vault and was accompanied by only minor increases in body size. Can we match this punctuated pattern of change in overall brain size to changes in behavior? Maybe, but the task is difficult. It is tempting, for instance, to argue that the increase in relative brain size at the origin of the australopithecines was related to toolmaking, but stone tools did not appear until 3–4 million years after that increase. Perhaps early australopithecines made other kinds of tools that did not fossilize (Panger et al., 2002) but, then again, chimpanzees also make non-stone tools (McGrew, 1992). Similarly, we cannot assert with confidence that australopithecines were more social than today's chimpanzees, since the latter are no slouches when it comes to sociality (Goodall, 1986; Stanford, 1999). Even more dubious would be the claim that australopithecines were "capable of producing language by arbitrary symbols" (Holloway, 1972). In fact, the earliest indicators of symbolic thought and/or communication (i.e., burial sites and body decorations) do not appear in the fossil record until about 1.3 million years after the australopithecines became extinct. Therefore, it is prudent simply to admit that we do not yet know what special feats australopithecines accomplished with their enlarged brains.

The phylogenetic enlargement of early *H. erectus* brains is somewhat simpler to explain. When *H. erectus* first evolved, the African forests were being replaced by grasslands, which probably led to an increased abundance of large herbivores and carnivores (Vrba, 1995; Leonard et al., 2003). In this environment, the physically large *H. erectus* would have been better equipped than early australopithecines to chase carnivores off carcasses (Blumenschine, 1987), hunt animals by running them to heat exhaustion (Carrier, 1984), and dig deeply for tubers (O'Connell et al., 1999). Since meat and tubers are more nutrient-dense than "grass" (i.e., herbaceous plants), the diet of *H. erectus* would have been better than that of the australopithecines. This is important because, without that increase in diet quality, *H. erectus* would probably have been unable to produce such large, metabolically expensive brains (Leonard and Robertson, 1994). Conversely, having bigger brains probably allowed *H. erectus* to invent new tools and tricks for obtaining even more nutritious foods, such as shellfish or eggs (Crawford, 1992; Cunnane and Crawford, 2003). The discovery that cooking foods frees up their nutrients and makes them more digestible was probably particularly important (Wrangham and Conklin-Brittain, 2003). In aggregate, these dietary improvements probably allowed *H. erectus* to evolve smaller teeth and intestines, which in turn would have allowed them to devote more metabolic energy to growing and maintaining larger brains (Aiello and Wheeler, 1995). In other words, early *H. erectus* (as well as *H. habilis* and *H. ergaster*) probably experienced strong, coordinated natural selection for both larger brains and better dietary quality. That would explain why brain size increased so rapidly when *H. erectus* first evolved, but why did it then plateau? Most likely, the benefits of improving diet quality no longer offset the costs of increasing brain size (see Chapter 7). Less technically, we might say that any novel food procurement tricks made possible by having larger brains no longer paid for themselves.

If this is true, then how do we explain the exponentially enlarging brains of late *H. erectus* and early *H. sapiens*? As far as I can tell, the most important factor here was not the struggle to extract food from the environment but altercations with fellow humans, for by then humans had probably become "so ecologically dominant that they in effect became their own principal hostile force of nature" (Alexander, 1990, p. 4). Instead of struggling to find natural resources and fight off predators, they became increasingly concerned with fighting one another, as evidenced by the unusually thick skulls of *H. erectus* (Boaz and Ciochon, 2004) and the near ubiquity of healed head injuries in early *H. sapiens* (Wolpoff, 1999). Those early humans probably competed both within their social groups (Humphrey, 1976; Miller, 2000) and with conspecific "foreigners" for access to females and environmental resources. This intraspecific competition hypothesis emphasizes the belligerent and devious (or "Machiavellian") aspects of human nature but also involves more positive aspects, since having true friends and allies makes competition easier. As Richard Alexander put it, early *Homo sapiens* probably "began to cooperate to compete, specifically against like groups of conspecifics" (Alexander, 1990, p. 4).

The most interesting aspect of this hypothesis is that it sets up a positive feedback loop. If individuals with larger brains are more likely to compete successfully for resources and mates, then brain size would tend to increase within the population; and, since the losers in this competition would be drawn from the same gene pool as the winners, they would always be just a step behind. Thus, an intraspecific "arms race" for bigger brains and better ways to outwit your opponents would have been set up (Rose, 1980). Of course, the resultant increases in absolute brain size would had to have been paid for by dietary improvements, but the winners of those competitions would have benefited from the "spoils of war." Thus, the intraspecific competition hypothesis provides a reasonable explanation for why brain size increased exponentially as early *H. sapiens* evolved.

This leads us to another fascinating question: Why did human brains cease expanding roughly 100,000 years ago? Again, the most likely answer is that the costs of increasing brain size outweighed the benefits. The costs in this case probably included the same metabolic and computational costs that I already discussed. However, *H. sapiens* probably also encountered an additional "childbirth constraint," which arose when fetal heads became so large that they had trouble passing through the mother's birth canal (Schultz, 1969; Leutenegger, 1974, 1987; Rosenberg, 1992). As you can see from Figure 9.9A, in chimpanzees and australopithecines, neonate heads fit comfortably through the birth canal, but in *H. erectus* that fit was tight. Neonate passage through the birth canal is likewise difficult in *H. sapiens*, which probably explains why so many women used to die during childbirth (and still do in some parts of the world). Childbirth is similarly difficult in many small monkeys (e.g., marmosets and squirrel monkeys; Leutenegger, 1982), but that is not relevant to our present discussion, for small creatures generally have proportionately larger brains and, therefore, larger heads (see Chapter 4). Crucial is that in all hominids except for *H. erectus* and *H. sapiens* the birth canal is significantly wider than a newborn's head. This implies that a severe childbirth constraint probably emerged during the reign of *Homo erectus*. If that is true, then how did *Homo sapiens* manage to increase its relative brain size beyond the level seen in *Homo erectus*? There are several possibilities.

One way around the childbirth constraint would have been for *H. sapiens* to evolve a wider birth canal. This seems not to have happened, at least not to a significant extent, for pelvic anatomy is relatively constant among hominins and was likely shaped more by the requirements of bipedal locomotion than by the need to accommodate larger fetal heads (McHenry, 1975; Abitbol, 1987). Alternatively, *H. sapiens* might have shortened its gestation period, giving birth to infants that were less developed and, hence, less bigheaded than those of *H. erectus*, australopithecines, or other apes (Martin, 1983). This hypothesis seems plausible because human newborns and infants are indeed remarkably helpless (see Smith and Tompkins, 1995). It also cannot be correct, however, because the gestation period of modern humans is actually slightly *longer* than we would expect in a primate of our body size (Harvey et al., 1987). How, then, did humans circumvent the childbirth constraint? Apparently, they increased brain growth *after* birth (Owen, 1859; Leutenegger, 1982). As you can see from graphs of brain growth versus body growth (see Figure 9.9B), *H. sapiens* differs from other primates in maintaining rapid brain growth well past birth (see also Figure 4.11). Thus, whereas most primate brains expand postnatally by a factor of approximately two, the brain of *H. sapiens* expands about threefold (Harvey et al., 1987). By postponing much of their brain growth to after birth, humans could increase adult brain size without increasing neonate skull size beyond the point where childbirth is impossible.

(A) Birthing in a chimpanzee, an australopithecine, and a modern human

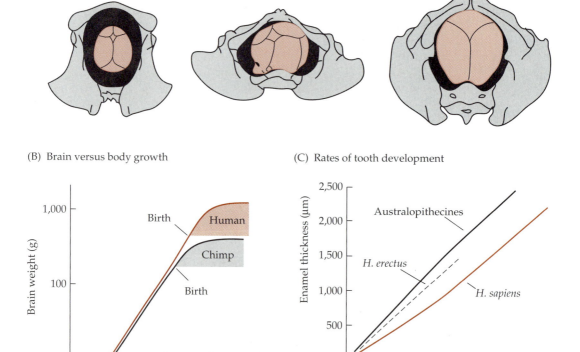

(B) Brain versus body growth

(C) Rates of tooth development

Figure 9.9 Hominin Childbirth and Development The enormous increase in brain size with *Homo sapiens* correlates with several changes in development. (A) In chimpanzees, an infant's skull passes easily through the birth canal. In australopithecines it probably did as well, but in *H. sapiens* the fit became so tight that further increases in relative brain size required changes in development. (B) Humans have circumvented this "childbirth constraint" at least in part by having their brain grow more after birth than it does in chimpanzees. This is evident when we plot brain versus body weight during development for both humans and chimpanzees (just as we did in Figure 4.11) and compare at what point in development both species give birth. (C) Detailed studies of enamel deposition in fossils suggest that the teeth of modern humans develop more slowly than those of early hominins. This finding suggests that the period of juvenile development lengthened as modern humans evolved. This prolongation of childhood probably increased the opportunities for social learning and, more specifically, socialization. (A after Tague and Lovejoy, 1986; B after Count, 1947; C after Dean et al., 2001.)

This much makes sense, but serious puzzles remain. Chief among them is that the brains of human neonates are fairly mature, passing many crucial landmarks of development well before birth (Clancy et al., 2001). This finding suggests that the extreme helplessness of human infants may be due, at least in part, to bodily awkwardness rather than neuronal immaturity. Specifically, one might point out that human neonates are undermuscled and unusually fat (Kuzawa, 1998; Pawlowski, 1998). Even after humans lose their baby fat, somatic development is slow (Leigh, 2001). Human teeth, for instance, grow more slowly than the teeth of *Homo erectus* or australopithecines (see Figure 9.9C). Thus, evolutionary changes in bodily development help to explain why human neonates are helpless and why human childhood is unusually long (if not entirely unique; see Bogin, 1997). However, if the brains of human newborns are fairly mature, what accounts for their extreme expansion after birth? We do not know for sure, but I suspect that human brains expand postnatally primarily because myriad glial cells continue to be "born" (especially in the neocortex), more synapses are formed, and many axons continue to grow and become ensheathed in myelin (see Chapter 3). Be that as it may, a safe bet is that at least some aspects of postnatal brain expansion have made human brains exceptionally responsive to environmental influence, thereby transforming human children into prolific learners. Since protracted learning generally helps social animals to bond and pick up skills, prolonging childhood was most likely beneficial for early *H. sapiens*. As any parent knows, however, there are serious costs to having helpless infants and dependent children (see Peccei, 2001). Moreover, there are probably limits to how much brains can expand postnatally (as long as they remain "fairly mature" prior to birth). Overall, then, I suspect that our human strategy for circumventing the childbirth constraint was largely exhausted by the time modern *H. sapiens* evolved. This would help explain why, in conjunction with the aforementioned metabolic and/or computational constraints, human brain size plateaued about 100,000 years ago.

Now, if absolute brain size plateaued during the last 100,000 years, why did human behavior change so dramatically in that same period? Perhaps the increase in relative brain size that was caused by the decrease in modern human body size freed up some neurons that were then reassigned to more "intelligent" functions. This idea comes up repeatedly in discussions of brain size and intelligence (Jerison, 1973) but has little empirical support (see Chapter 4). In fact, the decrease in modern human body size is most likely the result, rather than the cause, of increased human intelligence. The invention of clothing, for instance, surely reduced the need for having cold-adapted (i.e., big and robust) bodies, and the manufacturing of weapons allowed even gracile individuals to be lethal. But if the decrease in body size did not make modern humans more intelligent, what did? My favorite suspect here is language. As I reviewed earlier, the beginnings of symbolic thought date back 50,000–100,000 years, and some kind of human language probably emerged around that time. Perhaps it began with simple gestures or sounds that symbolized objects and events (Corballis, 2002); perhaps it emerged more fully formed. The true origins of language remain murky, and this is not the place for further stabs into that fog (see Deacon, 1997; Hudson, 1999; Dickins and Dickins, 2001; Hauser et al., 2002). However, it seems reasonable to speculate that, once human language had evolved, it increased the rate of changes in human behavior. Instead of having to learn by watching (e.g., Whiten et al., 1996), speaking humans could also learn by listening, and they could learn to solve their problems in more abstract ways. Thus, language probably increased the potency of "culture" to the point where it replaced the much slower process of natural selection as the major agent for behavioral change. If this idea is right, then it makes sense that absolute brain size plateaued after language came upon the scene. Once we had speech, we did not need perpetually larger brains; we just needed to improve our usage of the language faculty.

The above scenario may be simplistic in its exclusive emphasis on language, but it does return us forcefully to the question Huxley had struggled with more than 100 years ago: What evolutionary changes in the brain caused humans to be capable of learning how to speak? Perhaps it was just a numbers game; perhaps spoken language could only be learned after the total number of neurons in the human brain exceeded some unspecified threshold. Such explanations are what Ralph Holloway (1972) called "Rubicon models" of human brain and language evolution; they are elegant in their simplicity but also vague and probably naïve. More realistic explanations must be based on structurally and functionally specific changes in human brain anatomy and physiology. As we shall see below, such "reorganization models" of human language evolution also remain incomplete. However, we now have sufficient evidence to demonstrate that human brains have been reorganized in several ways. Moreover, at least some of those organizational changes can be linked to vocal dexterity and, from there, to speech.

Evolutionary Changes in Hominin Brain Organization

As hominin brains increased in size, not all brain regions became larger and/or more complex; some became smaller and/or simpler. The olfactory bulb, for instance, is smaller in humans than in chimpanzees (or other primates) and much simpler in its structural details (Stephan and Andy, 1970). Similarly, the magnocellular portion of the red nucleus, which projects mainly to the spinal cord (see Chapter 7), is so reduced in adult humans that I would call it vestigial (see Ulfig and Chan, 2001). An intriguing aspect of these and other simplifications in the human brain is that they are extensions of primate-wide trends. For instance, the dorsal cochlear nucleus, one of the auditory nuclei in the mammalian medulla, is highly laminated in prosimians, simpler in New World and Old World monkeys, and simplest in humans and

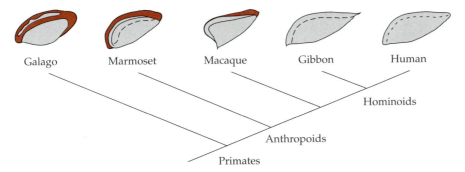

Figure 9.10 Simplification of the Dorsal Cochlear Nucleus in Primates The dorsal cochlear nucleus (DCN) contains several granule cell layers (red) in most mammals, including prosimians such as the galago. In New World monkeys, such as marmosets, granule cells are found only in the most superficial portion of the DCN, and in Old World Monkeys, such as macaques, this granular layer is very thin. In apes, including humans, the granular layer is essentially absent; only indistinct laminar boundaries (dashed lines) are evident. Given this phylogenetic distribution, it is most parsimonious to infer that the DCN was simplified as primate brains enlarged (see Figure 10.1). The functional significance of this simplification trend remains unclear. Perhaps some functions of the DCN were lost during phylogeny; perhaps they were subsumed by the neocortex. (After Moore, 1980.)

other apes (Figure 9.10) (Moore, 1980). Why do these primate-wide simplification trends exist? We do not know for sure, but a reasonable hypothesis is that they represent the flip side of the other major trend in primate phylogeny, namely, the trend of increasing neocortex size (see Figure 7.11). That is, those non-neocortical structures may have been simplified because some of their old functions were shifted to (or subsumed by) the neocortex. As I mentioned in Chapter 7, this idea of functional neocorticalization is controversial and alternative explanations remain on the table. The simplification of the olfactory bulb, for instance, may simply reflect a phylogenetic reduction in the human sense of smell (Rouquier et al., 2000). Similarly, the reduction in the size of the dorsal cochlear nucleus may be related to the fact that the larger primates tend to have less mobile ears (pinnae). At this point, very little is known about why some brain regions became smaller and/or simpler as primates evolved. Therefore, let us instead focus on the much better-studied trend of increasing neocortex size. Specifically, let us ask how the human neocortex differs from that of other primates.

As you might expect from the general correlation between absolute brain and neocortex size (discussed earlier; see also Figure 5.4), the aforementioned enlargement of modern human brains is mainly due to an expansion of the neocortex. This is most obvious when you consider that the ratio of neocortical gray matter to medulla is 30:1 in chimpanzees but 60:1 in humans

(Stephan et al., 1981; Frahm et al., 1982). Since the medulla is unlikely to have shrunk as *H. sapiens* evolved (even the dorsal cochlear nucleus is quite large, albeit simple, in humans), we can infer that the neocortex enlarged disproportionately as humans diverged from chimpanzees. But has it changed in aspects other than size? The answer is yes.

In the sections that follow, I first discuss hominin-specific changes in the projections from the neocortex to the brain stem and spinal cord; generally speaking, those projections became more extensive and direct as hominins evolved. Next, I address whether the human neocortex contains any "new" areas, whether any of its "old" areas are exceptionally large, and how those intracortical changes might have altered behavior. My basic conclusion is that the evolutionary changes in the human neocortex served mainly to make our behavior more flexible and unconventional, and allowed language to emerge. Finally, I review some possible downsides of increasing neocortex size. Specifically, I address the possibility that the human neocortex has become so large that its ability to compensate for damage has been reduced.

As we discussed in Chapter 7, Deacon's rule of "large equals well-connected" predicts that the human neocortex should have exceptionally extensive projections to the spinal cord. Indeed, we have already discovered that it does (Figures 7.11 and 7.13). What we have not yet discussed is that the human neocortex also invaded several parts of the medulla. Specifically, the human neocortex has apparently evolved an unprecedented level of direct access to the motor neurons innervating the muscles of our jaws, face, tongue, and vocal cords. This conclusion was originally derived from "old-fashioned" axonal degeneration studies (Kuypers, 1958; Iwatsubo et al., 1990), but more modern methods have generally supported that hypothesis. Modern axon-tracing studies, for example, have revealed that the strength of direct projections from the neocortex to the motor neurons for the muscles of the tongue correlates with absolute brain size (Jürgens and Alipour, 2002). Moreover, magnetic stimulation of the cortical tongue area in humans elicits the kind of short-latency tongue contractions you would expect from a direct pathway (Rödel et al., 2003). Similarly, magnetic stimulation of the human neocortex seems to cause direct (i.e., monosynaptic) activation of the motor neurons for the major muscles of the lips and jaw (Liscic et al., 1998; Pearce et al., 2003), just as the older anatomical studies had predicted. Finally, a recent study has confirmed that in monkeys the motor neurons that innervate the vocal folds (or chords) lack the kind of direct neocortical inputs that they apparently receive in humans (Iwatsubo et al., 1990; Simonyan and Jürgens, 2003). Overall, these data indicate that neocortical access to the motor neurons of the medulla and spinal cord increases steadily with neocortex size and is greatest in humans. Anatomically, this is explicable in terms of Deacon's rule, but functionally what does it mean?

Back in Chapter 7, we noted that the strength of the direct projections from the neocortex to the motor neurons of the hand correlates with manual dexterity. Indeed, humans are clear paragons when it comes to making finely controlled hand movements, and without the neocortex those movements evaporate. Given those observations, it seems likely that the increasingly direct cortical projections to the motor neurons for the mouth endowed humans with improved "oral dexterity" and that direct cortical projections to the motor neurons of the tongue and vocal cords greatly improved our "vocal dexterity" (Figure 9.11). In other words, the increased access of the neocortex to diverse lower motor neurons probably made humans more dextrous in a variety of realms, ranging from hand movements to vocalization. This increased dexterity hypothesis makes sense but leaves an open question: What aspects of the neocortex are responsible for that increased dexterity? What is so special about neocortical control?

We do not know, but one possibility is that neocortical neurons are more capable than other neurons of being molded by experience, of "learning" how to fire for an optimal result. Indeed, diverse evidence now indicates that even primary motor cortex can exhibit remarkable plasticity (Svensson et al., 2003). A second possibility is that the neocortex, with its multitude of intracortical connections (see Chapters 6 and 7), may provide an ideal substrate for coordinating the activities of all those neurons that are needed to perform highly skilled movements. After all, dexterity implies coordination as well as precision. Combining these two possibilities, we may surmise that neocortical control probably improves dexterity by modifying motor patterns until precise coordination is achieved. That kind of learned coordination is what makes us seem like "angels in action" (at least occasionally). But are we not more than a beautifully coordinated bunch of moving parts? Shakespeare clearly thought we are. Pursuant to that thought, let us search for other special attributes of the human neocortex.

Given the tight correlation between neocortex size and the number of neocortical areas across mammals (see Figure 6.16), we would expect humans to have at least a few new cortical areas. Direct evidence for human-specific cortical areas is, however, scant. One might have thought, for instance, that the two major "language areas," named after Paul Broca and Carl Wernicke, would be unique to human brains, but that is not the case. Cortical areas 44 and 45, which collectively comprise Broca's area in humans (Figure 9.12), have now been identified in both monkeys and prosimians (Preuss and Goldman-Rakic, 1991c; Rizzolatti and Arbib, 1998; Petrides and Pandya, 1999; Wu et al., 2000; Petrides and Pandya, 2001). Because monkeys clearly do not talk (notwithstanding evidence for limited "sign language" in some chimpanzees; see Terrace, 1979), we can infer that Broca's area in humans underwent a major change in function to become a "language area" (1979; Aboitiz and García, 1997; Jürgens, 2002). We shall return to that later. Despite this change in function, the topological and cytoarchitectural similarities between areas 44 and 45 in monkeys and Broca's area in humans are so striking that most authors do not doubt their homology; for instance, area 45 in both humans and macaques has clusters of large, deeply stained pyramidal cells in the

(A) Nonhuman primates

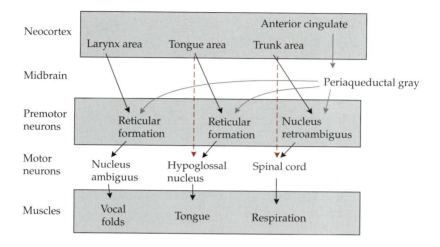

(B) Humans

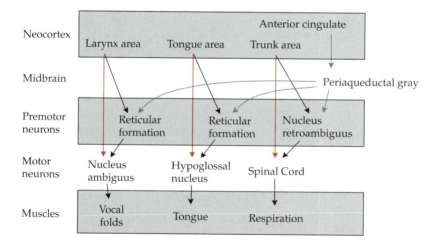

Figure 9.11 Neural Circuits for Vocal Production in Primates The neural circuitry for vocal production differs considerably across primates. (A) In nonhuman primates, the brain stem and spinal motor neurons that are involved in vocal production receive most of their inputs from other neurons in the midbrain and medulla. Direct projections from the neocortex to vocal motor neurons are relatively weak (dashed red arrows) or nonexistent. (B) Humans, in contrast, sport robust direct projections from the neocortex to all major vocal motor neuron groups (red arrows). This difference in connectivity fits well with the observation that humans generally exhibit more "vocal dexterity" than other primates. It may also explain why neocortical lesions cause no major vocalization deficits in nonhumans, but render humans virtually mute (though they can still laugh and cry). (After Jürgens, 2002; Jürgens and Alipour, 2002.)

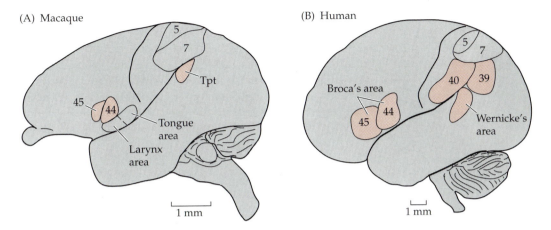

Figure 9.12 Enigmatic Areas in Macaque and Human Neocortices   Many, but perhaps not all, human neocortical areas have homologues in macaques. Neocortical areas 44 and 45 in macaques are probably homologous to Broca's area in humans, and the temporoparietal auditory area (Tpt) of macaques is apparently homologous to Wernicke's area in humans. Although many different brain regions clearly cooperate when humans speak, Broca's area and Wernicke's area are the two areas that are most widely known as "language areas." Since nonhumans do not talk, these areas most likely changed in function as humans evolved. Areas 39 and 40 of the human inferior parietal cortex seem to have no macaque homologues, but this conclusion remains tentative (they may be homologous to parts of macaque area 7). Just ventral and caudal to Broca's area lie the cortical larynx and tongue areas, which help produce speech sounds; in humans, those areas are hidden behind the temporal lobe.

lower part of layer III and a well-developed layer IV (see Petrides and Pandya, 2001). Similarly, neuroanatomical analyses suggest that Wernicke's area in humans is homologous to the "temporoparietal auditory area" of nonhuman primates (see Figure 9.12A)(Preuss and Goldman-Rakic, 1991a; Pandya, 1995). Thus, at least two cortical areas that one might have suspected to be uniquely human are not so. Of course, this does not prove that humans have no unique cortical areas at all. In fact, some authors do suspect that parts of the lateral prefrontal region, as well as areas 39 and 40 in the inferior parietal lobe (see Figure 9.12B), are unique to human brains (e.g., Karnath, 2001). However, the comparative data on those regions are so meager (see Vanduffel et al., 2002) that we cannot yet reject the null hypothesis that all human neocortical areas have nonhuman homologues.

Although most (if not all) neocortical areas are conserved between humans and other apes, the proportional sizes of those areas vary considerably across species. The primary visual or *striate* cortex, for instance, occupies 5% of the entire neocortex in chimpanzees, but only about 2% in humans (Stephan et al., 1981). That difference in proportional size is not entirely unexpected, since the fraction of the neocortex occupied by striate cortex generally decreases with increasing neocortex size (Passingham, 1973). However, when we plot striate cortex size against neocortex size in double logarithmic coordinates, we see that the human data point lies far below the best-fit allometric line (Figure 9.13A). Even after we account for the considerable intraspecific variation in human striate cortex size (e.g., Klekamp et al., 1994), it is fairly clear that human striate cortex is considerably smaller (on average) than we would expect in a primate with a neocortex that is as large as ours. Of course, there are two ways of looking at this: Either the visual cortex shrank as humans evolved, or some other neocortical areas ballooned. The former hypothesis is most likely false, since the human visual system is not generally degenerate. Our eyes are roughly as large as we would expect them to be given our body size, and when we look at how striate cortex size scales with body size, the human data point falls squarely on the allometric line (Figure 9.13B) (Passingham, 1973). In addition, human striate cortex is no smaller than we would expect from the size of its main thalamic input region, the lateral geniculate (Stevens, 2001). Therefore, striate cortex is unlikely to have shrunk as *H. sapiens* evolved. Combining this conclusion with the aforementioned finding that

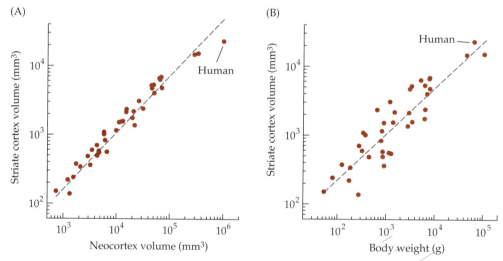

Figure 9.13 Primary Visual Cortex Scaling in Primates For most primates, the primary visual, or striate, cortex scales quite predictably with overall neocortex size (A). A notable exception are humans, for in them the striate cortex is smaller that we would expect given the absolute size of the human neocortex. This could mean either that the striate cortex was reduced in size during hominin phylogeny or that other neocortical regions ballooned. The latter possibility is supported by the finding that the human striate cortex is roughly as large as we would expect it to be given human body size (B). The dashed lines represent the best-fit lines for all nonhuman data points. (After Stephan et al., 1981.)

striate cortex is proportionately smaller in humans than in chimpanzees, we can deduce that human evolution must have involved a "great expansion in neocortical areas other than striate cortex" (Passingham, 1973). The crucial question is: What are those other areas?

Some of those other, expanded human cortical areas may lie within the temporal lobe (see Figure 9.5), since the human temporal lobe is larger than one would expect for apes with our absolute brain size (Rilling and Seligman, 2002). Given that the inferior temporal cortex, whose neurons respond selectively to faces, lies further dorsally in monkeys than in humans (Farah et al., 1999), I suspect that what expanded in the human temporal lobe was mainly its dorsal component, which processes auditory stimuli including speech. In the overall scheme of things, however, this human-specific expansion of the temporal lobe was relatively small, and across all primates the proportional size of the temporal lobe cortex has tended to decrease with increasing brain size (from 30% in squirrel monkeys to about 16% in humans and chimpanzees). Therefore, the great expansion of the human neocortex most likely involved at least some areas other than the temporal lobe. One promising candidate for human-specific expansion is the parietal lobe, which includes the somatosensory cortex and areas 5, 7, 39, and 40 (see Figures 9.5 and 9.12). Since areas 39 and 40 are virtually impossible to identify in nonhuman primates (Preuss and Goldman-Rakic, 1991a), we may infer that part of the parietal cortex probably ballooned during hominin phylogeny. Unfortunately, however, there have been no detailed comparative analyses of how the parietal lobe scales across primates. Therefore, we do not yet know to what extent the neocortical expansion in hominins was fueled by an increase in the size of the parietal cortex. Similarly, there have been no quantitative comparative studies of occipital lobe areas other than the striate cortex. That leaves us with the frontal lobe (see Figure 9.5).

Across primates in general, the frontal lobes increase in proportional size as absolute brain size goes up (Bush and Allman, 2004). It is not particularly surprising, therefore, that African apes have significantly larger frontal cortices than other primates with smaller brains (Figure 9.14A). Within the African apes, humans tend to have proportionately larger frontal lobes than chimpanzees, but the difference is small and statistically insignificant (Semendeferi, 2002). This is not the end of the story, however, for as I reviewed earlier, the frontal lobe is composed of several divisions, some of which do seem to be enlarged in humans. As Brodmann (see Chapter 2) reported back in 1912, the lateral prefrontal cortex (see Figure 9.5) occupies about 29% of the neocortex in humans, but only 17% in chimpanzees (Figure 9.14B). These numbers must be regarded cautiously since other authors obtained different results (see Passingham, 1973; Uylings and van Eden, 1990). However, Semendeferi et al. (2001) recently confirmed that at least one lateral prefrontal region, namely Area 10, is almost twice as large (percentagewise) in humans as in other apes. Area 13 of the orbitofrontal cortex, in

(A) Frontal lobe scaling according to Semendeferi et al. (1997 and 2002)

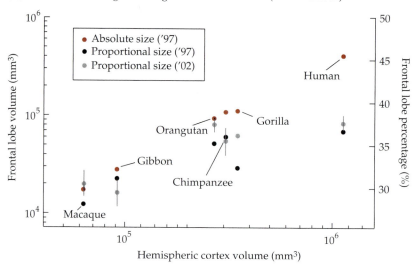

(B) Prefrontal cortex scaling according to Brodmann (1912)

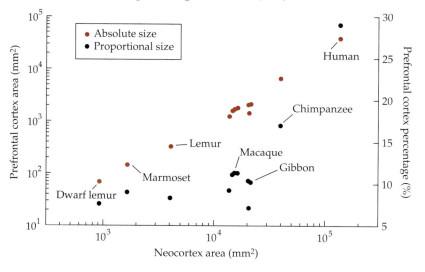

contrast, has roughly the same proportional size across all apes (Semendeferi et al., 1998). Collectively, these data reveal that the lateral prefrontal cortex, but not the entire frontal lobe, became disproportionately large as hominins evolved. This much is relatively clear. Less clear is whether human lateral prefrontal cortex is larger than one would expect, given that the proportional size of this neocortical region generally increases with increasing absolute brain

◀ **Figure 9.14 Frontal and Prefrontal Cortex Scaling in Primates**   Few topics in evolutionary neurobiology are as heavily debated as whether or not humans have disproportionately large frontal and/or prefrontal neocortices. (A) Katerina Semendeferi and her colleagues (1997, 2002) used magnetic resonance imaging to compare frontal lobe volume across various primates and found that, in terms of absolute size, the frontal lobe scales quite predictably across primates (red data points). In terms of proportional size, the data are less clear, mainly because there seems to be a lot of variance, both across and within studies. However, a safe conclusion is that the frontal lobes are proportionately larger in African apes than in macaques. (B) Back in 1912, Brodmann compared the size of the lateral prefrontal cortex across primates and found it proportionately larger in humans than in other primates. Is the human lateral prefrontal cortex larger than one would expect given human neocortex size? This question remains unresolved, but there does seem to be at least a weak primate-wide correlation between the proportional size of the lateral prefrontal cortex and the absolute size of the neocortex.

size (see Figure 9.14B). This question has not yet been resolved (see Deacon, 1997; Holloway, 2002; Passingham, 2002), but even if the human lateral prefrontal cortex is no larger than expected, it is still *disproportionately* large. That finding is important because changes in a region's proportional size may be functionally significant even if its relative size (i.e., its size relative to allometric expectations) has remained constant. We discussed this in Chapter 5 and shall come back to it later. But first let me place the lateral prefrontal cortex into its neuroanatomical context.

The lateral prefrontal cortex is densely interconnected with several other regions that are also selectively enlarged in human brains. Within the neocortex, its major partners are the posterior parietal cortex and the cortex of the temporal lobe (Preuss and Goldman-Rakic, 1991b; Petrides and Pandya, 1999). As we discussed earlier, in humans the former probably contains some unique areas, and the latter is slightly larger than expected. Outside the neocortex, the lateral prefrontal cortex interacts mainly with several cell groups in the dorsal thalamus, particularly the medial dorsal nucleus and the pulvinar. Both of these nuclei are disproportionately large in humans (compared to nuclei that lack prefrontal connections; Figure 9.15). The human pulvinar is especially intriguing because its enlargement is causally related to a major change in its embryogenesis: Only in humans does the pulvinar contain neurons that migrated into the thalamus from the telencephalon (Letinic and Rakic, 2001). Whether the pulvinar also contains telencephalic "immigrants" in chimpanzees or other apes remains unknown, but in macaques no such migration is in evidence. This suggests that pulvinar development was altered radically in apes or hominins. The other fascinating aspect of human pulvinar hypertrophy is that it involves mainly the dorsal pulvinar, which has strong reciprocal connections with the lateral prefrontal, parietal, and temporal cortices (Romanski et al., 1997; Gutierrez et al., 2000). This dorsal pulvinar

(A) Allometry of dorsal thalamic nuclei in apes

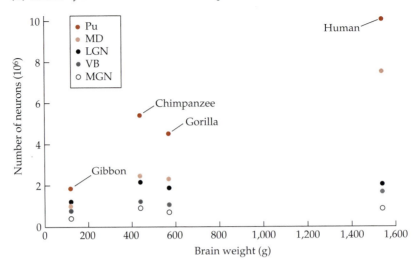

(B) Primate pulvinar hypertrophy

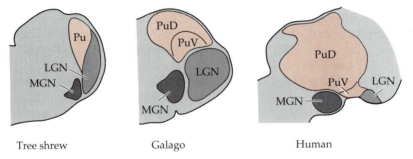

Tree shrew   Galago   Human

Figure 9.15 Evolutionary Changes in the Dorsal Thalamus   Two dorsal thalamic nuclei that are related to the lateral prefrontal cortex, namely the mediodorsal nucleus (MD) and the pulvinar (Pu), became disproportionately large in *Homo sapiens*. (A) Within apes, the number of neurons in MD and Pu increases rapidly with absolute brain size. Most other thalamic nuclei, such as, the ventrobasal (VB), and the medial and lateral geniculate nuclei (MGN, and LGN), scale with much lower allometric slopes. This implies that MD and Pu occupy a greater fraction of the dorsal thalamus in humans than they do in other primates. (B) The enlargement of the pulvinar involves mainly the dorsal pulvinar (PuD), which appears to have no homologue in nonprimates. Within the dorsal pulvinar, medial and lateral divisions are commonly recognized; of these, the medial pulvinar is remarkable for its lack of visual inputs and its massive connections with most higher-order cortices, including the lateral prefrontal cortex. The ventral pulvinar (PuV) receives mainly visual inputs from the superior colliculus and projects mainly to the visual cortex (see Figure 9.4). (A after Armstrong, 1982; B after Harting et al., 1972; Mai et al., 1997.)

is probably unique to primates (Harting et al., 1972) and separate from the ventral pulvinar, whose major function is to convey visual information from the midbrain to the telencephalon (see Figure 9.4). Collectively these data indicate that what enlarged in humans is not a motley group of areas and nuclei, but an entire circuit that includes the lateral prefrontal cortex and several "associates" in both the neocortex and the thalamus.

So, how did the enlargement of the lateral prefrontal cortex and its various associates impact human brain function? Given that increases in a region's proportional size generally increase its influence within the brain (see Chapter 5), we would expect the lateral prefrontal cortex to be more important in human than in nonhuman brains. According to the principle of "large equals well-connected," we would also expect the human lateral prefrontal cortex to have some novel connections. Both hypotheses are difficult to test because, as I mentioned already, we still know very little about the detailed connectivity (and functional interdependencies) of human brains. Still, it is interesting to note that in macaques the lateral prefrontal cortex has at least some connections that are not found in prosimians (Preuss and Goldman-Rakic, 1991b), and that in human brain imaging studies, the lateral prefrontal cortex is activated during many different tasks (see Duncan and Owen, 2000). Therefore, the available data are at least consistent with the notion that the lateral prefrontal cortex in humans exerts more control over other brain regions than it does in other primates. Assuming this is true, what might the nature of that control be? Earlier I mentioned that the lateral prefrontal cortex probably helps primates contemplate alternate scenarios of how to interact with external objects. Now, I would like to be more specific. This is challenging because the literature on lateral prefrontal cortex function is fractured into diverse schools of thought that are each complex in their own right and difficult to reconcile with one another. Nonetheless, I see a common theme in all this work and think it can be used to construct a general hypothesis of how the phylogenetic enlargement of the lateral prefrontal cortex has affected human behavior. Let me first state the hypothesis in general terms and then delve into the details and implications.

Basically, I propose that having an unusually large lateral prefrontal cortex has made humans exceptionally capable of doing things that are "unconventional" in the sense of representing novel solutions to behavioral problems. Alternatively, one might say that the large lateral prefrontal cortex allows humans an unprecedented capacity for voluntary acts, but the term "voluntary" is so enmeshed with issues like "free will" that the argument becomes philosophically too charged. Wise and Murray (2000) have used "arbitrary" to denote essentially what I here mean by "unconventional," but for most of us "arbitrary" is synonymous with "random," which is not what Wise and Murray had in mind. Similarly, saying that the lateral prefrontal cortex helps us make nonreflexive movements is unsatisfactory because "non-reflexive" is too easily confused with "learned," which is again not what I mean. Primates without lateral prefrontal cortices can still learn many things, but they become less flexible, losing the ability to choose from several alternative actions the one that is most likely to pay off (given the behavioral context). Essential to that flexibility is the ability to suppress reflexive responses (learned or not) in favor of the unconventional and nonreflexive ones. It is not surprising, therefore, that damage to the prefrontal cortex makes subjects act much more impulsively. This has long been know from patients with "prefrontal lobotomies" but is also a recurrent theme in recent work on prefrontal function (Miller and Cohen, 2001). For instance, it has now been shown that monkeys with damage to their prefrontal cortex have trouble reaching around transparent barriers for food; instead, they seem compelled to reach directly for the food and, consequently, bump into the barriers (Dias et al., 1996). Generally speaking, recent studies have confirmed that the primate prefrontal cortex plays a major role in "response inhibition," with different prefrontal subdivisions tending to inhibit different kinds of responses (Wallis et al., 2001; Aron et al., 2004). Hence I deem it fair to say that the disproportionate enlargement of the lateral prefrontal cortex probably made humans less impulsive and more capable of unconventional yet purposive behavior.

This hypothesis does not imply that the capacity for unconventional behavior is entirely unique to humans. Indeed, it probably is not. Some crows, for instance, can come up with unconventional solutions in a task that asks them to pull food out of narrow tubes. After trying repeatedly to get the food with straight lengths of wire, the crows stop, bend the wire into hooks, and retrieve the food efficiently (Weir et al., 2002). This behavior requires the inhibition of conventional behavior (poking with straight wires) and devising a new strategy (making a hook). Although these crows are known to make some hook tools in the wild (Hunt, 1996), they do not normally bend wires. Therefore, their wire-bending trick in the laboratory setting represents what I would call an unconventional solution to a behavioral problem. Now, you might be tempted to conclude from these data that the capacity for unconventional behavior is broadly conserved across the vertebrates, but that is not the case. Instead, it seems to be an exceedingly rare capacity that evolved independently in primates and some birds. Supporting this hypothesis is the discovery that birds apparently lack a "true" prefrontal cortex homologue but have modified one of their noncortical brain regions (part of the DVR; see Chapter 8) so that it is both structurally and functionally similar to the mammalian prefrontal cortex (Divac and Mogensen, 1985; Bast et al., 2002). Aside from birds, some other large-brained nonprimates (e.g., elephants and cetaceans) may also be capable of solving problems unconventionally. However, given that those species are but distant relatives of primates, crows, and one another, it is likely that they all became so clever independently. Such instances of convergence do not negate my basic claims, which are that (1) the invention of the lateral prefrontal cortex made primates more capable of being unconventional than their nonprimate ancestors; and that (2) within primates, the expansion of the lateral prefrontal cortex (and its associates) made humans the most unconventional primates of all. Having settled that, let us consider some details, beginning with the literature on primate eye movements.

Imagine you are in a darkened room, looking straight ahead, with bright lights flashing periodically off to the side. Your natural inclination is to move your eyes toward the lights, but what if you were instructed to look diametrically away from the flashing lights? Such anti-saccades, as they are called by specialists, require the more impulsive, positive saccades to be suppressed (see Everling and Fischer, 1998). Therefore, they represent what I call unconventional behavior. Although children have great trouble performing anti-saccades, adult humans generally succeed, and so do some adult nonhuman primates. In contrast, no nonprimates appear capable of making anti-saccades. In terms of neural mechanisms, we know that primates with damage in the "frontal eye field" region, which lies caudodorsally within the lateral prefrontal cortex, lose their capacity for making anti-saccades. Apparently, the neurons in that region actively suppress the midbrain and medullary neurons that normally drive positive saccades (Everling et al., 1999).

How do anti-saccades allow us to find "novel solutions to behavioral problems"? That question seems odd at first because we tend not to think of our eyes as problem-solving effectors. However, if you want to use your eyes for communicative purposes, then there is a problem to be solved, namely how to maximize the information content of those eye movements. Since information carrying capacity generally increases as a message becomes less predictable, making eye movements more unconventional tends to make them more informative (at least potentially). For instance, "rolling your eyes" can convey exasperation, and looking away from objects of interest can misinform competitors. Indeed, humans and chimpanzees, the two species with the largest lateral prefrontal cortices, are also unusually interested in the direction of another's gaze (Tomasello et al., 1999). Human eyes are further specialized in being considerably wider than tall and having a sclera that is white rather than brown (Figure 9.16). Those features must have made it easier for humans to determine one another's gaze direction (Kobayashi and Koshima, 2001). Thus, it seems fair to say that the expansion of the lateral prefrontal cortex, as well as changes in the eyes themselves, have expanded the capacity of humans to communicate through eye movements.

An analogous argument can be constructed for movements of the hands. Just as monkeys can learn to perform anti-saccades, they can learn to "anti-point." That is, they can learn to point diametrically away from stimulus objects. Under some conditions, they can even learn to point to the smaller of two food rewards, an act that requires inhibition of the urge to be greedy (Silberberg and Fujita, 1996). Moreover, as I mentioned earlier, many monkeys perform well on detour reaching tasks (see Santos et al., 1999). Collectively, these data indicate that simians in general have a well-developed capacity for

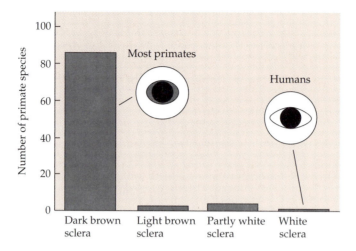

Figure 9.16 Human Eyes Are Quite Unusual   In most primates, the eyeball's outer covering (or sclera) is white except for where it can be seen from the outside; there it is generally dark brown. This scleral pigmentation makes it easier for an animal to hide the direction of its gaze. In humans, however, the sclera is completely white, which probably enhances the ability of humans to advertise what they are looking at. That is, it probably enhances the ability of humans to communicate with eye movements. Also interesting is that human eyes are considerably wider than tall, whereas most other primate eyes are round. This difference in eye shape may reflect the fact that humans often move their eyes along the horizon, but it also helps expose more of the white sclera, which should further facilitate communication via eye movements. (After Kobayashi and Koshima, 2001.)

making unconventional hand and arm movements. Among other animals, that ability is rare. Neurobiologically, the lateral prefrontal cortex is once again involved, as its neurons are active during anti-pointing (Connolly et al., 2000), and lesions there impair performance on the detour reaching task (Dias et al., 1996). Based on these data, I propose that the expansion of the lateral prefrontal cortex in humans is at least partly responsible for our superior ability to communicate through manual gestures and use our hands constructively. Of course, some apes make manual displays and some chimpanzees have learned elements of sign language (Terrace, 1979), but humans have expanded manual communication well beyond what is seen in other animals. Similarly, chimpanzees can make some simple tools (McGrew, 1992), but humans build machines! Particularly interesting is that early *H. sapiens* generally chipped away at rocks progressively, shaping them until the final tools emerged, while later humans took constructional detours along the way. That is, they sculpted "prepared cores" that looked very different from completed tools but could be used to produce several finished tools by simple and efficient strokes. Now, I admit that it sounds strange to call the prepared core technique "unconventional," since it became the norm. However, any individual who wants to learn the prepared core technique must first realize that taking the detour is more efficient than the obvious, direct approach. Think of the first human to invent the prepared core technique; he or she was clearly being unconventional.

On to the most fascinating human specialty, our ability to speak! I have already discussed the remarkable vocal dexterity of humans, but language is far more than the capacity to produce complex, precisely controlled sounds. It hinges on the still largely mysterious ability to represent objects symbolically. Consider a vervet monkey giving her "leopard" alarm call. When her fellow monkeys hear that call, they tend to run away and look for where the danger lurks (Cheney and Seyfarth, 1990). In contrast, when humans hear the word "leopard," they know that this need not imply immediate peril (at least in places where leopards are rare). In other words, humans generally use their words not just to signal what is present, but also to "talk about" things that may not be in evidence (Hartshorne and Weiss, 1931–58; Langer, 1942; Percy, 1975; Deacon, 1997). This uncoupling of the signifier from the signified (of the word from its object) is conceptually analogous to anti-pointing or anti-saccades insofar as it likewise involves the inhibition of direct, impulsive responses. Moreover, the link between a signifier and its signified is arbitrary in that the linkage is not necessary; the leopard's name could be "drapoel" instead. As in the aforementioned case of prepared core technology, it seems odd to designate symbolic speech as unconventional behavior because, after all, the names we give to objects or concepts all apply "by convention." Again, however, we must recall the innovators. Anyone who invents new symbolic names is, by definition, being unconventional! Neurologically, symbolic speech involves many different brain regions, but the lateral prefrontal cortex is once again a major player (e.g., Petersen et al., 1988), and the rostral part of Broca's area (area 45) clearly lies within the lateral prefrontal subdivision (Petrides and Pandya, 2001). Overall, these considerations lead me to suggest that the expansion of the lateral prefrontal cortex (and its associates) helped to give us human speech, just as it has endowed us with an unparalleled capacity for unconventional hand and eye movements.

Admittedly, this "prefrontal language hypothesis" remains vague and debatable. Perhaps the most important question is whether the lateral prefrontal cortex represents a set of truly parallel pathways that function separately in eye, hand, and speech control (as I have implied in the preceding discussion), or whether some of those pathways overlap. Most authors seem to agree that ocular control is separate from hand control, but some authors have argued that the neural substrates underlying manual control and speech are more or less coincident within the frontal lobe, diverging only in the premotor and motor cortices. Michael Arbib (in press) for instance, has proposed that the neural circuitry for speech evolved "atop of" the circuitry for

(A) Divergent circuits hypothesis

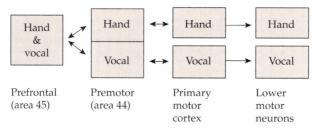

Prefrontal (area 45)   Premotor (area 44)   Primary motor cortex   Lower motor neurons

(B) Parallel circuits hypothesis

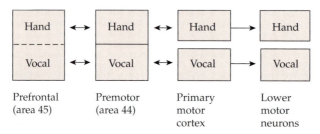

Prefrontal (area 45)   Premotor (area 44)   Primary motor cortex   Lower motor neurons

Figure 9.17 Two Hypotheses on Circuits through Broca's Area   Broca's area includes both rostral and caudal components, which are generally referred to as areas 45 and 44, respectively. Area 45 lies ventrally within the lateral prefrontal cortex, while area 44 is the ventralmost part of the premotor cortex. Most likely, both areas are reciprocally connected with one another, and area 44 has reciprocal connections with the primary motor cortex, which projects to the "lower" motor neurons in the brain stem and spinal cord. Clearly, the motor neurons that control hand movements are separate from those that control vocal production (see Figure 9.11). Within the primary motor cortex, regions that control the hands are also clearly separate from those that control the main vocal organs (i.e., the larynx, lips, and tongue), and the same kind of segregation is evident in the premotor cortex (though it has not yet been studied extensively). What we do not yet know, however, is whether hand and vocal control areas are segregated in the human prefrontal cortex. Therefore, we are left with two hypotheses: The cortical circuits for hand and speech control are either (A) divergent after a common "origin" in the prefrontal cortex or (B) parallel in their entirety. The former scenario implies that all of Broca's area is "old" and that its axons invaded the vocal motor system as humans evolved. The latter scenario implies that the ventral, speech-related part of area 45 is "new" in *Homo sapiens*.

manual gestures. Similarly Corballis (2003) argued that the region in charge of manual gestures was modified in humans such that it "added speech to its portfolio." According to this "divergent circuits hypothesis" (Figure 9.17A), the prefrontal neurons that control unconventional hand movements and vocalizations should be one and the same. Indeed, some data show that part of Broca's area is active during hand movements (Binkofski et al., 1999; Horwitz et al., 2003). However, those brain imaging studies lack the kind of spatial resolution we would need for a definite conclusion (see Rizzolatti and Arbib, 1998). Moreover, diverse other data indicate that the prefrontal cortex generally is part of a much larger system (the cortico-striato-pallido-thalamo-cortical system) that is well known to be divisible into many parallel pathways (Groenewegen et al., 1990; Middleton and Strick, 2002). In light of these considerations, I strongly suspect that the prefrontal circuits for manual gestures and speech are also organized in parallel (see Figure 9.17A), performing analogous functions but being spatially distinct. Further work is needed, however, to test this "parallel circuit hypothesis" (see Figure 9.17B). More generally, we must admit that Huxley's sense of mystery at how the human brain became exceptionally capable of speech remains largely in force.

Thus far, I have reviewed mainly those aspects of modern human brains that can be viewed as functional improvements over prior, smaller brains, but the great expansion of the human brain probably entailed some costs as well. As I mentioned already, the increases in hominin brain size probably made birthing difficult (see Figure 9.9) and required a more nutritious diet. In addition, there are network scaling costs. The most obvious of these network scaling costs is that, as brains increase in size, their axons must increase in length. Those increases in axon length tend to increase the average amount of time it takes for information to be exchanged between distant neurons, which probably makes it more difficult to synchronize neural activity in widely separated cortical regions, such as the left and right cerebral hemispheres (Ringo et al., 1994). This synchronization problem may be alleviated by making a few axons very thick (thereby increasing conduction speed; see Aboitiz, 2003), but it is potentially serious because neuronal synchrony probably plays a major role in neuronal computation and cognition (Singer and Gray, 1995; Varela, 1995). The second major network scaling cost stems from the fact that, as we discussed in Chapters 4 and 7, increases in absolute brain size generally require decreases in a brain's connection density or, more formally, its proportional connectivity (see Figures 4.16 and 7.15). That decrease in connection density tends to increase the degree of separation between individual neurons, making it more difficult to shuttle information to and fro within the brain (even if the networks are "small worlds"; see Chapter 7). Thus far we have discussed this network scaling cost in highly abstract terms, but for evolving hominins it must have been a rather concrete cost that required changes in the brain's internal wiring. Specifically, we can infer that, as brain size increased and connection density decreased, some neuronal connections must have been lost (see Chapter 7). What connections were eliminated and which ones were retained? How were large primate brains "rewired" to cope with network scaling costs?

In part, large primate brains have coped with the size-related decrease in connection density by moving away from a massively parallel network design with widely converging and diverging connections between individual neurons to a more serial design that includes several distinct "processing streams." Evidence for that hypothesis comes from the visual cortex, which is divisible into dorsal and ventral cortical streams in macaques but not in cats (Young, 1992; Scannell and Young, 1993). The same kind of circuit serialization probably occurred in the thalamocortical system. The LGN, for instance, projects almost exclusively to the primary visual cortex in macaques, but to various different cortical regions in cats (see Jones, 1985). Similarly, the main somatosensory component of the dorsal thalamus projects only to the primary somatosensory cortex in simians, but to multiple areas in nonprimates and prosimians (Kaas and Preuss, 2003). This move toward more serial thalamocortical and intracortical circuits minimizes wiring costs but implies a major change in computation strategy. Instead of processing information in a "parallel distributed" manner (Rumelhart et al., 1986), serial circuits are more hierarchical, with each node in the circuit processing the output from the prior step. This is an efficient way to run a brain.

However, a major disadvantage of the serial mode is that it leaves the system vulnerable to damage at low-level nodes. That is, serial circuits tend to have information bottlenecks, and taking out those bottleneck wreaks havoc with the whole system. Indeed, when the primary visual or somatosensory cortices are lesioned in macaques, the remaining cortices are deprived of the sensory inputs they need to function properly; as a result, the lesioned animals are severely impaired (Rodman and Moore, 1997; Preuss and Kaas, 1999; Preuss, in press). In most nonprimates, on the other hand, focal neocortical lesions tend to have less severe effects (Lashley, 1929). Given that even mild closed head injury (e.g., from concussions) can cause widespread brain damage (Heitger et al., 2004) and that early humans had a penchant for bashing heads, it is likely that those size-related increases in brain vulnerability amounted to nontrivial costs.

In addition, large-brained primates have addressed the problem of decreased connection density by making their brains less symmetrical. As we discussed in Chapter 7, brains tend generally to fractionate into functionally and anatomically distinct modules as they increase in size (even if they are "small worlds"). The most obvious manifestation of that increased fractionation is that the two cerebral hemispheres tend to be less densely interconnected and functionally more independent in larger brains. Within primates, the decrease in interhemispheric connectivity is evident from the fact that the corpus callosum becomes proportionately smaller as neocortex size increases (Figure 9.18) (Rilling and Insel, 1999a; see also Olivares et al., 2000). This decrease in connection density should make it more difficult for the two hemispheres to cooperate with one another and, to the extent that they cannot cooperate on common tasks, they might as well specialize for different tasks (see Ringo et al., 1994). Indeed, human brains are far more asymmetric, both structurally and functionally, than any other brains (Gannon et al., 1998; Buxho-

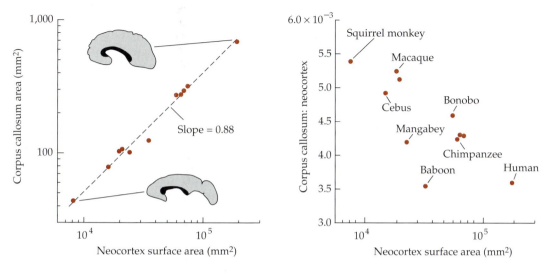

Figure 9.18 Corpus Callosum Scaling in Primates The corpus callosum, which interconnects the two cerebral hemispheres, gets proportionally smaller as the neocortex increases in size. This is evident from the fact that a cross-sectional area of the corpus callosum scales against neocortical surface area with a slope of less than one (left graph). It is also apparent when we plot the corpus callosum:neocortex ratio against neocortical surface area (right graph). The insets depict human (top) and squirrel monkey (bottom) neocortices from a medial perspective, with the corpus callosum shown in black. (After Rilling and Insel, 1999a; with raw data kindly provided by J. Riling.)

eveden and Casanova, 2000; Gilissen, 2001; Rogers and Andrews, 2002). Moreover, one-sided brain lesions tend to have more deleterious effects in humans than in other animals. For instance, lesions of the left inferior temporal lobe in humans cause serious deficits in face recognition, but in macaques bilateral lesions are required to yield comparable deficits (Heywood and Cowey, 1992). Similarly, right-sided lesions of the inferior parietal and superior temporal cortices in humans cause patients to neglect the left side of their external world, but monkeys seem to present similar symptoms only after bilateral lesions (see Karnath, 2001). This increase in brain asymmetry was probably beneficial insofar as it allowed for a greater degree of functional specialization, but it also increased our vulnerability to focal brain damage. Simply put, in functionally asymmetric adult brains, if one side is damaged, the other one cannot pick up the slack. Combining this insight with the above observations on increased circuit serialization in large primate brains, we see that increasing brain size was not "all good", but entailed some costs.

Overall, we can infer that human brains have been significantly reorganized, relative to chimpanzee and other primate brains. Most of those organizational changes were beneficial. They gave us more dexterity in various domains and more flexibility in terms of what to do in any given situation; they made us generally more intelligent (see Chapter 4) and more capable of exercising choice (free will). They also helped us learn to use symbolic language, which then opened up new possibilities for thought and cultural advance. Since most of these changes were at least indirectly linked to the enormous increase in hominin brain size, we can say that increasing absolute brain size has, in general, been good for us. At the same time, however, increasing brain size entailed various costs, including increased dietary needs, increased physical constraints on giving birth, and an enhanced vulnerability to brain lesions. Considering these costs and benefits, our modern human brains may well have reached a relatively stable equilibrium, where further increases in size are not "selected for" because they are offset by size-related costs. In concrete terms, this means that even if we ate a more nutritious diet (not likely here in the United States) and delivered all our babies by cesarean section (likewise improbable), the network scaling costs would tend to limit our absolute brain size. Of course, some human brains work well despite being exceptionally large (the German poet Schiller, for example, had a brain that weighed about 2 kg). Beyond that size, however, human brains are likely to become vulnerable and error-prone (Hofman, 2001). If this is true, and I admit the notion is adventurous, then large cetacean brains, weighing in at more than twice that size, must have evolved some other, novel ways to cope with network scaling costs. Those constraint-busting tricks remain unknown, but cetacean brains do harbor many mysteries (see Chapter 10), including the ability to "sleep" with one cerebral hemisphere at a time (Ridgway, 1986).

Conclusions

As you have seen, the story of human brain evolution is complex. Compared to the brains of cats or rats, human brains are larger, differ in the proportional size of their various sub-regions, and exhibit numerous unusual features. Conversely, rat and cat brains show some features that do not exist in human brains (though I have not stressed this here). All brains are special in some ways. The real difficulty in deciding what is special about human brains is that we have to isolate the uniquely human features from those that make all primates special in comparisons to rats, cats, or other mammals. This task is arduous because, as we have seen, primate brains differ from nonprimate brains in numerous respects, while the differences between human and non-human primate brains are much less discernible. Although human brains are 3–4 times as large as those of chimpanzees, they appear very similar in gross anatomy when we examine them with simple histological techniques. Thus, Huxley was certainly correct to claim that human brains are perfectly good primate brains and that, therefore, man must "take his place in the same order" with the other apes (Huxley, 1863; see Striedter, 2004).

But what about those aspects of brain structure that are not so evident? Could human brains differ from those of chimpanzees in terms of structural details? My answer here has been that, yes, human brains have been structurally "reorganized" compared to chimpanzees. They have proportionately larger neocortices, more direct projections from the neocortex to the medulla, and enlarged lateral prefrontal cortices, to name just the most striking differences (see also Preuss and Coleman, 2002). Most important, I have argued that most of these differences are predictable, given the enormous increase in human brain size and the size-related principles we discussed in Chapters 5–7. Essentially, my claim is that there might have been no way to make a primate brain as large as ours without also endowing it with most of our brains' unique organizational features.

This "forced move" hypothesis is disconcerting at first glance, because it seems akin to arguing that natural selection acted *only* on brain size, not on structural details. That conclusion would indeed be dubious, since brain function depends more on structural details than size per se. This whole chain of reasoning is flawed, however, because if there really are some causal scaling laws that govern what kind of organizational changes *must* occur when brains increase in size, then it makes no sense to argue that selection acted only on brain size. By way of illustration, consider polar bears and body size. No one doubts that the ability of polar bears to live in freezing temperatures is at least partly due to their large body size (since heat loss is proportional to surface area, which scales against body size with a power of about two-thirds). Yet it would be silly to proclaim that cold tolerance in polar bears is merely an accidental, non-selected by-product of increased body size. Natural selection cannot "see" one without the other. It must weigh all of the fitness-enhancing correlates of increased body size (e.g., cold tolerance and freedom from predators) against its costs (e.g., the need to eat more food). And so it is with human brains: their increase in size had many causal correlates, some of which were beneficial, others deleterious. The balance of those benefits and costs changed over time (which probably explains why hominin brain size increased in fits and starts, rather than monotonically) but natural selection never acted *solely* on brain size. It always acted on a complex web of attributes, many of which happened to be linked, either directly or indirectly, to absolute brain size.

The biggest challenge in comprehending how our brains evolved is to determine how the various changes in brain size and/or structure relate to changes in behavior. As I reviewed above, some of the major evolutionary increases in hominin brain size do correlate with major changes in behavior. Thus, the enormous increase in absolute and relative brain size near the origin of the genus *Homo* was apparently associated with more complicated tool-making, better nutrition, and extensive wandering within and out of Africa. Within *H. sapiens*, absolute brain size probably increased mainly because larger brains allowed for more sophisticated social interactions. That kind of "social intelligence" provides selective advantages both within a social group and between different groups (or species), since working with allies and knowing how to climb a social "ladder" tends to boost one's chances of making babies that prosper (Humphrey, 1976; Byrne & Whiten, 1988).

Turning to evolutionary changes in brain organization, we can infer that the proportional expansion of the neocortex and its increased invasion of the medulla and spinal cord probably helped to make modern humans more dextrous than their ancestors. This improved dexterity applies to hand and arm movements but also extends to movements of the eyes, jaws, tongue, and vocal chords. It may likewise explain why humans can control their breathing better than most other animals. The disproportionate enlargement of the lateral prefrontal cortex (and associates) probably enhanced the ability of humans to suppress stimulus-bound behaviors and to replace them with more arbitrary, voluntary acts. The combined effect of more dexterity and increased flexibility gave modern humans a profound competitive advantage over other species. It also lies at the core of our most uniquely human trait— our capacity for symbolic language. Without the large lateral prefrontal cortices and the massive neocortical projections to the medulla and spinal cord, humans probably could not have learned to speak. Once language did take hold, it opened new vistas and changed some longstanding rules of how brains generally evolve. Specifically, it allowed for immense changes in behavior that were not accompanied by corresponding changes in the brain. Even before human language came upon the scene, however, *H. sapiens* was a very special animal. It had managed to tame fire, manufacture diverse tools, and develop means of carrying those tools, as well as food and drink, over great distances. These and other tricks made humans quite successful in the Darwinian "struggle for existence" and allowed them to spread into diverse corners of the world. In some ways, language was just the icing on the cake of our success.

It is imperative to realize, however, that all this success came at the expense of other species and our own compatriots. Most likely, for instance, the success of early *H. erectus* contributed to the demise of the australopithecines, much as our own success today is at least a factor in the ongoing extinction of chimpanzees and other apes. Thus, our great success and dominance has always been morally ambiguous. As Shakespeare pointed out in this chapter's epigraph, we can be like angels in so many ways, yet are not beyond base treachery. We have plenty of disreputable impulses but, compared to other species, we can inhibit them far more effectively. This gives our minds freedom to contemplate the long-term consequences of our acts and to behave accordingly. All that freedom and perspective naturally comes at a price, for it also means we have become responsible for our deeds. This unprecedented power and responsibility of choice, makes us a terribly conflicted "paragon of animals."

10 *Reflections and Prospect*

When a new system is analyzed in animals, it is customary to recognize simi-larities. Workers sample the most divergent forms and, whenever possible, note the characteristics they share .... Although the discovery of similarity remains most important, sooner or later workers begin to characterize differ-ences on multiple levels and to attempt an explanation of their causes. This appears to be the point at which comparative neuroanatomy finds itself.
                                                                  —Carl Gans, 1969

As I reviewed in Chapter 2, evolutionary neuroscience has a long and fas-cinating history. When it first began, the field harbored obvious tensions between those who emphasized species similarities in brains (e.g., Huxley) and those who focused on the differences (Owen), but since then, evolution-ary neuroscience has been dominated by the search for similarities. Carl Gans, an eminent comparative morphologist, noted this emphasis on species simi-larities in 1969 (see epigraph), at the first major conference on comparative neuroanatomy. Remarkably, he also predicted at the time that this era of seek-ing species similarities in brains was drawing to a close. However, his pre-diction was premature by at least 30 years, for in 1969 new techniques for studying neuronal connections and histochemistry had just begun to reveal a treasure trove of unexpected similarities between species (see Chapter 3). Later, in the 1980s and 90s, comparative molecular studies revealed that most genes and proteins typically found in neurons are highly conserved across species (see Chapter 3). Collectively, these discoveries of broadly conserved neuronal features ensured that evolutionary neuroscience in the twentieth century would remain squarely focused on the search for species similarities. The turning point Gans prophesied may, however, be at hand.

Even as laboratory mice are rapidly replacing monkeys, cats, and even lab-oratory rats (Logan, 2001) as the most commonly used animal subjects in modern neuroscience, work continues on a wide variety of other species,

345

including various birds, frogs, fishes, and, of course, invertebrates. This work still brings novel species similarities to light, but as those similarities are clarified, species *differences* increasingly stand out. For instance, recent studies on gerbils have suggested that several aspects of the old "standard model" of vertebrate sound localization, which was based largely on extensive studies in owls (Konishi et al., 1988; Carr and Konishi, 1990), may be quite owl-specific (McAlpine and Grothe, 2003). Further studies are required to determine the full extent of variation in vertebrate sound localization mechanisms, but already the available data have shown that we cannot simply assume all brains to be alike (see also Bolker and Raff, 1997; Striedter, 2002). Although this insight is quite old and widely accepted, it is occasionally drowned out by calls for "species standardization" (Logan, 2002), which imply that cross-species extrapolations are safe and straightforward. In the last few years, however, evolutionary neuroscientists have once again begun to stress species differences in brain structure and function (Preuss, 1995b; Striedter, 2002; McAlpine and Grothe, 2003).

This resurgent interest in species differences is wonderful but also raises an important dilemma: How are we to explain the species differences we find? Merely cataloging them hardly seems worthwhile, but explaining species differences is considerably harder than explaining species similarities, which are easily attributed to common ancestry or convergence. We could say that species differences are due to "descent with modification," to use Darwin's phrase, but such a statement would not constitute an "explanation of their causes" (Gans, 1969). What we really need is a causal explanation of how and why a given modification occurred (Mayr, 1961). In this book I have reviewed numerous attempts to provide such causal explanations but, at this point, let us step back briefly and ask a general epistemological question: What kind of strategies can evolutionary neuroscientists use to explain species differences?

Explanatory Strategies in Evolutionary Neuroscience

One frequently mentioned strategy for explaining species differences is to ask what a given non-conserved feature might have been selected for. This "adaptationist programme" (Gould and Lewontin, 1979) is arduous for neuroscientists because our understanding of how brain anatomy and physiology relate to animal behavior is still so incomplete that, for most neuronal features, it is difficult to fathom what their adaptive function might have been. Furthermore, as I discussed in Chapter 1, it is extremely difficult to test experimentally hypotheses of adaptation for neuronal traits. A peacock's tail, for instance, can easily be cut or augmented to see how the peahens react (and how the peacock's fitness is altered), but how would you manipulate a peacock's brain? No simple matter, certainly, for making brain lesions is tedious and imprecise. Nor would it be trivial to find correlations between naturally occurring variation in brain structure and variation in behavior. Because of these limitations, hypotheses of neural adaptation are much easier to talk about than demonstrate.

One way to circumvent these limitations is to ask whether the neuronal feature in question has evolved repeatedly (i.e., convergently) in several different lineages and is consistently associated with a specific behavior. If it is, then the feature is likely to have evolved as an adaptation for that behavior. For example, the observation that rod-dominated retinas have evolved repeatedly in species that are active at night (or during twilight) suggests that rod-dominated retinas probably evolved as an adaptation for night vision (see Chapter 8). Of course, correlations cannot prove a causal link, for the correlation might always be due to some third variable, but when such correlations are combined with direct functional data, they become strong evidence that adaptation has occurred. Thus, the physiological finding that rods are more light-sensitive than cones strongly supports the hypothesis that rod-dominated retinas evolved *because* they provide superior light sensitivity for animals that are active at night. Thus, the "comparative method" of testing whether neuronal features are consistently associated with specific behaviors (Harvey and Pagel, 1991) can, in conjunction with some functional data, help us demonstrate what neuronal features might have been selected for.

A second, less frequently encountered, strategy for explaining species differences is to ask what mechanical factors or principles "forced" them to evolve (Gould and Lewontin, 1979). We encountered this idea in our discussion of the childbirth constraint in human evolution (see Figure 9.10), but the best example of a mechanical constraint in brain evolution is one we have not yet discussed, namely the evolution of highly folded neocortices. As you can see in Figure 10.1, the neocortices of very small primates are fairly smooth, but above an absolute brain size of 5–10 g, the neocortex becomes increasingly folded (von Bonin, 1941; Zilles et al., 1989). The best explanation for that variation in the degree of neocortical folding is, I think, the following: As absolute brain size increases, the neocortex remains relatively constant in thickness (Figure 10.2), whereas the telencephalic base thickens considerably (Figure 10.3). This causes the neocortex to expand in surface area more quickly than the base to which it is attached. That, in turn, forces the expanding neocortex either to balloon outward or to fold inward (see Figure 10.3). Since ballooning would create individuals with large, unwieldy heads, folding is the more viable option. In other words, the neocortex tends to fold with increasing brain size because the spatially most efficient means of expanding a sheet with relatively immobile edges is to throw it into folds.

This mechanical constraint hypothesis is supported by the observation that highly folded neocortices evolved not just in large-brained primates but also in large-brained representatives of many different lineages. Among monotremes, for instance, the large-brained echidna has a more highly folded

(A) Primate brains arranged by absolute brain size

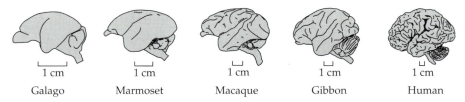

(B) Larger neocortices are more folded

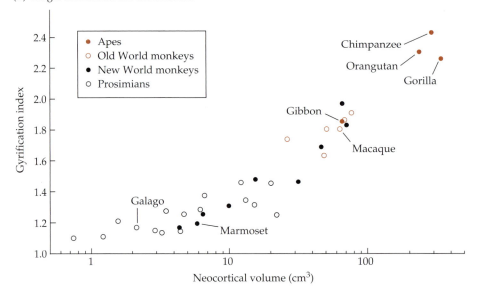

Figure 10.1 Neocortical Folding Correlates with Neocortex Size in Primates
Within primates, the degree of neocortical folding (or gyrification) correlates with neo-cortex size, which has increased in the lineage leading to *Homo sapiens*. (A) Although humans did not evolve from gibbons, nor macaques from marmosets, a cladistic analy-sis reveals that absolute brain and neocortex size probably did increase in the phyloge-netic branch that extends from early primates to modern humans (see Figure 9.1). Arranging extant primate brains by size (left to right), we see that the larger brains have more complexly folded neocortices. (B) Plotting the neocortical gyrification index against neocortical volume, we find that neocortical gyrification increases significantly once primate neocortices become larger than about 10 cm³. The data also show that, while neocortex size tends to increase from prosimians to New World monkeys, to Old World monkeys, and to apes, some of today's prosimians have larger neocortices than some New World monkeys, and some monkeys have neocortices that are larger than those of the smallest living apes (gibbons). These data suggest that absolute brain size probably increased not just within the lineage leading to humans, but independently within several primate lineages. However, since the degree of neocortical folding corre-lates more strongly with absolute neocortex size than with taxonomic affiliation, we may conclude that neocortical folding is most likely a consequence of increasing neocor-tex size. (A after images from the Comparative Mammalian Brain Collections website [http://brainmuseum.org] from the University of Wisconsin–Madison Brain Collection; data in B from Zilles et al., 1989.)

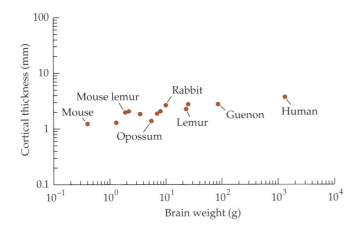

Figure 10.2 Neocortical Thickness Is Relatively Constant Across mammals, absolute brain size varies by more than three orders of magnitude, but neocortical thickness (measured here for motor cortex) varies only about 3-fold. In fact, as long as we compare homologous neocortical areas and allow for some exceptional species (see Figure 10.6), the number of neurons within a radial neocortical column is fairly constant across mammals (Rockel et al., 1980). (Data from Brodmann, 1909.)

neocortex than the platypus (see Figure 8.3). Even among rodents, which generally have smooth neocortices, the largest species (the capybara) has a neocortex with many well-developed folds (Welker, 1990). The most parsimonious interpretation of these data is that highly folded neocortices evolved repeatedly among mammals, whenever absolute brain weight increased above 5–10 g (that threshold varying slightly between taxonomic groups). This robust correlation between neocortical folding and absolute brain size suggests, in and of itself, that neocortical folding might be causally related to brain size (just as the correlation between rod-dominated retinas and night vision was suggestive of a causal link). Combined with the theoretical considerations I just discussed (and depicted in Figure 10.3), those correlative data strengthen the hypothesis that neocortical folding evolved *because* expanding sheets with (relatively) immobile edges must fold to conserve space.

It is important to note, however, that this mechanical constraint hypothesis is at best a partial explanation of neocortical folding. It does not, for instance, explain why neocortical thickness is so remarkably constant across mammals (see Figure 10.2). In fact, that constancy is poorly understood. We might hypothesize, for instance, that neocortical thickness is limited by the length of the radial glia that are critical for its development (because young neocortical neurons tend to migrate along those glia to their adult positions; see Rakic, 1972), but some radial glia are much longer than the cortex is thick (see Figure 8.8A). A more compelling explanation is that neocortical thickness is limited because, as the number of cellular laminae increases, the radially coursing axons and dendrites within the neocortex force the neocortical cell bodies to spread laterally (Figure 10.4), thereby decreasing overall cell density and forcing all tangential intracortical connections to increase in length. This hypothesis (see also Wright, 1934) requires further elaboration (e.g., calculations of how many laminae can be accommodated before lateral spread-

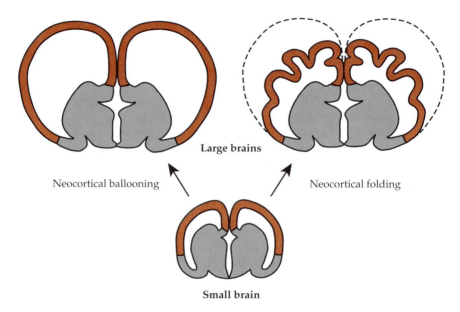

Large brains

Neocortical ballooning

Neocortical folding

Small brain

Figure 10.3 Neocortical Folding Saves Space These schematic cross sections illustrate why large mammalian brains (top) generally exhibit more neocortical gyrification than small brains (bottom). As absolute brain size increases (bottom to top), the neocortex (red) increases in surface area but does not thicken appreciably; in contrast, the telencephalic base (gray) thickens considerably but increases less in surface area. Because of this mismatch, the neocortex must expand either by ballooning (top left) or by folding (top right). The latter option yields brains that are more compact spatially (the dashed line indicates where the neocortical surface would be with the ballooning option). Individuals with compact brains, in turn, are likely to fare better in the Darwinian struggle for existence, mainly because their heads would be less cumbersome. (After Welker, 1990.)

ing is required) and may itself be incomplete (see van Essen, 1997). However, it does illustrate that a complete explanation of neocortical folding would probably entail mechanical constraints at several levels of analysis.

The third major strategy for explaining neural traits is to ask whether they evolved because of some constraining rule of brain development (Alberch, 1982; Maynard Smith et al., 1985; Amundson, 1994). The clearest example of that strategy is Finlay and Darlington's (1995) attempt to explain the observed correlations between absolute brain size and the proportional size of various brain regions in terms of when those brain regions were born during development (see Chapter 5). The constraining rule in this case is that neurogenetic schedules are generally stretched, rather than rearranged, to create larger brains. As long as this is true, late-born regions must become disproportionately large with increasing brain size. Another good example is Deacon's rule

(A) Two laminae

(B) Five laminae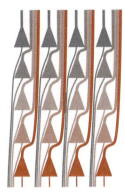

Figure 10.4 The Neocortical Thickness Constraint One likely explanation for the relatively constant thickness of the neocortex is that the radially coursing dendrites and axons (thick and thin lines, respectively) that are so characteristic of neocortical neurons must have room to pass between the neocortical cell bodies (triangles). As the number of neocortical layers increases (A versus B), this space dwindles until the radially coursing elements push the neocortical cell bodies laterally apart (note greater spacing between neocortical columns in B). This lateral spreading of the neocortical columns is probably deleterious because it decreases neuron density and forces the tangential intracortical connections to lengthen. In other words, mammalian neocortex might be relatively constant in thickness because it cannot thicken without reducing the effectiveness of its internal organization.

of large equals well-connected, which states that, as regions become disproportionately large in evolution, they develop more widespread connections with other brain regions (see Chapter 7). As you may recall, I used that principle to explain why large-brained primates, especially humans, develop such unusually extensive projections from the neocortex to the medulla and spinal cord (see Chapters 7 and 9).

Of these three explanatory strategies—adaptational, mechanical, and developmental—this book has emphasized the third, mainly because recent progress in our understanding of how brains develop has been so extensive, at both the cellular and the molecular levels of analysis, that we are now well-positioned to explore the causal linkages between brain development and evolution. It is important to note, however, that this emphasis on developmental rules of brain evolution does not imply that adaptation played no role in how brains have evolved, or that brain evolution is free of mechanical constraints. Instead, all three explanatory strategies are important and complementary. Indeed, as I spelled out in Chapter 1, one of this book's major aims is to show that brain evolution involves not just a single principle, or a single kind of principle, but a plethora of different principles. Integrating all these principles into a single theoretical framework remains an awesome challenge, but some steps toward this synthesis can already be taken.

Steps Toward Synthesis

In their widely cited critique of the "adaptationist paradigm," Gould and Lewontin (1979) presented explanations that entail developmental or mechanical constraints as *alternatives* to adaptational accounts, stating that the latter should be favored only if the former can be excluded. This recommendation resonated strongly with biologists because, after all, scientists should never just pursue a single "pet hypothesis" (Platt, 1964; O'Donohue and Buchanan, 2001). However, the literature on constraints in evolution then expanded to include many other kinds of constraints, including phylogenetic, genetic, and ecological constraints (Antonovics and van Tienderen, 1991). This left evolutionary biologists to grapple with an ever-increasing set of alternative hypotheses, none of which are easy to eliminate. Consequently, debates about the relative merits of various explanatory strategies (or paradigms) have become increasingly heated (e.g., Dennett, 1995; Mark, 1996; Andrews et al., 2002). These debates are largely misguided, however, because the various explanatory strategies I just discussed need not be alternatives.

For instance, if we define developmental constraints as "biases on the production of variant phenotypes" (Maynard Smith et al., 1985), meaning that the mechanisms of development restrict the kinds of offspring organisms can produce (Alberch, 1982; Amundson, 1994), then natural selection may be limited in what it can "choose from," but it can still guide or channel evolution by favoring specific variants. In other words, even constrained variants might well be adaptive. Furthermore, natural selection may itself be viewed as a set of causal principles that bias or constrain which variants come to predominate within a species (or larger taxonomic group). Thus, the course of evolution is really biased, constrained, or influenced by a broad variety of causal interactions—within the organism, between an organism and its environment, and within the environment itself (Oyama, 1985; Griffiths and Gray, 1994). A nice example of this multifactorial approach to evolutionary explanation is John Endler's (1995) work on guppies, which has demonstrated that the intraspecific variation in those small teleosts is due not to a single cause but to a great variety of different causal principles that interact (Figure 10.5). In such a scheme, constraints and adaptation are not alternatives but pieces of a large puzzle—or strands within a "tangled bank" (see Chapter 1).

Of course, the multifactorial approach to evolutionary explanation is not new. Darwin himself was well aware that evolutionary explanations may involve natural selection as well as other principles, most of which he collectively referred to as "laws of variation" (Darwin, 1859). In evolutionary neu-

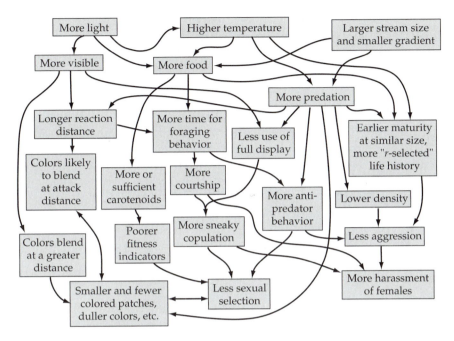

Figure 10.5 Complex Explanations in Biology Most explanations in biology amount to complex, tangled webs. Shown here is John Endler's diagrammatic explanation of the morphological and behavioral variation among guppies (*Poecilia reticulata*) that live in different kinds of habitats. Clearly, this explanation involves many different causal factors and relationships but, as Endler himself admits, a complete explanation would be even more complex. (After Endler, 1995.)

roscience, however, pluralistic explanations have historically been rare. Sven Ebbesson, for instance, presented parcellation as *the major*, if not exclusive, principle of brain evolution (Ebbesson, 1980, 1984). Similarly, Ariëns Kappers (1921) presented as the major law of vertebrate brain evolution his "law of neurobiotaxis," which states that in phylogeny neuronal cell bodies move toward their principal inputs. Those overarching "laws" are no longer widely accepted, but the hope of finding a single principle that can explain the myriad complexity of brains apparently persists (see Nieuwenhuys et al., 1998, quoted in Chapter 1). Even the debate about whether brain evolution is concerted *or* mosaic, adaptive *or* constrained (see Chapter 5) probably reflects, at least to some extent, a lingering desire for a single law to explain "everything." My approach here has been different (see also Deacon, 1990; Bass, 1998; Finlay et al., 2001): Instead of focusing on any single law or principle, I stress that comprehensive explanations of how brains evolved involve a diverse set of interlocking principles.

For example, my account of why the neocortex folds with increasing brain size (see Figure 10.1) emphasizes the role played by a mechanical constraint (that folding an expanding sheet saves space), but it simultaneously allows a major role for natural selection in eliminating the "unfit," namely the individuals with heads so large that they would droop or fail to pass through the birth canal. In addition, it is likely that natural selection favored neocortical folding because folding minimizes the lengths of intracortical connections, especially for axons coursing between areas that, in a folded neocortex, lie on opposing sides of a neocortical gyrus (van Essen, 1997). In large brains, that reduction in connection lengths would have saved significant amounts of metabolic energy and decreased intracortical conduction times considerably. Those savings would most likely have been "visible" to natural selection. Therefore, a full explanation of why neocortical folding evolved requires us to consider both mechanical constraints *and* natural selection.

Similarly, the principle of late equals large does not rule out a role for natural selection. As we discussed in Chapter 5, this principle implies that brain regions evolve concertedly, but even concerted changes in brain region size may be "selected for" if their benefits outweigh the costs. For example, the evolutionary increase in the proportional size of the human neocortex is largely explicable in terms of late equals large (Clancy et al., 2001), but that increase may nonetheless have been adaptive, for it probably was linked (through Deacon's rule) to an expansion of the neocortex's descending projections, which most likely helped humans become more dextrous and more flexible in several domains, including vocalization (see Chapter 9). This hypothesis is difficult to prove but, given that our unusual dexterity and behavioral flexibility are quite useful nowadays, we may surmise that they were probably selected for. In other words, those traits were probably adaptive for humans, *even though* they might have been "automatic" correlates of a change in absolute brain size.

That same meshing of apparently opposing principles also was evident in our discussion of Ebbesson's parcellation theory. At first glance, Ebbessonian parcellation seems diametrically opposed to Deacon's rule of large equals well-connected, since Deacon's rule involves the evolution of the kind of "unusual" connections that parcellation seemingly disallows. However, a closer look reveals that the two principles may coexist, for the weak version of Ebbesson's hypothesis states simply that a brain's *proportional* connectivity (or average connection density; see Figure 4.16) must decrease as neuron number increases—a contention that is not challenged by Deacon's rule (see Chapter 7). Indeed, I pointed out that the kind of connectional "invasion" predicted by Deacon's rule is actually required just to maintain a brain's *absolute* connectivity (the number of connections made by an average neuron) as neuron number increases. Thus, two supposedly opposing principles are reconciled.

Now, a major problem with this view of brain evolution as being governed by a spidery web of interacting principles is that it risks becoming overwhelmingly complex. When explanatory diagrams, such as the one in Figure 10.5, accumulate too many boxes and arrows, our ability to comprehend them fails and we must seek some sort of simplification. Probably the best strategy for achieving this simplification is to determine which of the web's many factors and interactions are central rather than peripheral, key "mediators" rather than mere "modulators" (Sanes and Lichtman, 1999). Therefore, let us ask: What are the central factors and/or principles that govern vertebrate brain evolution? My answer is the following.

If we are trying to explain species *similarities* in brain organization, then the most important principle is that of phylogenetic conservation, which states that closely related species tend to be similar because they have a common ancestor from which shared features were inherited. On the other hand, if we are trying to explain species *differences* in brains, then absolute brain size becomes a key explanatory variable, for it has changed repeatedly as brains evolved and can be linked to myriad aspects of brain structure and function. Both principles have formed the backbone of this book (see Chapter 1), but only phylogenetic conservation is widely appreciated, absolute brain size being barely mentioned in most prior books on brain evolution (e.g., Butler and Hodos, 1996; Nieuwenhuys et al., 1998; but see Allman, 1999). Therefore, let me briefly recap why absolute brain size is such a central variable in vertebrate brain evolution and review how it relates to the more widely appreciated measure of relative brain size.

Absolute and Relative Brain Size

As we discussed in Chapter 4, relative brain size has changed dramatically in many different lineages, with increases outnumbering decreases. These increases in relative brain size were generally accompanied by hefty increases in absolute brain size, since most lineages also underwent significant increases in body size. Therefore, one may well debate which variable was more important in brain evolution: relative or absolute brain size? Historically, most evolutionary neuroscientists have emphasized relative brain size (Jerison, 1973; Stephan et al., 1988; van Dongen, 1998), mainly because it "factors out" variation in absolute body size. I think, however, that this emphasis has been too exclusive, that variations in absolute brain size are far more important than most neuroscientists appreciate. As I have stressed repeatedly throughout this book, changes in absolute brain size are causally linked to a variety of changes in brain organization that are likely to affect an animal's behavior. In this section I briefly review those causal correlates of absolute brain size and then address the limitations of analyses in terms of absolute brain size, concluding that we really need to look at both: absolute *and* relative brain size.

Extant vertebrate brains range in size across 7 orders of magnitude, from about 1 mg to roughly 10 kg (see Figure 4.15). A parsimony analysis of this variation indicates that absolute brain size increased repeatedly in several different lineages (it also decreased occasionally). Those evolutionary increases in absolute brain size were generally associated with increased complexity, as measured by the number of distinct brain regions (see Chapter 6). They were also accompanied by law-like changes in brain region proportions, with late-born structures generally becoming disproportionately large (at least among mammals; see Chapter 5). Those size-related changes in brain region proportions probably caused major changes in neural connectivity, with disproportionately enlarged regions generally becoming "better connected" (see Chapter 7). That, in turn, most likely led to major changes in brain function, with the more widely connected regions, such as the neocortex of large-brained mammals, becoming disproportionately influential (see Chapters 5 and 7). In addition, the evolutionary increases in absolute brain size were generally accompanied by decreases in average connection density, causing larger brains to become structurally and functionally more modular (see Chapter 7). Therefore, the single variable of absolute brain size ties together many different structural and functional attributes of brains. Since those links are causal in nature, they have explanatory force.

It is important to stress, however, that the variable of absolute brain size does not capture or explain *all* of the variation that we see in brains. The evolutionary origin of the mammalian neocortex, for instance, is not explicable in terms of any change in absolute brain size (see Chapter 8). Clearly, some important changes in brain structure were causally independent of changes in absolute brain size. Indeed, I suspect that the origin of most major vertebrate lineages (e.g., mammals) involved some key neuronal innovations (such as the neocortex) that were causally unrelated to changes in absolute brain size but crucial to that lineage's overall "success" (Hunter, 1998). Nonetheless, I think those *key innovations* were relatively rare. Occasionally they "pushed" brain evolution onto novel "tracks," but within those tracks, brains varied mainly in absolute brain size (and its diverse correlates). Using this metaphor of "tracks," we can say, for instance, that human brains are firmly on the primate track but have become so large that they evolved a plethora of size-related specializations (see Chapter 9). This idea is clearly simplified, but it is a useful guiding principle.

Now, if different lineages tend to evolve along divergent tracks, then it makes little sense to compare the brains of *distant* relatives in terms of absolute brain size. For example, it is not very informative to know that the brains of some large cetaceans (whales) weigh roughly 5 times as much as human brains (Figure 10.6), because cetacean brains also differ from all primate brains in a multitude of other attributes (Kruger, 1959; Morgane and

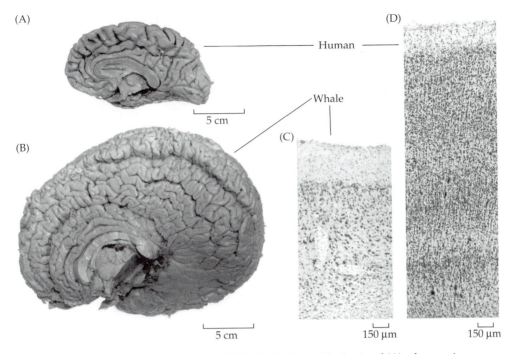

Figure 10.6 Comparison of Human and Whale Brains   The brain of (A) a human is considerably smaller than the brain of (B) a killer whale and exhibits fewer neocortical gyri. On the other hand, the neocortex is considerably thinner and less highly laminated in (C) a pygmy sperm whale (or most other cetaceans) than in (D) humans. This difference in neocortical thickness is but one of many features that distinguish human or other primate brains from those of cetaceans. It also explains why cetaceans have thinner and more numerous neocortical gyri: The thinner a sheet is, the more tightly it can be folded. (A and B from Ridgway, 1986; C and D from Preuss, 2001.)

Jacobs, 1972; Ridgway, 1986; Peichl et al., 2001). Most notably, the neocortex of all cetaceans is much thinner and less highly laminated than the neocortex of primates (see Figure 10.6 C,D). We do not yet know precisely how and why cetaceans evolved this unusually thin neocortex, but we can be fairly certain that this change was not a simple "automatic" consequence of a change in absolute brain size. Therefore, comparing human and whale brains in terms of absolute brain size is like comparing apples and grapefruits; the two differ in size-independent ways! This is not particularly surprising once you consider that primates and cetaceans are extremely distant relatives, with whales being closely related to hippopotami and other artiodactyls (Thewissen et al., 2001) (see Figure 2.11).

Whenever we compare distantly related taxonomic groups, such as humans and whales, *relative* brain size tends to be a more useful variable than absolute brain size. Thus, a relative brain size analysis immediately reveals that humans and toothed whales (porpoises, dolphins, orcas, and sperm whales) both have considerably larger brains than "average mammals" of their body size (see Figure 4.10). More specifically, it shows that both primates and toothed whales must have increased their absolute brain size beyond allometric expectations when they originally evolved from their respective ancestors. This finding is useful because it suggests that natural selection in both of these lineages favored increases in absolute brain size so strongly that their ancestral patterns of brain:body scaling were significantly modified. That insight, in turn, allows us to ask fascinating mechanistic questions about how development was modified to achieve those alterations in adult scaling (see Chapter 4). Moreover, once we have determined that those two lineages experienced selection for increased brain size, we can ask specific functional questions about the behavioral advantages conferred by larger brains. We can hypothesize, for instance, that both humans and toothed whales evolved larger brains to navigate more easily their complex social worlds and/or to forage more strategically (see Chapter 4). As I mentioned earlier, such hypotheses are difficult but not impossible to test. Therefore, in comparisons of distant relatives, relative brain size can be a useful variable.

Relative brain size can also be useful in comparisons between close relatives, when the specimens being compared vary in both brain and body size. This utility was evident at the beginning of this book, when we discussed the English sparrow data of Bumpus (1899). In that famous data set, comparisons of average absolute brain weight showed little difference between survivors and deceased (recall that roughly half the sparrows died in a snowstorm), but comparisons of relative brain size revealed that, at any given body size, the surviving sparrows had significantly larger brains (see Figure 1.1). In other words, the survivors exceeded the deceased in relative brain size but not in average absolute brain size. As discussed in the preceding paragraph, such a difference in relative brain size suggests that natural selection *might have* favored increases in absolute brain size. However, we should not dismiss the alternative hypothesis that natural selection in Bumpus's sparrows favored a reduction in body size rather than an increase in brain size (both would yield an increase in relative brain size!). As I discussed in Chapter 4, this kind of "increased encephalization by phylogenetic dwarfism" occurred only rarely if we consider vertebrate brain evolution overall. However, I suspect that it occurred more frequently in smaller taxonomic groups, especially within species. This idea of Bumpus's sparrows having experienced selection for reduced body size is interesting because it raises the following hypothetical question.

If Bumpus's sparrows became more encephalized by decreasing their body size, would the smaller, more encephalized sparrows have been "smarter" than their larger, less encephalized conspecifics? I do not think so, just as I do not believe that modern humans became more intelligent *because* their body size decreased during the last 100,000 years (see Chapter 9). In fact, I suspect the two groups of sparrows would have behaved nearly identically. Perhaps the smaller sparrows would have used fewer neurons to collect sensory inputs from their smaller body surface and fewer motor neurons to control their smaller limbs, but those differences would have produced no major changes in neuronal wiring or overall brain organization. Therefore, the sparrows' behavior would have remained largely unchanged. In contrast, if the sparrows had increased their absolute brain size while holding body size constant, then the larger brains would have had more neurons available to perform most computations. In addition, those larger brains might well have changed in terms of regional proportions (e.g., evolving proportionately larger telencephalons) and neuronal connections (e.g., more extensive telencephalic projections). Collectively, these correlates of increased absolute brain size would probably have caused measurable changes in behavior, including, perhaps, an increase in social and/or foraging intelligence (see Chapter 4). These considerations are admittedly speculative, but they illustrate my basic point, which is that, although analyses of relative brain size may be useful, differences in *absolute* brain and body size must at some point be factored in to complete our understanding of how brains evolved.

Perhaps the greatest obstacle to the widespread acceptance of absolute brain size as a key evolutionary variable is the widely held belief that only changes in relative brain size can be adaptive. As I discussed in Chapters 1, 5, and 9, this belief is logically flawed. Although changes in relative brain size do suggest that natural selection was at work, changes in brain size that are in line with allometric expectations may likewise be "selected for," because any major change in absolute brain size, whether "expected" or not, is likely to cause changes in an animal's behavior that are visible to natural selection even if it *also* acts on body size. For instance, the expected allometric increases in absolute brain size that accompanied the evolutionary increases in body size among nonhuman simians (see Figure 9.3) probably involved selection for both larger bodies *and* larger brains. Similarly, I suspect that the remarkable intelligence of large parrots (Pepperberg, 1981; Moore, 1992) is due at least in part to natural selection for large brains, even though the absolute brain size of large parrots is fully in line with allometric expectations (see Figure 5.12). Therefore, I conclude that evolutionary changes in absolute brain size are at least potentially adaptive, even if they are not associated with evolutionary changes in relative brain size.

Overall, then, I see no compelling reason to neglect absolute brain size in favor of relative brain size. Indeed, I think that in many comparisons, absolute brain size is a more informative, more interesting variable. Most importantly, comparisons of absolute brain size are valuable because they raise novel questions for research. For example, we can ask whether the enormous variation in absolute brain size *within cetaceans* is causally related to variation in cetacean brain organization and behavior. Preliminary data indicate already that for three cetacean species (harbor porpoise, white-sided dolphin, and sperm whale) the proportional size of the neocortex increases with absolute brain size (Mangold-Wirz, 1966), just as it does in primates and insectivores (see Chapter 4). In addition, it appears that absolute brain size in the toothed whales correlates with social and/or foraging complexity, but further studies are required to test this hypothesis. Finally, I wonder whether baleen whales, with their gigantic bodies and filter feeding mode of foraging, are as "dumb" as their low relative brain sizes would suggest (see Figure 4.9). The discovery of complex learned "songs" and clever fish corralling strategies in humpback whales (Payne and McVay, 1971; Sharpe and Dill, 1997; Rendell and Whitehead, 2001) certainly suggests that baleen whales are quite "intelligent." That conclusion would be consistent with the finding that the brains of baleen whales, while small in terms of relative brain size, are enormous in terms of absolute brain size.

Thus, my take-home message is that while absolute and relative brain size are both useful variables to analyze, comparisons of absolute brain size are far more interesting and meaningful than comparative neurobiologists historically assumed. Particularly in comparisons between closely related species, evolutionary changes in absolute brain size tend to have numerous important consequences for brain structure and, ultimately, animal behavior. Since prior books on vertebrate brain evolution have not emphasized these causal correlates of changing absolute brain size, this book will likely spawn a few debates and, hopefully, some new research.

Conclusions

Looking back over this endeavor and at the field of evolutionary neuroscience as a whole, I see vertebrate brains as wonderfully diverse but varying in ways that make some sense. I do not see a single, overarching "law" of brain evolution but a medley of interlocking causal principles. Several of these principles remain poorly understood, and they may never explain "all that has gone before" (Nieuwenhuys et al., 1998, p. 2189) but, with further work, they should allow us to explain much of the observed diversity in brains.

Such an understanding of how brains evolve can be invaluable. As most neuroscientists appreciate, the principle of phylogenetic conservation allows us to determine which neuronal features are likely to be conserved across which lineages, thereby facilitating our attempts to generalize from model species to humans. Less widely recognized is that evolutionary principles of *variation* can likewise yield useful predictions. For instance, according to Deacon's rule of large equals well-connected, the disproportionately large lateral prefrontal cortex of humans should have unusually extensive connections. To date, there is little data relevant to that hypothesis, but it could be used to guide further research. Most important, evolutionary neuroscience is required to interconnect the myriad bits of information we *already* have about the brains of diverse vertebrates. As the eminent geneticist Theodosius Dobzhanski once wrote: "Seen in the light of evolution, biology is, perhaps, intellectually the most satisfying and inspiring science. Without that light it becomes a pile of sundry facts, some of them interesting or curious but making no meaningful picture as a whole" (Dobzhansky, 1973, p. 129).

This unifying power of evolutionary theory is widely acknowledged, but its full impact has not yet been felt in neuroscience. Many neuroscientists do appreciate that evolutionary history unites the various species and that, therefore, it is useful to reconstruct that history. Less widely recognized, however, is that those facts of history are tied together by a web of diverse causal principles, of which natural selection is but one. Hopefully, this book will help to call attention to that point and, thereby, stimulate new work to comprehend those principles. With that aim in mind, let me end this book as it began, with a quote from Charles Darwin: "The Grand Question, which every naturalist ought to have before him, when dissecting a whale, or classifying a mite, a fungus, or an infusorian, is 'What are the laws of life'?" (Notebook B; quoted in Barret, 1987, p. 228).

Bibliography

Abitbol, M. M. 1987. Obstetrics and posture in pelvic anatomy. J Human Evol 16: 243–255.

Aboitiz, F. 1996. Does bigger mean better? Evolutionary determinants of brain size and structure. Brain Behav Evol 47: 225–245.

Aboitiz, F. 2003. Long distance communication in the human brain: timing constraints for inter-hemispheric synchrony and the origin of brain lateralization. Biol Res 36: 89–99.

Aboitiz, F., and García, R. 1997. The evolutionary origin of the language areas in the human brain. a neuroanatomical perspective. Brain Res Rev 25: 381–396.Aboitiz, F., Morales, D., and Montiel, J. 2003. The evolutionary origin of the mammalian isocortex: towards an integrated developmental and functional approach. Behav Brain Sci 26: 535–586.

Ackerly, D. D. 2000. Taxon sampling, correlated evolution, and independent contrasts. Evolution 54: 1480–1492.

Adrio, F., Anadón, R., and Rodríguez-Moldez, I. 2002. Distribution of tyrosine hydroxylase (TH) and dopamine beta-hydroxylase (DBH) immunoreactivity in the central nervous system of two chondrostean fishes (Acipenser baeri and Huso huso). J Comp Neurol 448: 280–297.

Ahnelt, P. K., and Kolb, H. 2000. The mammalian photoreceptor mosaic: adaptive design. Prog Ret Eye Res 19: 711–777.

Aiello, L. C., and Wheeler, P. 1995. The expensive-tissue hypothesis. Curr Anthrop 36: 199–221.

Airey, D. C., Castillo-Juarez H., Casella G., Pollak E. J., and deVoogd T. J. 2000. Variation in the volume of zebra finch song control nuclei is heritable: developmental and evolutionary implications. Proc R Soc Bio Sci B 267: 2099–2104.

Alberch, P. 1982. Developmental constraints in evolutionary processes. In: Bonner J. T., editor. Evolution and development. New York: Springer-Verlag, pp. 313–332.

Alberch, P. 1984. A return to the Bauplan. Behav Brain Sci 7: 332.

Alexander, R. D. 1990. How did humans evolve? Ann Arbor: University of Michigan Special Publications.

Allin, E. F., and Hopson, J. A. 1992. Evolution of the auditory system in Synapsida ("Mammal-like reptiles" and primitive mammals) as seen in the fossil record. In: Webster D. B., Fay R. R., Popper A. N., editors. The evolutionary biology of hearing. New York: Springer-Verlag, pp. 587–614.

Allman, J. 1990. The origin of the neocortex. Seminar in the Neurosciences 2: 257–262.

Allman, J., McLaughlin, T., and Hakeem, A. 1993. Brain weight and life-span in primate species. Proc Natl Acad Sci (USA) 90: 118–122.

Allman, J. M. 1999. Evolving brains. New York: Scientific American Library.

Allman, J. M., and Kaas, J. H. 1971. A representation of the visual field in the caudal third of the middle temporal gyrus of the owl monkey (Aotus trivirgatus). Brain Res 31: 85–105.

Allman, J. M., Hakeem A., Erwin J. M., Nimchinsky E., and Hof P. 2001. The anterior cingulate cortex: the evolution of an interface between emotion and cognition. Ann NY Acad Sci 935: 107–117.

Alonso, J.-M., Usrey, W. M., and Reid, R. C. 2001. Rules of connectivity between geniculate cells and simple cells in cat primary visual cortex. J Neurosci 21: 4002–4015.

Alroy, J. 1998. Cope's rule and the dynamics of body mass evolution in North American fossil mammals. Science 280: 731–734.

Alstermark, B., Isa, T., Ohki, Y., and Saito, Y. 1999. Disynaptic pyramidal excitation in forelimb motoneurons mediated via C3-C4 propriospinal neurons in the Macaca fuscata. J Neurophysiol 82: 3580–3585.

Alvarez-Bolado, G., Rosenfeld, M. G., and Swanson, L. W. 1995. Model of forebrain regionalization based on spatiotemporal patterns of POU-III homeobox gene expression, birthdates, and morphological features. J Comp Neurol 355: 237–295.

Amundson, R. 1994. Two concepts of constraint: adaptationism and the challenge from developmental biology. Philosophy of Science 61: 556–578.

Anderson, P. A. V., and Greenberg, R. M. 2001. Phylogeny of ion channels: clues to structure and function. Comp Biochem and Physiol B 129: 17–28.

Andrews, P. W., Gangestad, S. W., and Matthews, D. 2002. Adaptationism — how to carry out an exaptationist program. Behav Brain Sci 25: 489–553.

Antonovics, J., and van Tienderen, P. H. 1991. Ontoecogenophyloconstraints? The chaos of constraint terminology. Trends Ecol Evol 6: 166–168.

Appel, T. A. 1987. The Cuvier-Geoffroy debate. New York: Oxford University Press.

Arbib, M. A. In press. From monkey-like action recognition to human language: an evolutionary framework for neurolinguistics. Behav Brain Sci.

Arends, J. J. A., and Zeigler, H. P. 1986. Anatomical identification of an auditory pathway from a nucleus of the lateral lemniscal system to the frontal telencephalon nucleus basalis of the pigeon. Brain Res 398: 375–381.

Arensburg, B. 1994. Middle Paleolithic speech capabilities: a response to Dr. Lieberman. Am J Phys Anthrop 94: 279–280.

Ariëns Kappers, C. U. 1910. The migrations of the motor cells of the bulbar trigeminus, abducens and facialis in the series of vertebrates, and the differences in the course of their root-fibers. Verh Konikl Akad Wetensch Amsterdam 16(4):3–195.

Ariëns Kappers, C. U. 1921. On structural laws in the nervous system: the principles of neurobiotaxis. Brain 44: 125–149.

Ariëns Kappers, C. U., Huber, C. G., and Crosby, E. C. 1936. Comparative anatomy of the nervous system of vertebrates, including man. New York: Macmillan.

Armstrong, E. 1982. Mosaic evolution in the primate brain: differences and similarities in the hominid thalamus. In: Armstrong, E., Falk, D., editors. Primate brain evolution. New York: Plenum Press, pp. 131–161.

Armstrong, E. 1985. Relative brain size in monkeys and prosimians. Am J Phys Anthrop 66: 263–273.

Aron, A. R., Robbins, T. W., Poldrack, R. A. 2004. Inhibition and the right inferior frontal cortex. Trends Cog Sci 8: 170–177.

Asanuma, C., and Stanfield, B. B. 1990. Induction of somatic sensory inputs to the lateral geniculate nucleus in congenitally blind mice and in congenitally normal mice. Neuroscience 39: 533–545.

Atoji, Y., Wild, J. M., Yamamoto, Y., and Suzuki, Y. 2002. Intratelencephalic connections of the hippocampus in pigeons (*Columba livia*). J Comp Neurol 447: 177–199.

Aubín, D., and Dalmedico, A. D. 2002. Writing the history of dynamical systems and chaos: *Longue durée* and revolution, disciplines and cultures. Historia Mathematica 29: 273–339.

Bachy, I., Berthon, J., and Rétaux, S. 2002. Defining pallial and subpallial divisions in the developing *Xenopus* forebrain. Mech Dev 117: 163–172.

Balaban, C. D., and Ulinski, P. S. 1981. Organization of thalamic afferents to anterior dorsal ventricular ridge in turtles. I. Projections of thalamic nuclei. J Comp Neurol 200: 95–129.

Balaban, E., Teillet, M.-A., and Le Douarin, N. 1988. Application of the quail-chick chimera system to the study of brain development and behavior. Science 241: 1339–1342.

Barkow, J. H., Cosmides, L., and Tooby, J., editors. 1992. The adapted mind. New York: Oxford University Press.

Barlow, H. B. 1986. Why have multiple cortical areas? Vision Res 26: 81–90.

Baron, G., Stephan, H., and Frahm, H. D. 1996. Comparative neurobiology in chiroptera. Basel: Birkhäuser.

Barret, P. H., editor. 1987. Charles Darwin's notebooks, 1836–1844. Ithaca: Cornell University Press.

Barton, R. A. 2001. The coordinated structure of mosaic brain evolution. Behav Brain Sci 24: 281–282.

Barton, R. A., and Dunbar, R. I. M. 1997. Evolution of the social brain. In: Whiten, A., Byrne, R. W., editors. Machiavellian Intelligence II. Cambridge: Cambridge University Press, pp. 240–263.

Barton, R. A., and Harvey, P. H. 2000. Mosaic evolution of brain structure in mammals. Nature 405: 1055–1058.

Barton, R. A., Aggleton, J. P., and Grenyer, R. 2003. Evolutionary coherence of the mammalian amygdala. Proc R Soc Lond B 270: 539–543.

Basil, J. A., Kamil, A. C., Balda, R. P., and Fite, K. V. 1996. Differences in hippocampal volume among food storing corvids. Brain Behav Evol 47: 156–164.

Bass, A. H. 1998. Behavioral and evolutionary neurobiology: a pluralistic approach. Am Zool 38: 97–107.

Bast, T., Diekamp, B., Thiel, C., Schwarting, R. K. W., and Güntürkün, O. 2002. Functional aspects of dopamine metabolism in the putative prefrontal cortex analogue and striatum of pigeons (*Columba livia*). J Comp Neurol 446: 58–67.

Bastian, J. 1986. Electrolocation. In: Bullock T. H., Heiligenberg W., editors. Electroreception. New York: Wiley, pp. 577–612.

Bastian, J. 1998. Plasticity in an electrosensory system. III. Contrasting properties of spatially segregated dendritic inputs. J Neurophysiol 79: 1839–1857.

Bateson, W. 1892. On numerical variation in teeth, with a discussion of the conception of homology. Proc Zool Soc Lond (no vol. #): 102–115.

Bauchot, R., Platel, R., and Ridet, J.-M. 1976. Brain-body weight relationships in Selachii. Copeia 2: 305–310.

Bauchot, R., Thireau, M., and Diagne, M. 1983. Relations pondérales encéphalo-somatiques interspecifiques chez les amphibiens Anoures. Bull Mus Nat Hist Paris 4: 383–398.

Bauchot, R., Randall, J. E., Ridet, J.-M., and Bauchot, M.-L. 1989. Encephalization in tropical teleost fishes and comparison with their mode of life. J Hirnforsch 30: 645–669.

Bauchot, R., Platel, R., Ridet, J.-M., Diagne, M., and Delfini, C. 1995. Encéphalisation et adaptations écobiologiques chez les chondrichtyens. Cybium 19: 153–165.

Bazzaz, A. A. 1993. Cluster analysis of dorsal root ganglia (DRG) neurons of the Mongolian gerbil *Meriones unguiculatus*. Dirasat Series B Pure and Applied Sciences 19(4): 201–220.

Beatty, J. 1995. The evolutionary contingency thesis. In: Wolters, G., Lennox, J. G., editors. Concepts, theories, and rationality in the biological sciences. Pittsburgh: University of Pittsburgh Press, pp. 45–81.

Beatty, J. 1997. Why do biologists argue like they do? Philosophy of Science 64: S432–S443.

Beck, P. D., Popsichal, M. W., and Kaas, J. H. 1996. Topography, architecture, and connections of somatosensory cortex in opossums: evidence for five somatosensory areas. J Comp Neurol 366: 109–133.

Bekkers, J. M., and Stevens, C. F. 1970. Two different ways evolution makes neurons larger. Prog Brain Res 83: 37–45.

Bell, C. C., and Szabo, T. 1986. Electroreception in mormyrid fish: central anatomy. In: Bullock, T. H., Heiligenberg, W., editors. Electroreception. New York: John Wiley & Sons, pp. 375–421.

Bell, E., Ensini, E., Gulisano, M., and Lumsden, A. 2001. Dynamic domains of gene expression in the early avian forebrain. Dev Biol 236: 76–88.

Bennett, A. F. 1991. The evolution of activity capacity. J Exp Biol 160: 1–23.

Bennet, M. R., Gibson, W. G., and Lemon, G. 2002. Neuronal cell death, nerve growth factor and neurotrophic models: 50 years on. Autonomic Neuroscience: Basic and Clinical 95: 1–23.

Bennet, P. M., and Harvey, P. H. 1985a. Brain size, development and metabolism in birds and mammals. J Zool Lond (A) 207: 491–509.

Bennet, P. M., and Harvey, P. H. 1985b. Relative brain size and ecology in birds. J Zool Lond (A) 207: 151–169.

Bergquist, H. 1952. Transversal bands and migration areas in *Lepidochelys olivacea*. Kgl Fysiogr Sällsk Lund Handl N F 63: 13.

Bergquist, H., and Källén, B. 1953a. On the development of neuromeres to migration areas in the vertebrate cerebral tube. Acta Anat (Basel) 18: 65–73.

Bergquist, H., and Källén, B. 1953b. Studies on the topography of the migration areas in the vertebrate brain. Acta Anat (Basel) 17: 353–359.

Bergquist, H., and Källén, B. 1954. Notes on the early histogenesis and morphogenesis of the central nervous system in vertebrates. J Comp Neurol 100: 627–659.

Berman, N. L., and Maler, L. 1999. Neural architecture of the electrosensory lateral line lobe: adaptations for coincidence detection, a sensory searchlight and frequency-dependent adaptive filtering. J Exp Biol 202: 1243–1253.

Beutel, R. G., and Haas, A. 1998. Larval head morphology of *Hydroscapha natans* (Coleoptera, Myxophaga) with reference to miniaturization and the systematic position of Hydroscaphidae. Zoomorphol 118: 103–116.

Bhide, P. G., and Frost, D. O. 1992. Axon substitution in the reorganization of developing neural connections. Proc Natl Acad Sci (USA) 89: 11847–11851.

Binkofski, F., Buccino, G., Posse, S., Seitz, R. J., Rizzolatti, G., and Freund, H.-J. 1999. A fronto-parietal circuit for object manipulation in man: evidence from an fMRI-study. Eur J Neurosci 11: 3276–3286.

Bishop, G. H. 1959. The relation between nerve fiber size and sensory modality: phylogenetic implications of the afferent innervation of cortex. J Nerv Mental Dis 128: 89–114.

Bitterman, M. E. 1965. The evolution of intelligence. Sci Am 212: 92–100.

Blob, R. W. 2001. Evolution of hindlimb posture in nonmammalian therapsids: biomechanical tests of paleontological hypotheses. Paleobiology 27: 14–38.

Blumenschine, R. J. 1987. Characteristics of an early hominid scavenging niche. Curr Anthrop 28: 383–407.

Boaz, N. T., and Ciochon, R. L. 2004. Headstrong hominids. Natural History: 28–34.

Bock, W. 1999. Functional and evolutionary explanations in morphology. Neth J Zoology 49: 45–65.

Bogin, B. 1997. Evolutionary hypotheses for human childhood. Yearbook Phys Anthorpol 40: 63–89.

Bohm, D. 1957. Causality and chance in modern physics. Philadelphia: University of Pennsylvania Press.

Bok, S. T. 1959. Histonomy of the cerebral cortex. Amsterdam: Elsevier.

Bolker, J. A., and Raff, R. A. 1997. Beyond worms, flies, and mice: it's time to widen the scope of developmental biology. J NIH Res 9: 35–39.

Bond, J., Roberts, E., Mochida, G. H., Hampshire, D. J., Scott, S., Askham, J. M., Springell, K., Mahadevan, M., Crow, Y. J., Markham, A. F., Walsh, C. A., and Woods, C. G. 2002. ASPM is a major determinant of cerebral cortical size. Nature Genetics 32: 316–320.

Bone, Q. 1959. The central nervous system in amphioxus. J Comp Neurol 115: 27–64.

Bowler, J. P. 1988. The non-Darwinian revolution: reinterpreting a historical myth. Baltimore: Johns Hopkins University Press.

Braford, M. R. J. 1986. *De gustibus non est disputandum*: a spiral center for taste in the brain of the teleost fish, *Heterotis niloticus*. Science 232: 489–491.

Braford, M. R. J., and Northcutt, R. G. 1983. Organization of the diencephalon and pretectum of the ray-finned fishes. In: Davis, R. E., Northcutt, R. G., editors. Fish neurobiology, Vol. 2: Higher brain areas and function. Ann Arbor: University of Michigan Press, pp. 117–163.

Brandstätter, R., and Kotrschal, K. 1990. Brain growth patterns in four European cyprinid fish species (Cyprinidae, Teleostei): roach (*Rutilus rutilus*), bream (*Abramis braama*), common carp (*Cyprinus carpia*) and sabre carp (*Pelecus cultratus*). Brain Behav Evol 35: 195–211.

Brandt, S. A., Brocke, J., Röricht, S., Ploner, C. J., Villringer, A., and Meyer, B.-U. 2001. *In vivo* assessment of human visual system connectivity with transcranial electrical stimulation during functional magnetic resonance imaging. NeuroImage 14: 366–375.

Breathnach, A. S. 1960. The cetacean nervous system. Biol Rev 35: 187–230.

Brochu, C. A. 2000. A digitally-rendered endocast of *Tyrannosaurus rex*. J Vert Paleontol 20: 1–6.

Brodmann, K. 1909. Vergleichende Lokalisationslehre der Grosshirnrinde in ihren Prinzipien dargestellt auf Grund des Zellenbaues. Leipzig: Barth.

Bronchti, G., Heil, P., Sadka, R., Hess, A., Scheich, H., and Wollberg, Z. 2002. Auditory activation of 'visual' cortical areas in the blind mole rat (*Spalax ehrenbergi*). Eur J Neurosci 16: 311–329.

Brox, A., Puelles, L., Ferreiro, B., and Medina, L. 2003. Expression of the genes *GAD67* and *Distal-less-4* in the forebrain of *Xenopus laevis* confirms a common pattern in tetrapods. J Comp Neurol 461: 370–393.

Brox, A., Puelles, L., Ferreiro, B., and Medina, L. 2004. Expression of the genes *Emx1, Tbr1*, and *Eomes* (*Tbr2*) in the telencephalon of *Xenopus laevis* confirms the existence of a ventral pallial division in all tetrapods. J Comp Neurol 474: 562–577.

Bruce, L. L., and Butler, A. B. 1984. Telencephalic connections in lizards. I. Projections to cortex. J Comp Neurol 229: 585–601.

Bruce, L. L., and Neary, T. J. 1995. The limbic systems of tetrapods: a comparative analysis of cortical and amygdalar projections. Brain Behav Evol 46: 224–234.

Brunet, M., Guy, F., Pilbeam, D., Mackaye, H. T., Likius, A., Ahountas, D., Beauvilain, A., Blondel, C., Bocherens, H., Boisserie, J. R., De Bonis, L., Coppens, Y., Dejax, J., Denys, C., Duringer, P., Eisenmann, V. R., Fanone, G., Fronty, P., Geraads, D., Lehmann, T., Lihoreau, F., Louchart, A., Mahamat, A., Merceron, G., Mouchelin, G., Otero, O., Campomanes, P. P., De Leon, M. P., Rage, J. C., Sapanet, M., Schuster, M., Sudre, J., Tassy, P., Valentin, X., Vignaud, P., Viriot, L., Zazzo, A., and Zollikofer, C. 2002. A new hominid from the upper Miocene of Chad, central Africa. Nature 418: 145–151.

Brunso-Bechtold, J. K., and Casagrande, V. A. 1981. Effect of bilateral enucleation on the development of layers in the dorsal lateral geniculate nucleus. Neuroscience 2: 589–597.

Buchholtz, E. A., and Seyfarth, E.-A. 1999. The gospel of the fossil brain: Tilly Edinger and the science of paleoneurology. Brain Res Bull 48: 351–361.

Buffon, G. L. L. 1753. Histoire naturelle, générale et particuliére, Vol. 4.

Buhl, E. H., and Oelschlager, H. A. 1986. Ontogenetic development of the nervus terminalis in toothed whales - evidence for its non-olfactory nature. Anat Embryol 173: 285–294.

Bulfone, A., Puelles, L., Porteus, M. H., Frohman, M. A., Martin, G. R., and Rubenstein, J. L. R. 1993. Spatially restricted expression of *Dlx-1*, *Dlx-2* (*Tes-1*), *Gbx-2*, and *Wnt-3* in the embryonic day 12.5 mouse forebrain defines potential transverse and longitudinal boundaries. J Neurosci 13: 3155–3172.

Bullock, T. H. 2002. Grades in neural complexity: how large is the span. Integ Comp Biol 42: 757–761.

Bullock, T. H., and Heiligenberg, W., editors. 1986. Electroreception. New York: Wiley.

Bullock, T. H., and Horridge, G. A. 1965. Structure and function in the nervous systems of invertebrates. San Francisco: W. H. Freeman.

Bumpus, H. C. 1899. The elimination of the unfit as illustrated by the introduced sparrow, *Passer domesticus*. Woods Hole, MA: Biological lectures, Marine Biological Laboratory. pp. 209–226.

Burke A. C., Nelson, C. E., Morgan, B. A., and Tabin, C. 1989. Development of the turtle carapace: implications for the evolution of a novel bauplan. J Morphol 199: 363–378.

Bush, E. C., and Allman J. M. 2004. The scaling of frontal cortex in primates and carnivores. Proc Natl Acad Sci (USA) 101: 3962–3966.

Butler, A. 1995. The dorsal thalamus of jawed vertebrates: a comparative viewpoint. Brain Behav Evol 1995: 209–223.

Butler, A. B. 1994a. The evolution of the dorsal pallium in the telencephalon of amniotes: cladistic analysis and a new hypothesis. Brain Res Rev 19: 66–101.

Butler, A. B. 1994b. The evolution of the dorsal thalamus of jawed vertebrates. Brain Res Rev 19: 29–65.

Butler, A. B., and Hodos, W. 1996. Comparative vertebrate neuroanatomy. New York: Wiley-Liss.

Butler, A. B., Molnar, Z., and Manger, P. R. 2002. Apparent absence of claustrum in monotremes: implications for forebrain evolution in amniotes. Brain Behav Evol 60: 230–240.

Buxhoeveden D., and Casanova M. 2000. Comparative lateralisation patterns in the language area of human, chimpanzee, and rhesus monkey brains. Laterality 5: 315–330.

Byrne, R. W. 1997. Machiavellian intelligence. Evol Anthrop 5: 172–180.

Byrne, R. W., and Whiten A., editors. 1988. Machiavellian intelligence: social expertise and the evolution of intellect in monkeys, apes, and humans. Oxford: Clarendon Press.

Cáceres, L. J., Zapala, M. A., Redmond, J. C., Kudo, L., Geschwind, D. H., Lockhart, D. J., Preuss, T. M., and Barlow, C. 2004. Elevated gene expression levels distinguish human from non-human primate brains. Proc Natl Acad Sci (USA) 100: 13030–13035.

Calder, W. A., III. 1984. Size, function, and life history. Cambridge, MA: Harvard University Press.

Campbell, C. B. G., and Hodos, W. 1970. The concept of homology and the evolution of the nervous system. Brain Behav Evol 3: 353–367.

Cannon, W. 1945. The way of an investigator. New York: W. W. Norton.

Caramelli, D., Lalueza-Fox, C., Vernesi, C., Lari, M., Casoli, A., Mallegni, F., Chiarelli, B., Dupanloup, I., Bertranpetit, J., Barbujani, G., and Bertorelle, G. 2003. Evidence for a genetic discontinuity between Neandertals and 24,000 year-old anatomically modern Europeans. Proc Natl Acad Sci (USA) 100: 6593–6597.

Carr, C. E., and Konishi, M. 1990. A circuit for detection of interaural time differences in the brain stem of the barn owl. J Neurosci 10: 3227–3246.

Carr, C. E., and Maler, L. 1986. Electroreception in gymnotiform fish. In: Bullock, T. H., Heiligenberg, W., editors. Electroreception. New York: Wiley, pp. 319–373.

Carr, C. E., Maler, L., Heiligenberg, W., and Sas, E. 1981. Laminar organization of the afferent and efferent systems of the torus semicircularis of gymnotiform fish: morphological substrates for parallel processing in the electrosensory system. J Comp Neurol 203: 649–670.

Carrier, D. R. 1984. The energetic paradox of human running and hominid evolution. Curr Anthrop 25: 483–495.

Carrier, D. R. 1987. The evolution of locomotor stamina in tetrapods: circumventing a mechanical constraint. Paleobiology 13: 326–341.

Carroll, R. L. 1988. Vertebrate paleontology and evolution. New York: Freeman.

Carroll, S. B. 2003. Genetics and the making of *Homo sapiens*. Nature 422: 849–857.

Cartmill, M. 1990. Human uniqueness and theoretical content in paleoanthropology. Int J Primatol 11: 173–192.

Cartmill, M. 1992. New views on primate origins. Evol Anthrop 1: 105–111.

Cartwright, N. 2002. In favour of laws that are not *ceteris paribus* after all. Erkenntnis 57: 425–439.

Casagrande, V. A., and Norton, T. T. 1991. Lateral geniculate nucleus: a review of its physiology and function. In: Leventhal A. G., editor. The neural basis of visual function. London: Macmillan Press, pp. 41–84.

Catania, K. C., Lyon, D. C., Mock, O. B., and Kaas, J. H. 1999a. Cortical organization in shrews: evidence from five species. J Comp Neurol 410: 55–72.

Catania, K. C., Northcutt, R. G., and Kaas, J. H. 1999b. The development of a biological novelty: a different way to make appendages as revealed in the snout of the star-nosed mole *Condylura cristata*. J Exp Biol 202: 2719–2726.

Catania, K. C., Collins, C. E., and Kaas, J. H. 2000. Organization of sensory cortex in the East African hedgehog (*Atelerix albiventris*). J Comp Neurol 421: 256–274.

Changizi, M. A. 2001. Principles underlying mammalian neocortical scaling. Biol Cybern 84: 207–215.

Chen F. C., and Li, W. H. 2001. Genomic divergences between humans and other hominoids and the effective population size of the common ancestor of humans and chimpanzees. Am J Hum Genet 68: 444–456.

Cheney, D. L., and Seyfarth, R. M. 1990. How monkeys see the world. Chicago: University of Chicago Press.

Cherniak, C. 1995. Neural component placement. Trends Neurosci 18: 522–527.

Cheung, D., Darlington, R. B., and Finlay, B. L. In press. Unfolding developmental rules in an enlarging brain. In: Munakata Y., Johnson M., editors. Attention and performance XXI: Processes of change in brain and cognitive development. Oxford: Oxford University Press.

Chiang, A.-S., Lin, W.-Y., Liu, H.-P., Pszvzolkowski, M. A., Fu, T.-F., and Chiu, S.-L. 2002. Insect NMDA receptors mediate juvenile hormone biosynthesis. Proc Natl Acad Sci (USA) 99: 37–42.

Chisholm, A., and Tessier-Lavigne, M. 1999. Conservation and divergence of axon guidance mechanisms. Curr Opin Neurobio 9: 603–615.

Chklovskii, D. B., Schikorski, T., and Stevens, C. F. 2002. Wiring optimization in cortical circuits. Neuron 34: 341–347.

Christie, R. V. 1987. Galen on Erasistratus. Persp Biol and Med 30: 440–449.

Clancy, B., Darlington, R. B., and Finlay, B. L. 2001. Translating developmental time across mammalian species. Neuroscience 105: 7–17.

Clark, D. A., Mitra, P. P., and Wang, S.-H. 2001. Scalable architecture in mammalian brains. Nature 411: 189–193.

Clarke, E., and O'Malley, C. D. 1986. The human brain and spinal cord: a historical study illustrated by writings from antiquity to the twentieth century. Berkeley, CA: University of California Press.

Clayton, N. S. 1998. Memory and the hippocampus in food-storing birds: a comparative approach. Neuropharm 37: 441–452.

Clayton, N. S., and Dickinson, A. 1998. Episodic-like memory during cache recovery by scrub jays. Nature 395: 272–274.

Clayton, N. S., Griffiths, D. P., Emery, N. J., and Dickinson, A. 2001. Elements of episodic-like memory in animals. Phil Trans R Soc London B 356: 1483–1491.

Clayton, N. S., Bussey, T. J., and Dickinson, A. 2003. Can animals recall the past and plan for the future? Nat Rev Neurosci 4: 685–691.

Clutton-Brock, T. H., and Harvey, P. H. 1980. Primates, brains, and ecology. J Zool Lond (A) 190: 309–323.

Cobb, M. 2000. Reading and writing The Book of Nature: Jan Swammerdam (1637–1680). Endeavour 24: 122–128.

Cobos, I., Shimamura, K., Rubenstein, J. L. R., Martínez, S., and Puelles, L. 2001. Fate map of the avian anterior forebrain at the 4 somite stage, based on the analysis of chick-quail chimeras. Dev Biol 239: 46–67.

Colombo, M., and Broadbent, N. 2000. Is the avian hippocampus a functional homologue of the mammalian hippocampus? Neurosci Biobehav Rev 24: 465–484.

Combe, G. 1836. A system of phrenology. Edinburgh: Maclachlan & Stewart.

Compagno, L. J. V. 1999. Systematics and body form. In: Hamlett, W. C., editor. Sharks, skates, and rays. Baltimore: Johns Hopkins University Press, pp. 1–42.

Conley, M. 1988. Laminar organization of geniculostriate projections. Brain Behav Evol 32: 187–192.

Conlon, J. M. 2002. The origin and evolution of peptide YY (PYY) and pancreatic polypeptide (PP). Peptides 23: 269–278.

Connolly, J. D., Goodale, M. A., DeSouza, J. F. X., Menon, R. S., and Villis, T. 2000. A comparison of frontoparietal fMRI activation during anti-saccades and anti-pointing. J Neurophysiol 84: 1645–1655.

Connors, B. W., and Kriegstein, A. R. 1986. Cellular physiology of the turtle visual cortex: distinctive properties of pyramidal and stellate neurons. J Neurosci 6: 164–177.

Conrad, M. 1990. The geometry of evolution. BioSystems 24: 61–81.

Conroy, G. C. 2002. Speciosity in the early Homo lineage: too many, too few, or just about right? J Human Evol 43: 759–766.

Conroy, G. C., Weber, G. W., Seidler, H., Recheis, W., Nedden, D. Z., and Mariam, J. H. 2000. Endocranial capacity of the Bodo cranium determined from three-dimensional computed tomography. Am J Phys Anthrop 113: 111–118.

Conturo, T. E., Lori, N. F., Cull, T. S., Akbudak, E., Snyder, A. Z., Shimony, J. S., McKinstry, R. C., Burton, H., and Raichle, M. E. 1999. Tracking neuronal fiber pathways in the living human brain. Proc Natl Acad Sci (USA) 96: 10422–10427.

Conway Morris, S. Dec. 13. 1991. Rerunning the tape. Times Lit Suppl: 6.

Cook, E. P., and Johnston, D. 1999. Voltage-dependent properties of dendrites that eliminate location-dependent variability of synaptic input. J Neurophysiol 81: 535–543.

Cooper, H. M., Herbin, M., and Nevo, E. 1993a. Ocular regression conceals adaptive progression of the visual system in a blind subterranean mammal. Nature 361: 156–159.

Cooper, H. M., Herbin, M., and Nevo, E. 1993b. Visual system of a naturally microphthalmic mammal: the blind mole rat, Spalax ehrenbergi. J Comp Neurol 328: 313–350.

Corballis, M. C. 2002. From hand to mouth: the origins of language. Princeton: Princeton University Press.

Corballis, M. C. 2003. From mouth to hand: gesture, speech, and the evolution of right-handedness. Behav Brain Sci 26: 199–208.

Cosans, C. 1994. Anatomy, metaphysics, and values: the ape brain debate reconsidered. Biology and Philosophy 9: 129–165.

Count, F. W. 1947. Brain and body weight in man: their antecedents in growth and evolution. Ann NY Acad Sci 46: 993–1122.

Courtillot, V. 1999. Evolutionary catastrophes: the science of mass extinction. Cambridge: Cambridge University Press.

Cowey, A. 1981. Why are there so many visual areas? In: Schmidt, F. O., Warden, F. G., Adelman, G., Dennis, S. G., editors. The organization of the cerebral cortex. Cambridge, MA: MIT Press, pp. 395–413.

Crawford, M. A. 1992. The role of dietary fatty acids in biology: their place in the evolution of the human brain. Nutrition Rev 50: 3–11.

Crile, G., and Quiring, D. 1940. A record of the body weight and certain organ and gland weights of 3690 animals. Ohio J Sci 40: 219–259.

Cuadrado, M. I. 1987. The cytoarchitecture of the torus semicircularis in the teleost *Barbus meridionalis*. J Morphol 191: 233–245.

Cullen, M. J., and Kaiserman-Abramof, I. R. 1976. Cytological organization of dorsal lateral geniculate nuclei in mutant anopthalmic and postnatally enucleated mice. J Neurocytol 5: 407–424.

Cunnane, S. C., and Crawford, M. A. 2003. Survival of the fattest: fat babies were the key to evolution of the large human brain. Comp Biochem and Physiol A 136: 17–26.

Cuvier, G. 1805–1845. Lecons d'anatomie comparée. Paris: Baudouin.

Dallos, P., and Evans, B. N. 1995. High-frequency motility of outer hair cells and the cochlear amplifier. Science 267: 2006–2009.

Damasio, A. R. 1994. Descartes' error. New York: Putnam.

Darwin, C. 1871. The descent of man, and selection in relation to sex. London: J. Murray.

Darwin, C. R. 1859. On the origin of species by means of natural selection. London: J. Murray.

Davidson, R. S. 1966. Operant stimulus control applied to maze behavior: heat escape conditioning and discrimination reversal in *Alligator mississippiensis*. J Exp Anal Behav 9: 671–676.

da Vinci, L. 1911–16. Quaderni d'anatomia. Christiania: Jacob Dybwad.

de Beer, G. 1958. Embryos and ancestors. London: Oxford University Press.

De Miguel C., Henneberg, M. 2001. Variation in hominid brain size: how much is due to method? Homo 52: 3–58.

de Winter, W., and Oxnard, C. E. 2001. Evolutionary radiations and convergences in the structural organization of mammalian brains. Nature 409: 710–714.

Deacon, T. W. 1990a. Fallacies of progression in theories of brain-size evolution. Int J Primatol 11: 193–236.

Deacon, T. W. 1990b. Problems of ontogeny and phylogeny in brain size evolution. Int J Primatol 11: 237–282.

Deacon, T. W. 1990c. Rethinking mammalian brain evolution. Am Zool 30: 629–705.

Deacon, T. W. 1997. The symbolic species. London: Penguin.

Dean, C., Leakey, M. G., Reid, D., Schrenk, F., Schwartz, G. T., Stringer, C., and Walker, A. 2001. Growth processes in teeth distinguish modern humans from *Homo erectus* and earlier hominins. Nature 414: 628–631.

Dehay, C., Giroud, P., Berland, M., Killackey, H. P., and Kennedy, H. 1996a. Phenotypic characterisation of respecified visual cortex subsequent to prenatal enucleation in the monkey: development of acetylcholinesterase and cytochrome oxidase patterns. J Comp Neurol 376: 386–402.

Dehay, C., Gourd, P., Berland, M., Killackey, H. P., and Kennedy, H. 1996b. Contribution of thalamic input to the specification of cytoarchitectonic cortical fields in the primate: effects of bilateral enucleation in the fetal monkey on the boundaries, dimensions, and gyrification of striate and extrastriate cortex. J Comp Neurol 367: 70–89.

Demski, L. S., and Northcutt, R. G. 1996. The brain and cranial nerves of the white shark: an evolutionary perspective. In: Klimley, A. P., Ainley, D. G., editors. Great white sharks. San Diego: Academic Press, pp. 121–130.

Dennett, D. C. 1995. Darwin's dangerous idea: evolution and the meanings of life. New York: Simon & Schuster.

Derrington, A. M., Krauskopf, J., and Lennie, P. 1984. Chromatic mechanisms in lateral geniculate nucleus of macaque. J Physiol (London) 357: 241–265.

Desan, P. H. 1988. Organization of cerebral cortex in turtle. In: Smeets, W. J. A. J., editor. The forebrain of reptiles. Basel: Karger, pp. 1–11.

Desimone, R., and Duncan, J. 1995. Neural mechanisms of selective visual attention. Ann Rev Neurosci 18: 193–222.

Desmond, A. 1994. Huxley: from Devil's disciple to evolution's high priest. Reading, MA: Addison-Wesley.

Dexler, H., and Eger O. 1911. Beiträge zur Anatomie des Säugerrückenmarkes. I. *Halicore dugong*. Erxl Morphol Jahrb 43: 107–207.

Diamond, M. E., and Ebner, F. F. 1990. Emergence of radial and modular units in neocortex. In: Finlay, B. L., editor. The neocortex. New York: Plenum Press, pp. 159–171.

Dias, R., Robbins, T. W., and Roberts, A. C. 1996. Dissociation in prefrontal cortex of affective and attentional shifts. Nature 380: 69–72.

Dickins, T. E., and Dickins, D. W. 2001. Symbols, stimulus equivalence and the origins of language. Behavior and Philosophy 29: 221–244.

Di Ferdinando, A., Calabretta, R., and Parisi, D. 2001. Evolving modular architectures for neural networks. In: French, R., Sougné, J., editors. Proceedings of the Sixth Neural Computation and Psychology Workshop: evolution, learning, and development. London: Springer Verlag, pp. 253–262.

Dingus, L., and Rowe, T. 1997. The mistaken extinction. New York: W. H. Freeman & Co.

Disbrow, E., Roberts, T., and Krubitzer, L. 2000. Somatotopic organization of cortical fields in the lateral sulcus of *Homo sapiens*: evidence for SII and PV. J Comp Neurol 418: 1–21.

Divac, I., and Mogensen, J. 1985. The prefrontal cortex in the pigeon: catecholamine histofluorescence. Neuroscience 15: 677–682.

Dobzhansky, T. 1973. Nothing in biology makes sense except in the light of evolution. Am Biol Teach 35: 125–129.

Dominy, N. J., and Lucas, P. W. 2001. The ecological importance of trichromatic colour vision in primates. Nature 410: 363–366.

Donoghue, J. P., and Wise, S. P. 1982. The motor cortex of the rat: cytoarchitecture and microstimulation mapping. J Comp Neurol 212: 76–88.

Dryer, L. 2000. Evolution of odorant receptors. Bioessays 22: 803–810.

Dubbeldam, J. L., Brauch, C. S. M., and Don, A. 1981. Studies on the somatotopy of the trigeminal system in the mallard, *Anas platyrhynchos* L. III. Afferents and organization of the nucleus basalis. J Comp Neurol 196: 391–405.

Dubois, E. A., Zandbergen, M. A., Peute, J., and Goos, H. J. T. 2002. Evolutionary development of three gonadotropin-releasing hormone (GnRH) systems in vertebrates. Brain Res Bull 57: 413–418.

Duchaine, B., Cosmides, L., and Tooby, J. 2001. Evolutionary psychology and the brain. Curr Opin Neurobio 11: 225–230.

Dulai, K. S., von Dornum, M., Mollon, J. D., and Hunt, D. M. 1999. The evolution of trichromatic color vision by opsin gene duplication in New World and Old World primates. Genome Res 9: 629–638.

Dunbar, R. I. M. 1998. The social brain hypothesis. Evol Anthrop 6: 178–190.

Duncan, J., and Owen, A. M. 2000. Common regions of the human frontal lobe recruited by diverse cognitive demands. Trends Neurosci 23: 475–483.

Durbin, R., and Mitchison, G. 1990. A dimension reduction framework for understanding cortical maps. Nature 343: 644–647.

Earman, J., and Roberts, J. 1999. *Ceteris paribus*, there is no problem of provisos. Synthese 118: 439–478.

Ebbesson, S. O. E. 1980. The parcellation theory and its relation to interspecific variability in brain organization, evolutionary and ontogenetic development, and neuronal plasticity. Cell Tiss Res 213: 179–212.

Ebbesson, S. O. E. 1984. Evolution and ontogeny of neural circuits. Behav Brain Sci 7: 321–366.

Ebbesson, S. O. E., editor. 1980. Comparative neurology of the telencephalon. New York: Plenum.

Ebbesson, S. O. E., and Heimer, L. 1970. Projections of the olfactory tract fibers in the nurse shark, *Gynglimostoma cirratum*. Brain Res 17: 47–55.

Ebinger, P., and Röhrs, M. 1995. Volumetric analysis of brain structures, especially of the visual system in wild and domestic turkeys (*Meleagris gallopavo*). J Brain Res 36: 219–228.

Ebinger, P., Wächtler, K., and Stähler, S. 1983. Allometrical studies in the brain of cyclostomes. J Hirnforsch 24: 545–550.

Economos, A. C. 1980. Brain-life span conjecture: a re-evaluation of the evidence. Gerontology 26: 82–89.

Edinger, L. 1908a. The relations of comparative anatomy to comparative psychology. J Comp Neurol Psychol 18: 437–457.

Edinger, L. 1908b. Vorlesungen über den Bau der Nervösen Zentralorgane des Menschen und der Tiere - für Ärzte und Studierende. Leipzig: Vogel Verlag.

Edinger, T. 1929. Die fossilen Gehirne. Ergebn Anatomie Entwicklungsgesch 28: 1–249.

Eisenberg, J. F. 1981. The mammalian radiations. Chicago: University of Chicago Press.

Eisenberg, J. F., and Wilson, D. 1978. Relative brain size and feeding strategies in the Chiroptera. Evolution 32: 740–751.

Eldredge, N., and Cracraft, J. 1980. Phylogenetic patterns and the evolutionary process. New York: Columbia University Press.

Eldredge, N., and Gould, S. J. 1972. Punctuated equilibria: an alternative to phyletic gradualism. In: Schopf, T. J. M., editor. Models in Paleobiology. San Francisco: Freeman, Cooper and Company, pp. 82–115.

Elliott, T., and Shadbolt, N. R. 1999. A neurotrophic model of the development of the retinogeniculocortical pathway induced by spontaneous retinal waves. J Neurosci 15: 7951–7970.

Enard, W., Khaitovich, P., Klose, J., Zöllner, S., Heissig, F., Giavalisco, P., Nieselt-Struwe, K., Muchmore, E., Varki, A., Ravid, R., Doxiadis, G. M., Bontrop, R. E., and Pääbo, S. 2002. Intra- and interspecific variation in primate gene expression patterns. Science 296: 340–343.

Endler, J. A. 1986. Natural selection in the wild. Princeton: Princeton University Press.

Endler, J. A. 1995. Multiple-trait coevolution and environmental gradients in guppies. Trends Ecol Evol 10: 22–29.

Everling, S., and Fischer, B. 1998. The antisaccade: a review of basic research and clinical studies. Neuropsychologia 36: 885–899.

Everling, S., Dorris, M. C., Klein, R. M., and Munoz, D. P. 1999. Role of primate superior colliculus in preparation and execution of anti-saccades and pro-saccades. J Neurosci 19: 2740–2754.

Falck, B., Hillarp, N. A., Thieme, G., and Thorp, A. 1962. Fluorescence of catecholamines and related compounds condensed with formaldehyde. J Histochem Cytochem 10: 348–354.

Farah, M., Humphreys, G. W., and Rodman, H. R. 1999. Object and face recognition. In: Zigmond, M. J., Bloom, F. E., Landis, S. C., Roberts, J. L., Squire, L. R., editors. Fundamental Neuroscience. San Diego: Academic Press, pp. 1339–1361.

Farrell, W. J., Bottger, B., Ahmadi, F., and Finger, T. E. 2002. Distribution of cholecystokinin, calcitonin gene-related peptide, neuropeptide Y., and galanin in the primary gustatory nuclei of the goldfish. J Comp Neurol 450: 103–114.

Farries, M. A. 2001. The oscine song system considered in the context of the avian brain: lessons from comparative neurobiology. Brain Behav Evol 58: 80–100.

Felsenstein, J. 1985. Phylogenies and the comparative method. Am Nat 125: 1–15.

Fernandez, A., Pieau, C., Reperant, J., Boncinelli, E., and Wassef, M. 1998. Expression of the *Emx-1* and *Dlx-1* homeobox genes define three molecularly distinct domains in the telencephalon of mouse, chick, turtle and frog embryos: implications for the evolution of telencephalic subdivisions in amniotes. Development 125: 2099–2111.

Ferrier, D. 1886. The functions of the brain. New York: G. P. Putnam's Sons.

Finger, T. E., Bell, C. C., and Russel, C. J. 1981. Electrosensory pathways to the valvula cerebelli in mormyrid fish. Exp Brain Res 42: 23–33.

Finger, S. 2000. Minds behind the brain. Oxford: Oxford University Press.

Finger, T. E. 1986. Electroreception in catfish: behavior, anatomy, and physiology. In: Bullock, T. H., Heiligenberg, W., editors. Electroreception. New York: Wiley & Sons, pp. 287–318.

Finger, T. E. 1988. Sensorimotor mapping and oropharyngeal reflexes in goldfish, *Carassius auratus*. Brain Behav Evol 31: 17–24.

Finger, T. E. 1997. Feeding patterns and brain evolution in ostariophysean fishes. Acta Physiol Scand 161 Suppl: 59–66.

Finger, T. E., Bell, C. C., and Russel, C. J. 1981. Electrosensory pathways to the valvula cerebelli in mormyrid fish. Exp Brain Res 42: 23–33.

Finger, T. E., Bell, C. C., and Carr, C. E. 1986. Comparsions among electroreceptive teleosts: why are electrosensory systems so similar? In: Bullock, T. H., Heiligenberg, W., editors. Electroreception. New York: John Wiley & Sons, pp. 465–481.

Fink, R. P., and Heimer, L. 1967. Two methods for selective silver impregnation of degenerating axons and their synaptic endings in the central nervous system. Brain Res 4: 369–374.

Finlay, B. L., and Darlington, R. B. 1995. Linked regularities in the development and evolution of mammalian brains. Science 268: 1578–1584.

Finlay, B. L., Sengelaub, D. R., and Berian, C. A. 1986. Control of cell number in the developing visual system. I. Effects of monocular enucleation. Dev Brain Res 28: 1–10.

Finlay, B. L., Wikler, K. C., and Sengelaub, D. R. 1987. Regressive events in brain development and scenarios for vertebrate brain evolution. Brain Behav Evol 30: 102–117.

Finlay, B. L., Darlington, R. B., and Nicastro, N. 2001. Developmental structure in brain evolution. Behav Brain Sci 24: 263–308.

Fitch, W. T. 2000. The evolution of speech: a comparative review. Trends Cogn Sci 4: 258–267.

Fleming, T. H., Heithaus, E. R., and Sawyer, W. B. 1977. An experimental analysis of the food location behavior of frugivorous bats. Ecology 58: 619–697.

Fox, J. H., and Wilczynski, W. 1986. Allometry of major CNS divisions: towards a reevaluation of somatic brain-body scaling. Brain Behav Evol 28: 157–169.

Frahm, H. D., Stephan, H., and Stephan, M. 1982. Comparison of brain structure volumes in insectivora and primates. I. Neocortex. J Hirnforsch 23: 375–389.

Fraser, S., Keynes, R., and Lumsden, A. 1990. Segmentation in the chick embryo hindbrain is defined by cell lineage restrictions. Nature 344: 431–435.

Freitag, J., Ludwig, G., Andreini, I., Rössler, P., and Breer, H. 1998. Olfactory receptors in aquatic and terrestrial vertebrates. J Comp Physiol A 183: 635–650.

Fritzsch, B., Nikundiwe, A. M., and Will, U. 1984. Projection patterns afferents in anurans: a comparative HRP study. J Comp Neurol 229: 451–469.

Fritzsch, B., Ryan, M. J., Wilczynski, W., Hetherington, T. E., and Walkowiak, W., editors. 1988. The evolution of the amphibian auditory system. New York: Wiley & Sons.

Frost, S. B., and Masterton, R. B. 1992. Origin of auditory cortex. In: Webster, B., Fay, R. R., Popper, A. N., editors. The evolutionary biology of hearing. New York: Springer-Verlag, pp. 655–671.

Frost, S. B., and Masterton, R. B. 1994. Hearing in primitive mammals: *Monodelphis domestica* and *Marmosa elegans*. Hearing Res 76: 67–72.

Frost, S. B., Milliken, G. W., Plautz, E. J., Masterton, R. B., and Nudo, R. J. 2000. Somatosensory and motor representations in cerebral cortex of a primitive mammal (*Monodelphis domestica*): a window into the early evolution of sensorimotor cortex. J Comp Neurol 421: 29–51.

Fryxell, K. J. 1995. The evolutionary divergence of neurotransmitter receptors and second-messenger pathways. J Molec Evol 41: 85–97.

Fukuchi-Shimogori, T., and Grove, E. A. 2001. Neocortex patterning by the secreted signaling molecule FGF8. Science 294: 1071–1074.

Gall, F. J. 1798. Schreiben über seinen bereits geendigten Prodromus über die Verrichtungen des Gehirns der Menschen und der Thiere an Herrn Jos. Fr. von Retzer. Der neue Teutsche Merkur 3: 311–332.

Gannon, P. J., Holloway, R. L., Broadfield, D. C., and Braun, A. R. 1998. Asymmetry of chimpanzee planum temporale: humanlike pattern of Wernicke's brain language area homolog. Science 279: 220–222.

Gans, C. 1969. Discussion: some questions and problems in morphological comparison. Ann NY Acad Sci 167: 506–513.

Garland, T. J., Harvey, P. H., and Ives, A. R. 1992. Procedures for the analysis of comparative data using phylogenetically independent contrasts. Syst Biol 41: 18–32.

Gatesy, J., and O'Leary, M. A. 2001. Deciphering whale origins with molecules and fossils. Trends Ecol Evol 16: 562–570.

Gauthier J., Kluge, A. G., and Rowe, T. 1988. Amniote phylogeny and the importance of fossils. Cladistics 4: 105–209.

Geoffroy Saint-Hilaire, E. 1807. Considérations sur les pièces de la tête osseuse de animaux vertebrés, et particulièrment sur celles du crêne des oiseaux. Ann Mus Hist Nat 10: 342–343.

Gilbert, S. F., Opitz, J. M., and Raff, R. A. 1996. Resynthesizing evolutionary and developmental biology. Dev Biol 173: 357–372.

Gilissen, E. 2001. Structural symmetries and asymmetries in human and chimpanzee brains. In: Falk D., Gibson K. R., editors. Evolutionary anatomy of the primate cerebral cortex. Cambridge: Cambridge University Press, pp. 187–215.

Gilland, E., and Baker, R. 1993. Conservation of neuroepithelial and mesodermal segments in the embryonic vertebrate head. Acta Anat 148: 110–123.

Glendenning, K. K., and Masterton, R. B. 1998. Comparative morphometry of mammalian central auditory systems: variation in nuclei and form of the ascending system. Brain Behav Evol 51: 59–89.

Glezer, I. I., Jacobs, M. S., and Morgane, P. J. 1988. Implications of the "initial brain" concept for brain evolution in Cetacea. Behav Brain Sci 11: 75–116.

Glusman, G., Bahar, A., Sharon, D., Pilpel, Y., White, J., and Lancet, D. 2000. The olfactory receptor gene superfamily: data mining, classification, and nomenclature. Mamm Genome 11: 1016–1023.

Goethe, J. W. 1981. Goethes Werke. Berlin: Aufbau-Verlag.

Goodall, J. 1986. The chimpanzees of Gombe: patterns of behavior. Cambridge, MA: Harvard University Press.

Goodhill, G. J., and Richards, L. J. 1999. Retinotectal maps: molecules, models and misplaced data. Trends Neurosci 22: 529–534.

Gould, S. J. 1970. Dollo on Dollo's law: irreversibility and the status of evolutionary laws. J History Biol 3: 189–212.

Gould, S. J. 1975. Allometry in primates, with emphasis on scaling and the evolution of the brain. Contrib Primatol 5: 244–292.

Gould, S. J. 1977. Ontogeny and phylogeny. Cambridge: Harvard University Press.

Gould S. J., and Eldredge, N. 1993. Punctuated equilibrium comes of age. Nature 366: 223–227.

Gould, S. J., and Lewontin, R. C. 1979. The spandrels of San Marco and the Panglossian paradigm: a critique of the adaptationist program. Proc R Soc Lond 205: 581–598.

Grant, B. R., and Grant, P. R. 1989. Natural selection in a population of Darwin's finches. Am Nat 133: 377–393.

Grant, P. R. 1972. Centripetal selection and the house sparrow. Syst Zool 21: 23–30.

Gregory, W. K. 1935. Reduplication in evolution. Quart Rev Biol 10: 272–290.

Griffiths, P. E., and Gray, R. D. 1994. Developmental systems and evolutionary explanation. J Philosophy 91: 277–304.

Groenewegen, H. J., Berndse, H. W., Wolters, J. G., and Lohman, A. H. M. 1990. The anatomical relationship of the prefrontal cortex with the striatopallidal system, the thalamus and the amygdala - evidence for a parallel organization. Prog Brain Res 85: 95–118.

Guillery, R. W. 1966. A study of Golgi preparations from the dorsal lateral geniculate nucleus of the adult cat. J Comp Neurol 128: 21–50.

Guirado, S., Dávila, J. C., Real, M. Á., and Medina, L. 2000. Light and electron microscopic evidence for projections from the thalamic reticular nucleus rotundus to targets in the basal ganglia, the dorsal ventricular ridge, and the amygdaloid complex in a lizard. J Comp Neurol 424: 216–232.

Gutierrez, C., Cola, M. G., Seltzer B., and Cusick C. 2000. Neurochemical and connectional organization of the dorsal pulvinar complex in monkeys. J Comp Neurol 419: 61–86.

Haeckel, E. 1889. Natürliche Schöpfungsgeschichte. Berlin: Georg Reimer.

Haeckel, E. 1893. The evolution of man. London: Kegan, Paul, Trench and Co.

Haight, J. R., and Neylon, L. 1978. The organization of neocortical projections from the ventroposterior thalamic complex in the marsupial brush-tailed possum, Trichosurus vulpecula: a horseradish peroxidase study. J Anat 126: 459–485.

Haight, J. R., and Neylon, L. 1979. The organization of neocortical projections from the ventrolateral thalamic nucleus in the brush-tailed possum, Trichosurus vulpecula, and the problem of motor and somatic sensory convergence within the mammalian brain. J Anat 129: 673–694.

Hall, B. K. 2000. Evo-devo or devo-evo - does it matter? Evol Development 2: 177–178.

Hamburger, V., 1975. Cell death in the development of the lateral motor column of the chick embryo. J Comp Neurol 160: 535–546.

Hamrick, M. W., 2001. Primate origins: evolutionary change in digital ray patterning and segmentation. J Human Evol 40: 339–351.

Harping, J. E., Pearson, J. C., Norris, J. R., Mann, B. L. 1985. Subclassification of neurons in the ventrobasal complex of the dog: quantitative Golgi study using principal components analysis. J Comp Neurol 242: 230–246.

Harris-Warrick, R. 2000. Ion channels and receptors: molecular targets for behavioral evolution. J Comp Physiol A 186: 605–616.

Harting, J. K., Hall, W. C., Diamond, I. T. 1972. Evolution of the pulvinar. Brain Behav Evol 6: 424–452.

Hartshorne, C., Weiss, P., editors. 1931–58. Collected papers of Charles Sanders Peirce. Cambridge, MA: Harvard University Press.

Harvey, P. H., and Krebs, J. H. 1990. Comparing brains. Science 249: 140–146.

Harvey, P. H., and Pagel, M. D. 1991. The comparative method in evolutionary biology. Oxford: Oxford University Press.

Harvey, P. M., Martin, R. D., and Clutton-Brock, T. H. 1987. Life histories in comparative perspective. In: Smuts, B., Cheney, D. L., Seyfarth, R. M., Wrangham, R., Struhsaker, T., editors. Primate societies. Chicago: University of Chicago Press, pp. 181–196.

Haug, H. 1987. Brain sizes, surfaces, and neuronal sizes of the cortex cerebri: a stereological investigation of man and his variability and a comparison with some mammals (primates, whales, marsupials, insectivores, and one elephant). Am J Anat 180: 126–142.

Hauptmann, G., and Gerster, T. 2000. Regulatory gene expression patterns reveal transverse and longitudinal subdivisions of the embryonic zebrafish forebrain. Mech Dev 91: 105–118.

Hauser, M. 2000. Wild minds: what animals really think. New York: Henry Holt and Co.

Hauser, M. D., Chomsky, N., and Fitch, W. T. 2002. The faculty of language: what is it, who has it, and how did it evolve? Science 298: 1569–1579.

Hawks, J., Hunley, K., Lee, S-H., and Wolpoff, M. H. 2000. Population bottlenecks and Pleistocene human evolution. Mol Biol Evol 17: 2–22.

Healy, S., and Guilford, T. 1990. Olfactory-bulb size and nocturnality in birds. Evolution 44: 339–346.

Healy, S. D., and Krebs, J. R. 1996. Food storing and the hippocampus in Paridae. Brain Behav Evol 47: 195–199.

Hedges, S. B. 1994. Molecular evidence for the origin of birds. Proc Natl Acad Sci 91: 2621–2624.

Heesey, C. P., and Ross, C. F. 2001. Evolution of activity patterns and chromatic vision in primates: morphometrics, genetics and cladistics. J Human Evol 40: 111–149.

Heffner, H. E., and Heffner, R. S. 1986. Hearing loss in Japanese macaques following bilateral auditory cortex lesions. J Neurophysiol 55: 256–271.

Heffner, R., and Masterton, B. 1975. Variation in form of the pyramidal tract and its relationship to digital dexterity. Brain Behav Evol 12: 161–200.

Heffner, R. S., Masterton, R. B. 1983. The role of the corticospinal tract in the evolution of human digital dexterity. Brain Behav Evol 23: 165–183.

Heiligenberg, W. 1986. Jamming avoidance responses. In: Bullock, T. H., Heiligenberg, W., editors. Electroreception. New York: Wiley, pp. 613–649.

Heiligenberg, W. 1991. Neural nets in electric fish. Cambridge, MA: MIT Press.

Heitger, M. H., Anderson, T. J., Jones, R. D., Dalrymple-Alford, J. C., Frampton, C. M., and Ardagh, M. W. 2004. Eye movement and visuomotor arm movement deficits following mild closed head injury. Brain 127: 575–590.

Heller, S. B., and Ulinski, P. S. 1987. Morphology of geniculocortical axons in turtles of the genera *Pseudemys* and *Chrysemys*. Anat Embryol 175: 505–515.

Hemmi, J. M. 1999. dichromatic colour vision in an australian marsupial, the tammar wallaby. J Comp Physiol A 185: 509–515.

Hempel, C. G. 1942. The function of general laws in history. J Philosophy 39: 35–48.

Hennig, W. 1950. Grundzüge einer Theorie der Phylogenetischen Systematik. Berlin: Deutscher Zentralverlag.

Hennig, W. 1966. Phylogenetic Systematics. Urbana: University of Illinois Press.

Henshilwood, C., d'Errico, F., Vanhaeren, M., van Niekirk, K., and Jacobs, Z. 2004. Shell beads from South African cave show modern human behaviours 75,000 years ago. Science 304: 404.

Herrick, C. J. 1933. Morphogenesis and the brain. J Morphol 54: 233–258.

Herrick, C. J. 1948. The brain of the tiger salamander *Amblystoma tigrinum*. Chicago: University of Chicago Press.

Heywood, C. A., and Cowey, A. 1992. The role of the "face cell" area in the discrimination and recognition of faces by monkeys. Phil Trans R Soc London B 335: 31–37.

Hill, W. G., and Mbaga, S. H. 1998. Mutation and conflicts between artificial and natural selection for quantitative traits. Genetica 102–103: 171–181.

Hille, B. 1984. Ionic channels of excitable membranes. Sunderland, MA: Sinauer.

Hillenius, W. J. 1994. Turbinates in therapsids: evidence for late Permian origins of mammalian endothermy. Evolution 48: 207–229.

His, W. 1904. Die Entwicklung des menschlichen Gehirns während der ersten Monate. Leipzig: Hirzel.

Hodos, W., and Butler, A. B. 1997. Evolution of sensory pathways in vertebrates. Brain Behav Evol 50: 189–197.

Hodos, W., and Campbell, C. B. G. 1969. *Scala naturae*: why there is no theory in comparative psychology. Psychol Review 76: 337–350.

Hofman, M. A. 2001. Brain evolution in hominids: are we at the end of the road? In: Falk, D., Gibson, K., editors. Evolutionary anatomy of the primate cerebral cortex. Cambridge: Cambridge University Press, pp. 113-127.

Hofman, M. H. 1983. Encephalization in hominids: evidence for the model of punctualism. Brain Behav Evol 22: 102–117.

Holdefer, R. N., and Norton, T. T. 1995. Laminar organization of receptive-field properties in the dorsal lateral geniculate nucleus of the tree shrew (*Tupaia glis* Belangeri). J Comp Neurol 358: 401–413.

Holloway, R. L. 1972. Australopithecine endocasts, brain evolution in the hominoidea, and a model of hominid evolution. In: Tuttle, R. H., editor. The functional and evolutionary biology of primates, pp. 185–203.

Holloway, R. L. 1974. On the meaning of brain size. Science 184: 677–679.

Holloway, R. L. 2002. How much larger is the relative volume of Area 10 of the prefrontal cortex in humans? Am J Phys Anthrop 118: 399–401.

Hollyday, M., and Hamburger, V. 1976. Reduction of the naturally occurring motor neuron loss by enlargement of the periphery. J Comp Neurol 170: 311–320.

Holmgren, N. 1922. Points of view concerning forebrain morphology in lower vertebrates. J Comp Neurol 34: 391–459.

Holmgren, N. 1925. Points of view concerning forebrain morphology in higher vertebrates. Acta Zool Stockh 6: 413–477.

Holmgren, S., and Jensen, J. 2001. Evolution of vertebrate neuropeptides. Brain Res Bull 55: 723–735.

Hoogland, P. V., and Vermeulen-Vanderzee, E. 1995. Efferent connections of the lateral cortex of the lizard *Gekko gekko*: evidence for separate origins of medial and lateral pathways from the lateral cortex to the hypothalamus. J Comp Neurol 352: 469–480.

Hopson, J. A. 1979. Paleoneurology. In: Gans, C., Northcutt, R. G., Ulinski, P., editors. Biology of the reptilia. London: Academic Press, pp. 39–146.

Horwitz, B., Amunts, K., Bhattacharyya, R., Patkin, D., Jeffries, K., and Braun, A. R. 2003. Activation of Broca's area during the production of spoken and signed language: a combined cytoarchitectonic mapping and PET analysis. Neuropsychologia 41: 1868–1876.

Hough, G. E., II., Pang, K. C. H., and Bingman, V. P. 2002. Intrahippocampal connections in the pigeon (*Columba livia*) as revealed by stimulation-evoked field potentials. J Comp Neurol 452: 297–309.

Hoyle, C. H.V. 1999. Neuropeptide families and their receptors: evolutionary perspectives. Brain Res 848: 1–25.

Huberman, A. D., Stellwagen, D., and Chapman, B. 2002. Decoupling eye-specific segregation from lamination in the lateral geniculate nucleus. J Neurosci 22: 9419–9429.

Hublin, J.-J., Spoor, F., Braun, M., Zonneveld, F., and Condemi, S. 1996. A later Neanderthal associated with Upper Paleolithic artefacts. Nature 381: 224–226.

Hudson, R. 1999. Review of Terrence Deacon, the Symbolic Species. J Pragmatics 33: 129–135.

Huffman, K. J., Molnar, Z., Van Dellen, A., Kahn, D. M., Blakemore, C., and Krubitzer, L. 1999. Formation of cortical fields on a reduced cortical sheet. J Neurosci 19: 9939–9952.

Huisman, A. M., Kuypers, H. G. J. M., and Verburgh, C. A. 1981. Quantitative differences in collateralization of the descending spinal pathways from red nucleus and other brainstem cell groups in rat as demonstrated with the multiple fluorescent retrograde tracer technique. Brain Res 209: 271–286.

Huisman, A. M., Kuypers, H. G. J. M., and Verburgh, C. A. 1982. Differences in collateralization of the descending spinal pathways from red nucleus and other brain stem cell groups in cat and monkey. Prog Brain Res 57: 185–217.

Hull, D. L. 1988. Science as a process. Chicago: University of Chicago Press.

Humphrey, N. K. 1976. The social function of intellect. In: Bateson, P. P. G., Hinde, R. A., editors. Growing points in ethology. Cambridge: Cambridge University Press, pp. 303–317.

Hunt, G. R. 1996. Manufacture and use of hook tools by New Caledonian crows. Nature 379: 249–251.

Hunt, G. R., Corballis, M. C., and Gray, R. D. 2001. Laterality in tool manufacture by crows. Nature 414: 707.

Hunter, J. P. 1998. Key innovations and the ecology of macroevolution. Trends Ecol Evol 13: 31–36.

Husband, S. A., and Shimizu, T. 1999. Efferent projections of the ectostriatum in the pigeon (*Columba livia*). J Comp Neurol 406: 329–345.

Hutcheon, J. M., Kirsch, J. A. W., and Garland, T. J. 2002. A comparative analysis of brain size in relation to foraging ecology and phylogeny in the Chiroptera. Brain Behav Evol 60: 165–180.

Huxley, J. S. 1932. Problems of relative growth. London: Methuen & Co. Ltd.

Huxley, J. S., and de Beer, G. R. 1934. The mosaic style of differentiation. In: Barcroft J., Saunders J. T., editors. The elements of experimental embryology. London: Cambridge University Press, pp. 195–270.

Huxley, T. H. 1861. On the zoological relations of man with the lower animals. Nat Hist Rev 1: 67–84.

Huxley, T. H. 1863. Man's place in Nature. Ann Arbor: University of Michigan Press.

Ichida, J. M., and Casagrande, V. A. 2002. Organization of the feedback pathway from striate cortex (V1) to the lateral geniculate nucleus (LGN) in the owl monkey (*Aotus trivirgatus*). J Comp Neurol 454: 272–283.

Iglesia, J. A. L., and Lopez-Garcia, C. 1997. A Golgi study of the principal projection neurons of the medial cortex of the lizard *Podarcis hispanica*. J Comp Neurol 385: 528–564.

Innocenti, G. M. 1981. Growth and reshaping of axons in the establishment of visual callosal connections. Science 218: 824–827.

Inoue, T., Nakamura, S., and Osumi, N. 2000. Fate mapping of the mouse prosencephalic neural plate. Dev Biol 219: 373–383.

Insel, T. R., Winslow, J. T., Wang, Z., and Young, L. J. 1998. Oxytocin, vasopressin, and the neuroendocrine basis of pair bond formation. Adv Exp Med Biol 449: 215–224.

Ito, H., and Vanegas, H. 1983. Cytoarchitecture and ultrastructure of nucleus prethalamicus, with special reference to degenerating afferents from the optic tectum of teleosts (*Holocentrus ascensonis*). J Comp Neurol 221: 401–415.

Ito, H., Yoshimoto, M., and Somiya, H. 1999. External brain form and cranial nerves of the megamouth shark, *Megachasma pelagios*. Copeia 1999(1): 210–213.

Ito, H., Naoyuki, Y., Yoshimoto, M., Sawai, N., Yang, C.-H., Xue, H.-G., and Imura, K. 2003. Fiber connections of the torus longitudinalis in a teleost: *Cyprinus carpio* re-examined. J Comp Neurol 457: 202–211.

Iwaniuk, A. N., and Nelson, J. E. 2003. Developmental differences are correlated with relative brain size in birds: a comparative analysis. Can J Zool 81: 1913–1928.

Iwaniuk, A. N., Pellis, S. M., and Whishaw, I. Q. 1999. Is digital dexterity really related to corticospinal projections? A re-analysis of the Heffner and Masterton data set using modern comparative statistics. Behav Brain Res 101: 173–187.

Iwaniuk, A. N., Nelson, J. E., and Pellis, S. M. 2001. Do big-brained animals play more? Comparative analyses of play and relative brain size in mammals. J Comp Psychol 115: 29–41.

Iwaniuk, A. N., Dean, C., and Nelson, J. E. 2004. A mosaic pattern characterizes the evolution of the avian brain. Proc R Soc Lond B (Suppl) 271: S148–S151.

Iwatsubo, T., Kuzuhara, S., Kanemitsu, A., Shimada, H., and Toyokura, Y. 1990. Corticofugal projections to the motor nuclei of the brainstem and spinal cord in humans. Neurology 40: 309–312.

Jacob, F. 1977. Evolution and tinkering. Science 196: 1161–1166.

Jacobs, G. H. 1993. The distribution and nature of colour vision among the mammals. Biol Rev Camb Philosoph Soc 68: 413–471.

Jacobs, R. A., and Jordan, M. I. 1992. Computational consequences of a bias toward short connections. J Cognitive Neurosci 4: 323–336.

Jeong, H., Mason, S. P., Barabási, A-L., and Oltvai, Z. N. 2001. Lethality and centrality in protein networks. Nature 411: 41–42.

Jerison, H. J. 1973. Evolution of the brain and intelligence. New York: Academic Press.

Jerison, H. J. 1976. Paleoneurology and the evolution of mind. Sci Am 234: 90–101.

Jerison, H. J. 1989. Brain size and the evolution of mind. New York: American Museum of Natural History.

Jerison, H. J. 2001. The study of primate brain evolution: where do we go from here? In: Falk D., Gibson K. R., editors. Evolutionary anatomy of the primate cerebral cortex. Cambridge: Cambridge University Press, pp. 305–337.

Jeserich, G., Strelau, J., and Lanwert, C. 1997. Partial characterization of the 5'-flanking region of trout I. P.: a Po-like gene containing a PLP-like promoter. J Neurosci Res 50: 781–790.

Jiao, Y., Medina, L., Veenman, C. L., Toledo, C., Puelles, L., and Reiner, A. 2000. Identification of the anterior nucleus of the ansa lenticularis in birds as the homolog of the mammalian subthalamic nucleus. J Neurosci 20: 6998–7010.

Johnston, J. B. 1923. Further contributions of the study of the evolution of the forebrain. J Comp Neurol 36: 143–192.

Jones, E. G. 1985. The thalamus. New York: Plenum Press.

Jukes, T. H., and Osawa, S. 1993. Evolutionary changes in the genetic code. Comp Biochem Physiol B 106: 489–494.

Julian, G. E., and Gronenberg, W. 2002. Reduction of brain volume correlates with behavioral changes in queen ants. Brain Behav Evol 60: 152–164.

Jungbluth, S., Larsen, C., Wizenmann, A., and Lumsden, A. 2001. Cell mixing between the embryonic midbrain and hindbrain. Curr Biol 11: 204–207.

Jürgens, U. 1979. Neural control of vocalization in non-human primates. In: Steklis H. D., Raleigh M. J., editors. Neurobiology of social communication in primates. New York: Academic Press, pp. 11–44.

Jürgens, U. 2002. Neural pathways underlying vocal control. Neurosci Biobehav Rev 26: 235–258.

Jürgens, U., and Alipour, M. 2002. A comparative study on the cortico-hypoglossal connections in primates, using biotin dextranamine. Neurosci Lett 328: 245–248.

Kaas, J. H. 1982. The segregation of function in the nervous system: why do sensory systems have so many subdivisions? Contrib Sensory Physiology 7: 201–240.

Kaas, J. H. 1983. What, if anything, is S-I? The organization of the "first somatosensory area" of cortex. Physiol Rev 63: 206–231.

Kaas, J. H. 1984. Duplication of brain parts in evolution. Behav Brain Sci 7: 342–343.

Kaas, J. H. 1987. The organization of neocortex in mammals: implications for theories of brain function. Ann Rev Psychol 38: 129–151.

Kaas, J. H. 1999. The transformation of association cortex into sensory cortex. Brain Res Bull 50: 425.

Kaas, J. H. 2002. Convergences in the modular and areal organization of the forebrain of mammals: implications for the reconstruction of forebrain evolution. Brain Behav Evol 59: 262–272.

Kaas, J. H., and Collins, C. E. 2001. Variability in the sizes of brain parts. Behav Brain Sci 24: 288–289.

Kaas, J. H., and Huerta, M. F. 1988. The subcortical visual system of primates. In: Steklis H. D., Erwin J., editors. Comparative primate biology, Vol. 4: neurosciences. New York: Alan Liss, pp. 327–391.

Kaas, J. H., and Lyon, D. C. 2001. Visual cortex organization in primates: theories of V3 and adjoining visual areas. Prog Brain Res 134: 285–295.

Kaas, J. H., and Pons, T. P. 1988. The somatosensory system of primates. Comarative primate biology, Vol. 4: Neurosciences, pp. 421–468.

Kaas, J. H., and Preuss, T. M. 2003. Human brain evolution. In: Squire, L. R., McConnell, S. K., Roberts, J. L., Spitzer, N. C., Zigmond, M. J., editors. Fundamental neuroscience (Second Edition). San Diego: Academic Press, pp. 1147–1166.

Kaas, J. H., Hall, W. C., and Diamond, I. T. 1970. Cortical visual area I and II in the hedgehog: the relation between evoked potential maps and achitectonic subdivisions. J Neurophysiol 33: 595–615.

Kaas, J. H., Guillery, R. W., and Allman, J. M. 1972. Some principles of organization in the dorsal lateral geniculate nucleus. Brain Behav Evol 6: 253–299.

Kaas, J. H., Huerta, M. F., Weber, J. T., and Harting, J. K. 1978. Patterns of retinal terminations and laminar organization of the lateral geniculate nucleus of primates. J Comp Neurol 182: 517–554.

Kahn, D. M., and Krubitzer, L. 2002a. Massive cross-modal cortical plasticity and the emergence of a new cortical area in developmentally blind mammals. Proc Natl Acad Sci (USA) 99: 11429–11434.

Kahn, D. M., and Krubitzer, L. 2002b. Retinofugal projections in the short-tailed opossum (*Monodelphis domestica*). J Comp Neurol 447: 114–127.

Kahn, M. C., Hough, G. E., II., ten Eyck, G. R., and Bingman, V. P. 2003. Internal connectivity of the homing pigeon (*Columba livia*) hippocampal formation: an anterograde and retrograde tracer study. J Comp Neurol 459: 127–141.

Källén, B. 1953. On the nuclear differentiation during ontogenesis in the avian forebrain and some notes on the amniote strio-amygdaloid complex. Acta Anat 17: 72–84.

Kamiya, T., and Pirlot, P. 1988. The brain of the Malayan bear (*Helarctos malayanus*). Z Zool Syst Evolutionsforsch 26: 225–235.

Kanwal, J. S., and Finger, T. E. 1992. Central representation and projections of gustatory systems. In: Hara, T. J., editor. Fish chemoreception. London: Chapman and Hall, pp. 79–103.

Kappelman, J. 1996. The evolution of body mass and relative brain size in fossil hominids. J Human Evol 30: 243–276.

Karnath, H. O. 2001. New insights into the functions of the superior temporal cortex. Nat Rev Neurosci 2: 568–576.

Karten, H. J. 1968. The organization of the ascending auditory pathway in the pigeon (*Columba livia*). II. Telencephalic projections of the nucleus ovoidalis thalami. Brain Res 11: 134–153.

Karten, H. J. 1969. The organization of the avian telencephalon and some speculations on the phylogeny of the amniote telencephalon. Ann NY Acad Sci 167: 164–179.

Karten, H. J., and Shimizu, T. 1989. The origins of neocortex: connections and lamination as distinct events in evolution. J Cog Neurosci 1: 291–301.

Karten, H. J., Hodos, W., Nauta, W. J. H., and Revzin, A. M. 1973. Neural connections of the "visual Wulst" of the avian telencephalon. Experimental studies in the pigeon (*Columba livia*) and owl (*Speotyto cunicularia*). J Comp Neurol 150: 253–277.

Kaslin, J., and Panula, P. 2001. Comparative anatomy of the histaminergic and other aminergic systems in zebrafish (*Danio rerio*). J Comp Neurol 440: 342–377.

Katz, M. J., and Lasek, R. L. 1978. Evolution of the nervous system: role of ontogenetic buffer mechanisms in the evolution of matching populations. Proc Natl Acad Sci (USA) 75: 1349–1352.

Katz, M. J., Lasek, R. L., and Kaiserman-Abramof, I. R. 1981. Ontophyletics of the nervous system: eyeless mutants illustrate how ontogenetic buffer mechanisms channel evolution. Proc Natl Acad Sci (USA) 78: 397–401.

Kauffman, S. A. 1993. The origins of order: self-organization and selection in evolution. New York: Oxford University Press.

Kaufman, J. A. 2003. On the expensive tissue hypothesis: independent support from highly encephalized fish. Curr Anthrop 5: 705–707.

Kawamura, T. 1947. Science of birdsong (reprinted in 1974, in Japanese). Tokyo: Chuokoronsha.

Kawasaki, M. 1997. Sensory hyperacuity in the jamming avoidance response of weakly electric fish. Curr Opin Neurobio 7: 473–479.

Keating, C., and Lloyd, P. E. 1999. Differential modulation of motor neurons that innervate the same muscle but use different excitatory transmitters in *Aplysia*. J Neurophysiol 82: 1759–1767.

Kenigfest, N., Martiníez-Marcos, A., Belekhova, M., Font, C., Lanuza, E., Desfilis, E., and Martínez-García, F. 1997. A lacertilian dorsal retinorecipient thalamus: a re-investigation in the old-world lizard *Podarcis hispanica*. Brain Behav Evol 50: 313–334.

Kermack, D. M., and Kermack, K. A. 1984. The evolution of mammalian characters. Washington, D.C.: Kapitan Szabo Publishers.

Kermack, K. A., and Mussett, F. 1983. The ear in mammal-like reptiles and early mammals. Acta Palaeontol Pol 28: 147–158.

Kicliter, E., and Northcutt, R. G. 1975. Ascending afferents to the telencephalon of ranid frogs: an anterograde degeneration study. J Comp Neurol 161: 239–254.

Killackey, H. P. 1994. Evolution of the human brain: a neuroanatomical perspective. In: Gazzaniga M., editor. The cognitive neurosciences. Boston: MIT Press, pp. 1243–1253.

Killackey, H. P., and Ebner, F. 1973. Convergent projection of three separate thalamic nuclei on to a single cortical area. Science 179: 283–285.

Kimmel, C. B. 1993. Patterning the brain of the zebrafish embryo. Annu Rev Neurosci 16: 707–732.

King, J. A., and Millar, R. P. 1995. Evolutionary aspects of gonadotropin-releasing hormone and its receptor. Cell Mol Neurobiol 15: 5–23.

Kirsch, J. A. W., and Johnson, J. I. 1983. Phylogeny through brain traits: trees generated by neural characters. Brain Behav Evol 22: 60–69.

Kirsch, J. A. W., Johnson, J. I., and Switzer, R. C. 1983. Phylogeny through brain traits: the mammalian family tree. Brain Behav Evol 22: 70–74.

Kirsch, J. A. W., Lapointe, F.-J., and Springer, M. S. 1997. DNA-hybridization studies of marsupials and their implications for metatherian classification. Austral J Zool 45: 211–280.

Kirsche, K., and Kirsche, W. 1964. Experimental study on the influence of olfactory nerve regeneration on forebrain regeneration of *Ambystoma mexicanum*. J Hirnforsch 7: 315–333.

Kitching, I. J., Forey, P. L., Humphries, C. J., and Williams, D. M. 1998. Cladistics. Oxford: Oxford University Press.

Kittler, R., Kayser, M., and Stoneking, M. 2003. Molecular evolution of *Pediculus humanus* and the origin of clothing. Curr Biol 13: 1414–1417.

Kitzes, L. M., Kageyama, G. H., Semple, M. N., and Kil, J. 1995. Development of ectopic projections from the ventral cochlear nucleus to the superior olivary complex induced by neonatal ablation of the contralateral cochlea. J Comp Neurol 353: 341–363.

Klein, R. G. 2000. Archeology and the evolution of human behavior. Evol Anthrop 9: 17–36.

Klekamp, J., Riedel, A., Harper, C., and Kretschmann, H.-J. 1994. Morphometric study on the postnatal growth of the visual cortex of Australian aborigines and Caucasians. J Hirnforsch 35: 541–548.

Kobayashi, H., and Koshima, S. 2001. Unique morphology of the human eye and its adaptive meaning: comparative studies on external morphology of the primate eye. J Human Evol 40: 419–435.

Konishi, M., Takahashi, T. T., Wagner, H., Sullivan, W. E., and Carr, C. E. 1988. Neurophysiological and anatomical sub-strates of sound localization in the owl. In: Edelman, G. M., Gan, W. E., Cowan, W. M., editors. Auditory function: neurobiological bases of hearing. New York: Wiley, pp. 721–745.

Kooy, F. H. 1917. The inferior olive in vertebrates. Folia Neurobiol 10: 205–369.

Kornack, D. R., and Rakic, P. 1998. Changes in cell-cycle kinetics during the development and evolution of primate neocortex. Proc Natl Acad Sci (USA) 95: 1242–1246.

Korneliussen, H. K. 1968. Comments on the cerebellum and its division. Brain Res 8: 229–236.

Kotrschal, K., Van Staden, M. J., and Huber, R. 1998. Fish brains: evolution and environmental relationships. Rev Fish Biol Fisheries 8: 373–408.

Kraemer, M., Zilles, K., Schleicher, A., Gebhard, R., Robbins, T. W., Everitt, B. J., and Divac, I. 1995. Quantitative receptor autoradiography of eight different transmitter-binding sites in the hippocampus of the common marmoset, *Callithrix jacchus*. Anat Embryol 191: 213–225.

Kramer, A., Crummett, T. L., and Wolpoff, M. H. 2001. Out of Africa and into the Levant: replacement or admixture in Western Asia? Quarternary Intl 75: 51–63.

Krawczyk, D. C. 2002. Contributions of the prefrontal cortex to the neural basis of human decision making. Neurosci Biobehav Rev 26: 631–664.

Krayniak, P. F., and Siegel, A. 1978. Efferent connections of the hippocampus and adjacent regions in the pigeon. Brain Behav Evol 15: 372–388.

Krebs, J. R., Sherry, D. F., Healy, S. D., Perry, H., and Vaccarino, A. L. 1989. Hippocampal specialization of food-storing birds. Proc Natl Acad Sci (USA) 86: 1388–1392.

Krubitzer, L. 1995. The organization of neocortex in mammals: are species differences really so different? Trends Neurosci 18: 408–417.

Kruger, L. 1959. The thalamus of the dolphin (*Tursiops truncatus*) and comparison with other mammals. J Comp Neurol 111: 133–194.

Kruska, D. 1988. Mammalian domestication and its effect on brain structure and behavior. In: Jerison H. J., Jerison I., editors. Intelligence and evolutionary biology. Berlin: Springer Verlag, pp. 211–250.

Kudo, H., and Dunbar, R. I. M. 2001. Neocortex size and social network size in primates. Anim Behav 62: 711–722.

Kuhlenbeck, H. 1967–1978 The central nervous system of vertebrates; a general survey of its comparative anatomy with an introduction to the pertinent fundamental biologic and logical concepts. Basel: S. Karger A.G.

Kuida, K., Hauydar, T. F., Kuan, C.-Y., Yang, D., Karasuyama, H., Rakic, P., and Flavell, R. A. 1998. Reduced apoptosis and cytochrome C-mediated caspase activation in mice lacking caspase 9. Cell 94: 325-337.

Kusuma, A., ten Donkelaar, H. J., and Nieuwenhuys, R. 1979. Intrinsic organization of the spinal cord. In: Gans, C., Northcutt, R. G., Ulinski, P., editors. Biology of the reptilia, Vol. 10: Neurology B. London: Academic Press.

Kuypers, H. G. J. M. 1958. Corticobulbar connexions to the pons and lower brain-stem in man. An anatomical study. Brain 10: 371–375.

Kuzawa, C. W. 1998. Adipose tissue in human infancy and childhood: an evolutionary perspective. Yearbook Phys Anthorpol 41: 177–209.

Lacalli, T. 1996. Frontal eye circuitry, rostral sensory pathways, and brain organization in amphioxus larvae: evidence from 3D reconstructions. Philos Trans R Soc Lond B 344: 165–185.

Lacalli, T. C., Holland, N. D., and West, J. E. 1994. Landmarks in the anterior central nervous system of amphioxus larvae. Philos Trans R Soc Lond B 344: 165–184.

Laing, D. G., Doty, R. L., and Breipohl, W., editors. 1991. The human sense of smell. Berlin: Springer-Verlag.

Land, M. F., and Osorio, D. C. 2003. Colour vision: colouring the dark. Curr Biol 13: R83–R85.

Lange, M. 2002. Who's afraid of *ceteris-paribus* laws? Or: how I learned to stop worrying and love them. Erkenntnis 57: 407–423.

Lange, W. 1975. Cell number and cell density in the cerebellar cortex of man and other mammals. Cell Tiss Res 157: 115–124.

Langer, S. K. 1942. Philosophy in a new key: a study in the symbolism of reason, rite, and art. Cambridge, MA: Harvard University Press.

Langworthy, O. R. 1967. A study of the brain of the porpoise, *Tursiops truncatus*. Brain 31: 225–235.

Lanuza, E., Belekhova, M., Martínez-Marcos, A., Font, C., and Martínez-García, F. 1998. Identification of the reptilian basolateral amygdala: an anatomical investigation of the afferents to the posterior dorsal ventricular ridge of the lizard *Podarcis hispanica*. Eur J Neurosci 10: 3517–3534.

Lapointe, F.-F., Baron, G., and Legendre, P. 1999. Encephalization, adaptation and evolution of chiroptera: a statistical analysis with further evidence for bat monophyly. Brain Behav Evol 54: 110–121.

Larsen, C. W., Zeltser, L. M., Lumsden, A. 2001. Boundary formation and compartition in the avian diencephalon. J Neurosci 21: 4699–4711.

Lasek, R. J., Joseph, B. S., and Whitlock, D. G. 1968. Evaluation of a radioautographic neuroanatomical tracing method. Brain Res 8: 319–336.

Lashley, K. S. 1929. Brain mechanisms and intelligence: a quantitative study of injuries to the brain. Chicago: University of Chicago Press.

Lashley, K. S., and Clark, G. 1946. The cytoarchitecture of the cerebral cortex of *Ateles*: a critical examination of architectonic studies. J Comp Neurol 85: 223–305.

Laska, M., and Seibt, A. 2002. Olfactory sensitivity for aliphatic alcohols in squirrel monkey and pigtail macaques. J Exp Biol 205: 1633–1643.

Latora, V., and Marchiori, M. 2001. Efficient behavior of small-world networks. Phys Rev Lett 87: Art. No. 198701.

Lauder, G. V. 1981. Form and function: structural analysis in evolutionary morphology. Paleobiology 7: 430–442.

Lauder, G. V., and Liem, K. F. 1989. The role of historical factors in the evolution of complex organismal functions. In: Wake, D. B., Roth, G., editors. Complex organismal functions: integration and evolution in vertebrates. New York: Wiley & Sons, pp. 63–78.

LaVail, J. H., and LaVail, M. M. 1972. Retrograde axonal transport in the central nervous system. Science 176: 1415–1417.

Lavdas, A. A., Grigoriou, M., Pachnis, V., and Parnavelas, J. G. 1999. The medial ganglionic eminence gives rise to a population of early neurons in the developing cerebral cortex. J Neurosci 99: 7881–7888.

Lee, S.-H., and Wolpoff, M. H. 2003. The pattern of evolution in Pleistocene human brain size. Paleobiology 29: 186–196.

Lefebvre, L., Whittle, P., Lascaris, E., and Finkelstein, A. 1997. Feeding innovations and forebrain size in birds. Anim Behav 53: 549–560.

Lefebvre, L., Nicolakakis, N., and Boire, D. 2002. Tools and brains in birds. Behaviour 139: 939–973.

Lehman, M. N., Lesauter, J., Kim, C., Berriman, S. J., Tresco, P. A., and Silver, R. 1995. How do fetal grafts of the suprachiasmatic nucleus communicate with the host brain? Cell Transplant 4: 75–81.

Leigh, S. R. 2001. Evolution of human growth. Evol Anthrop 10: 223–236.

Leise, E. M. 1990. Modular construction of nervous systems: a basic principle of design for invertebrates and vertebrates. Brain Res Rev 15: 1–23.

Lende, R. A. 1963. Cerebral cortex: a sensorimotor amalgam in the Marsupialia. Science 141: 730–732.

Lende, R. A. 1969. A comparative approach to the neocortex: localization in monotremes, marsupials and insectivores. Ann NY Acad Sci 167: 262–275.

Leonard, R. B., and Willis, W. D. 1979. The organization of the electromotor nucleus and extraocular motor nuclei in the stargazer (*Astroscopus y-graecum*). J Comp Neurol 183: 397–414.

Leonard, W. R., and Robertson, M. L. 1994. Evolutionary perspectives on human nutrition: the influence of brain and body size on diet and metabolism. Am J Hum Biol 6: 77–88.

Leonard, W. R., Robertson, M. L., Snodgrass, J. J., and Kuzawa, C. W. 2003. Metabolic correlates of hominid evolution. Comp Biochem and Physiol A 136: 5–15.

Letinic, K., and Rakic, P. 2001. Telencephalic origin of human thalamic GABAergic neurons. Nat Neurosci 4: 931–936.

Leutenegger, W. 1974. Functional aspects of pelvic morphology in simian primates. J Human Evol 3: 207–222.

Leutenegger, W. 1982. Encephalization and obstetrics in primates with particular reference to human evolution. In: Armstrong, E., Falk, D., editors. Primate brain evolution:

methods and concepts. New York: Plenum, pp. 85–95.

Leutenegger, W. 1987. Neonatal brain size and neurocranial dimensions in Pliocene hominids: implications for obstetrics. J Human Evol 16: 291–296.

LeVay, S., and Nelson, S. B. 1991. Columnar organization of the visual cortex. In: Leventhal, A. G., editor. The neural basis of visual function. London: MacMillan Press, pp. 266–315.

Levitt, J. B., Schumer, R. A., Sherman, S. M., and Spear, P. D. 2001. Visual response properties of neurons in the LGN of normally reared and visually deprived macaque monkeys. J Neurophysiol 85: 2111–2129.

Li, Q., and Martin, J. H. 2002. Postnatal development of connectional specificity of corticospinal terminals in the cat. J Comp Neurol 447: 57–71.

Liang, F., Moret, M., Wiesendanger, M., Rouiller, E. M. 1991. Corticomotoneuronal connections in the rat: evidence from double-labeling of motoneurons and corticospinal axon arborizations. J Comp Neurol 311: 356–366.

Lin, X.-W., Otto, C. J., and Peter, R. E. 1998. Evolution of neuroendocrine peptide systems: gonadotropin-releasing hormone and somatostatin. Comp Biochem and Physiol C 119: 375–388.

Linden, R. 1994. The survival of developing neurons: a review of afferent control. Neuroscience 58: 671–682.

Liscic, R. M., Zidar, J., and Mihelin, M. 1998. Evidence of direct connection of corticobulbar fibers to orofacial muscles in man: electromyographic study of individual motor unit responses. Muscle and Nerve 21: 561–566.

Liu, F.-G. R., Miyamoto, M. M., Freire, N. P., Ong, P. Q., Tennant, M. R., Young, T. S., and Gugel, K. F. 2001. Molecular and morphological supertrees for eutherian mammals. Science 291: 1786–1789.

Liu, G. B., and Pettigrew, J. D. 2003. Orientation mosaic in barn owl's visual Wulst revealed by optical imaging: comparison with cat and monkey striate and extra-striate areas. Brain Res 961: 153–158.

Llinás, R., Walton, K. D., and Lang, E. J. 2004. Cerebellum. In: Shepherd, G. M., editor. The synaptic organization of the brain. Oxford: Oxford University Press, pp. 271–309.

Lloll, R. 1512. Liber de ascensu, et descensu, intellectus. Republished by Plama de Mallorca in 1744.

Logan, C. A. 2001. "[A]re Norway rats . . . things?": Diversity versus generality in the use of albino rats in experiments on development and sexuality. J History Biol 34: 287–314.

Logan, C. A. 2002. Before there were standards: the role of test animals in the production of empirical generality in physiology. J History Biol 35: 329–363.

Lopreato, G. F., Lu, Y., Southwell, A., Atkinson, N. S., Hillis, D. M., Wilcox, T. P., and Zakon, H. H. 2001. Evolution and divergence of sodium channel genes in vertebrates. Proc Natl Acad Sci (USA) 98: 7588–7592.

Lorch, P. D., and Eadie, J. M. 1999. Power of concentrated changes test for correlated evolution. Syst Biol 48: 170–191.

Luksch, H., Cox, K., and Karten, H. J. 1998. Bottlebrush dendritic endings and large dendritic fields: motion-detecting neurons in the tectofugal pathway. J Comp Neurol 396: 399–414.

Lumsden, A. 1990. The cellular basis of segmentation in the developing hindbrain. Trends Neurosci 13: 329–335.

Luo, Z.-X., Crompton, A. W., and Sun, A.-L. 2001. A new mammaliaform from the early Jurassic and evolution of mammalian characteristics. Science 292: 1535–1540.

Luo, Z.-X., Kielan-Jaworowska, Z., and Cifellie, R. L. 2002. In quest for a phylogeny of Mesozoic mammals. Acta Palaeontol Pol 47: 1–78.

Luu, P., and Posner, M. I. 2003. Anterior cingulate cortex regulation of sympathetic activity. Brain 126: 2119–2120.

Lyon, D. C., and Kaas, J. H. 2002. Connectional evidence for dorsal and ventral V3, and other extrastriate areas in the prosimian primate, *Galago garnetti*. Brain Behav Evol 59: 114–129.

Ma, P. M. 1994. Catecholaminergic systems in the zebrafish. II. Projection pathways and pattern of termination of the locus coeruleus. J Comp Neurol 344: 256–269.

Ma, P. M. 1997. Catecholaminergic systems in the zebrafish. III. Organization and projection pattern of medullary dopaminergic and noradrenergic neurons. J Comp Neurol 381: 411–427.

Mace, G. M., Harvey, P. H., and Clutton-Brock, T. H. 1981. Brain size and ecology in small mammals. J Zool, Lond 193: 333–354.

MacLarnon, A. 1996. The scaling of gross dimensions of the spinal cord in primates and other species. J Human Evol 30: 71–87.

MacLean, P. D. 1972. Cerebral evolution and emotional processes: new findings on the striatal complex. Ann NY Acad Sci 193: 137–149.

MacLean, P. D. 1973. A triune concept of the brain and behaviour. Toronto: University of Toronto Press.

Macphail, E. M. 1982. Brain and intelligence in vertebrates. Oxford: Clarendon Press.

Maddison, W. P., Donoghue, M. J., and Maddison, D. R. 1984. Outgroup analysis and parsimony. Systematic Zoology 33: 83–103.

Madge, S., and Burn, H. 1994. Crows and jays. Boston: Houghton Mifflin Co.

Madsen, O., Scally, M., Douady, C. J., Kao, D. J., DeBry, R. W., Adkins, R., Amrine, H. M., Stanhope, M. J., de Jong, W. W., and Springer, M. S. 2001. Parallel adaptive radiations in two major clades of placental mammals. Nature 409: 610–614.

Maguire, E. A., Gadian, D. G., Johnsrude,, I. S., Good, C. D., Ashburner, J., Frackowiak, R. S. J., and Frith, C. D. 2000. Navigation-related structural change in the hippocampi of taxi drivers. Proc Natl Sci (USA) 97: 4398–4403.

Mai, J. K., Assheuer, J., Paxinos, G. 1997. Atlas of the human brain. San Diego: Academic Press.

Maier, N. R. F., and Schneirla, T. C. 1935. Principles of animal psychology. New York: McGraw-Hill.

Manger, P. R., Slutsky, D. A., and Molnár, Z. 2002. Visual subdivisions of the dorsal ventricular ridge of the iguana (*Iguana iguana*) as determined by electrophysiologic mapping. J Comp Neurol 453: 226–246.

Mangold-Wirz, K. 1966. Cerebralisation und Ontogenesemodus bei Eutherien. Acta Anat 63: 449–508.

Manley, G. A. 2000. Cochlear mechanisms from a phylogenetic viewpoint. Proc Natl Acad Sci (USA) 97: 11736–11743.

Marín F., and Puelles, L. 1995. Morphological fate of rhombomeres in quail/chick chimeras: a segmental analysis of hindbrain nuclei. Eur J Neurosci 7: 1714–1738.

Marín, O., and Rubenstein, J. L. R. 2001. A long, remarkable journey: tangential migration in the telencephalon. Nat Rev Neurosci 2: 780–790.

Marín, O., Smeets, W. J. A. J., and González, A. 1998. Basal ganglia organization in amphibians: evidence for a common pattern in tetrapods. Prog Neurobiol 55: 363–397.

Marino, L., Rilling, J. K., Lin, S. K., and Ridgway, S. H. 2000. Relative volume of the cerebellum in dolphins and comparison with anthropoid primates. Brain Behav Evol 56: 204–211.

Mark, R. 1996. Architecture and evolution. Am Scientist (July–August): 383–389.

Markl, H. 1985. Manipulation, modulation, information, cognition: Some of the riddles of communication. In: Hölldobler, B., Lindauer, M., editors. Experimental behavioral ecology and sociobiology. Sunderland, MA: Sinauer Assoc, pp. 163-194.

Marquis, D. G. 1934. Phylogenetic interpretation of the functions of the visual cortex. Arch Neurol Psychiatry 33: 807–812.

Marshall, C. R., Raff, E. C., and Raff, R. A. 1994. Dollo's law and the death and resurrection of genes. Proc Natl Acad Sci (USA) 91: 12283–12287.

Martin, G. F. 1983. Anatomical demonstration of the location and collateralization of rubral neurons which project to the spinal cord, lateral brainstem and inferior olive in the North American opossum. Brain Behav Evol 23: 93–109.

Martin, G. F., Cabana, T., and Humbertson, A. O. J. 1981. Evidence for a lack of distinct rubrospinal somatotopy in the North American opossum and for collateral innervation of the cervical and lumbar enlargements by single rubral neurons. J Comp Neurol 201: 255–263.

Martin, J. H., and Lee, S. J. 1999. Activity-dependent competition between developing corticospinal terminations. NeuroReport 10: 2277–2282.

Martin, R. D. 1981. Relative brain size and basal metabolic rate in terrestrial vertebrates. Nature 293: 57–60.

Martin, R. D. 1983. Human brain evolution in an ecological context. New York: American Museum of Natural History.

Martins, E. P. 2000. Adaptation and the comparative method. Trends Ecol Evol 15: 296–299.

Martins, E. P., Diniz-Filho, J. A. F., and Housworth, E. A. 2002. Adaptive constraints and the phylogenetic comparative method: a computer simulation test. Evolution 56: 1–13.

Maynard Smith, J., Burian, R., Kauffman, S., Alberch, P., Campbell, J., Goodwin, B., Lande, R., Raup, D., and Wolpert, L. 1985. Developmental constraints and evolution. Quart Rev Biol 60: 265–287.

Mayr, E. 1961. Cause and effect in biology. Science 134: 1501–1506.

McAllister, A. K. 2000. Cellular and molecular mechanisms of dendrite growth. Cereb Cortex 10: 963–973.

McAlpine, D., and Grothe, B. 2003. Sound localization and delay lines - do mammals fit the model? Trends Neurosci 26: 347–350.

McBrearty, S., and Brooks, A. S. 2000. The revolution that wasn't: a new interpretation of the origin of modern human behavior. J Human Evol 39: 453–563.

McCormick, C. A. 1989. Central lateral line mechanosensory pathways in bony fish. In: Coombs, S., Görner, P., Münz, H., editors. The mechanosensory lateral line: neurobiology and evolution. New York: Springer-Verlag, pp. 341–364.

McGrew, W. C. 1992. Chimpanzee material culture. Cambridge: Cambridge University Press.

McHenry, H. M. 1975. Biomechanical interpretation of the early hominid hip. J Human Evol 4: 343–356.

McIlwain, J. T. 1995. Lateral geniculate lamination and the corticogeniculate projection: a potential role in binocular vision in the quadrants. J theor Biol 172: 329–333.

McIntyre, L. 1997. Gould on laws in biological science. Bio Philos 12: 357–367.

McNab, B. K. 1989. Brain size and its relation to the rate of metabolism in mammals. Am Nat 133: 157–167.

McShea, D. W. 2000. Functional complexity in organisms: parts as proxies. Biol Philosophy 15: 641–668.

Medina, L., and Reiner, A. 1994. Distribution of choline-acetyltransferase immunoreactivity in the pigeon brain. J Comp Neurol 342: 497–537.

Medina, L., and Reiner, A. 2000. Do birds possess homologues of mammalian primary visual, somatosensory and motor cortices? Trends Neurosci 23: 1–12.

Medina, L., Smeets, W. J. A. J., Hoogland, P. V., and Puelles, L. 1993. Distribution of choline-acetyltransferase immunoreactivity in the brain of the lizard *Gallotia galloti*. J Comp Neurol 331: 261–285.

Medina, L., Veenman, C. L., and Reiner, A. 1997. Evidence for a possible avian dorsal thalamic region comparable to the mammalian ventral anterior, ventral lateral, and oral ventroposterolateral nuclei. J Comp Neurol 384: 86–108.

Meek, J., and Nieuwenhuys, R. 1991. Palisade pattern of mormyrid Purkinje cells: a correlated light and electron microscopic study. J Comp Neurol 306: 156–192.

Meek, J., Nieuwenhuys, R., and Elsevier, D. 1986a. Afferent and efferent connections of cerebellar lobe C1 of the mormyrid fish *Gnathonemus petersi*: an HRP study. J Comp Neurol 245: 319–341.

Meek, J., Nieuwenhuys, R., and Elsevier, D. 1986b. Afferent and efferent connections of cerebellar lobe C3 of the mormyrid fish *Gnathonemus petersi*: an HRP study. J Comp Neurol 245: 342–358.

Meek, J., Hafmans, T. G. M., Maler, L., and Hawkes, R. 1992. The distribution of zebrin II in the gigantocerebellum of the mormyrid fish *Gnathonemus petersii* compared with other teleosts. J Comp Neurol 316: 17–31.

Meinhardt, H., and Gierer, A. 2000. Pattern formation by local self-activation and lateral inhibition. Bioessays 22: 753–760.

Mel, B. W. 1994. Information-processing in dendritic trees. Neural Computation 6: 1031–1085.

Meltzoff, A. N. 1996. The human infant as imitative generalist: a 20-year progress report on infant imitation with implications for comparative psychology. In: Heyes, C. M., Galef, B. G., Jr., editors. Social Learning. San Diego: Academic Press, pp. 347–370.

Meng, J., and Wyss, A. R. 1995. Monotreme affinities and low-frequency hearing suggested by multituberculate ear. Nature 377: 141–144.

Messenger, J. B. 1996. Neurotransmitters of cephalopods. Invert Neurosci 2: 95–114.

Metcalfe, W. K., Mendelson, B., and Kimmel, C. B. 1986. Segmental homologies among reticulospinal neurons in the hindbrain of the zebrafish larva. J Comp Neurol 251: 147–159.

Meyer, A., and Dolven, S. I. 1992. Molecules, fossils, and the origin of tetrapods. J Molec Evol 35: 102–113.

Mezler, M., Fleischer, J., and Breer, H. 2001. Characteristic features and ligand specificity of the two olfactory receptor classes from *Xenopus laevis*. J Exp Biol 204: 2987–2997.

Middleton, F. A., and Strick, P. L. 2002. Basal ganglia 'projections' to the prefrontal cortex of the primate. Cerebral Cortex 12: 926–635.

Miller, E. K., and Cohen, J. D. 2001. An integrative theory of prefrontal cortex function. Annu Rev Neurosci 24: 167–202.

Miller, G. 2000. The mating mind. New York: Doubleday.

Millot, J., and Anthony, J. 1965. Anatomie de *Latimeria chalumnae*. Paris: Centre national de la recherche scientifique.

Mink, J. W., Blumenshine, R. J., and Adams, D. B. 1981. Ratio of central nervous system to body metabolism in vertebrates: its constance and functional basis. Am J Physiol 241: R203–R212.

Mish, F. C., editor. 1993. Merriam-Webster's collegiate dictionary (Tenth Edition). Springfield, MA: Merriam-Webster Inc.

Misson, J.-P., Austin, C. P., Takahashi, T., Cepko, C. L., and Caviness, V. S. J. 1991. The alignment of migrating neural cells in relation to the murine neopallial radial glial fiber system. Cereb Cortex 1: 221–229.

Möller, H. 1973. Zur Evolutionshöhe des Marsupialgehirns. Zool Jahrb Anat 91: 434–448.

Molnár, Z. 1998. Development and evolution of thalamocortical interactions. Eur J Morphol 38: 313–320.

Montagnese, C. M., Krebs, J. R., and Meyer, G. 1996. The dorsomedial and dorsolateral forebrain of the zebra finch, *Taeniopygia guttata*: a Golgi study. Cell Tiss Res 283: 263–282.

Moodie, R. L. 1915. A new fish brain from the coal measures of Kansas, with a review of other fossil brains. J Comp Neurol 25: 135–181.

Moore, B. R. 1992. Avian movement imitation and a new form of mimicry - tracing the evolution of a complex form of learning. Behaviour 122: 231–263.

Moore, J. K. 1980. The primate cochlear nuclei: loss of lamination as a phylogenetic process. J Comp Neurol 193: 609–629.

Morgan, C. L. 1894. An introduction to comparative psychology. London: Walter Scott.

Morgane, P. J., and Jacobs, L. F. 1972. Comparative anatomy of the cetacean nervous system. In: Harrison, R. J., editor. Functional anatomy of marine mammals. New York: Academic Press, pp. 117–244.

Morita, Y., and Finger, T. E. 1985. Topographic and laminar organization of the vagal gustatory system in the goldfish, *Carassius auratus*. J Comp Neurol 238: 187–201.

Morita, Y., Murakami, T., and Ito, H. 1983. Cytoarchitecture and topographic projections of the gustatory centers in a teleost, *Carassius carassius*. J Comp Neurol 218: 378–394.

Morris, S. C. 2004. Life's solution: inevitable humans in a lonely universe. Cambridge: Cambridge University Press.

Mueller, T., and Wullimann, M. F. 2002. BrdU-, *neuroD (nrd-)* and Hu-studies reveal unusual non-ventricular neurogenesis in the postembryonic zebrafish forebrain. Mech Dev 117: 123–135.

Müller, A. E., Thalmann, U. 2000. Origin and evolution of primate social organisation: a reconstruction. Biol Rev 75: 405–435.

Müller, G. B., Newman, S. A., editors. 2003. Origination of organismal form: beyond the gene in developmental and evolutionary biology. Cambridge, MA: MIT Press.

Mulligan, K. A., and Ulinski, P. S. 1990. Organization of geniculocortical projections in turtles: isoazimuth lamellae in the visual cortex. J Comp Neurol 296: 531–547.

Murphy, W. J., Eizirik, E., Johnson, W. E., Zhang, Y. P., Ryder, O. A., and O'Brien, S. J. 2001. Molecular phylogenetics and the origins of placental mammals. Nature 409: 614-618.

Murre, J. M. J., and Sturdy, D. P. F. 1995. The connectivity of the brain: multi-level quantitative analysis. Biol Cybern 73: 529–545.

Muske, L. E. 1993. Evolution of gonadotropin-releasing hormone (GnRH) neuronal systems. Brain Behav Evol 42: 215–230.

Myojin, M., Ueki, T., Sugahara, F., Murakami, Y., Shigetani, Y., Aizawa, S., Hirano, S., and Kuratani, S. 2001. Isolation of *Dlx* and *Emx* gene cognates in an agnathan species, *Lampetra japonica*, and their expression patterns during embryonic and larval development: conserved and diversified regulatory patterns of homeobox genes in vertebrate head evolution. J Exp Zool 291: 68–84.

Nakajima, K., Maier, M. A., Kirkwood, P. A., and Lemon, R. N. 2000. Striking differences in transmission of corticospinal excitation to upper limb motoneurons in two primate species. J Neurophysiol 84: 698–709.

Nauta, W. J. H. 1950. Über die sogenannte terminale Degeneration im Zentralnervensystem und ihre Darstellung durch Silberimprägnation. Arch Neurol Psychiat 66: 353–376.

Nauta, W. J. H. 1971. The problem of the frontal lobe: a reinterpretation. J Psychiat Res 8: 167–187.

Nauta, W. J. H., and Gygax, P. A. 1954. Silver impregnation of degenerating axons in the central nervous system: a modified technique. Stain Technol 29: 91–93.

Nauta, W. J. H., and Karten, H. J. 1970. A general profile of the vertebrate brain, with sidelights on the ancestry of cerebral cortex. In: Schmitt, F. O., editor. The neurosciences: second study program. New York: Rockefeller University Press, pp. 7–26.

Neary, T. J. 1990. The pallium of anuran amphibians. In: Jones, E. G., Peters, A., editors. Cerebral cortex, Vol. 8A. New York: Plenum, pp. 107–138.

Nelson, M. E., and Bower, J. M. 1990. Brain maps and parallel computers. Trends Neurosci 13: 403–408.

Newman, M. E. J., and Watts, D. J. 1999. Scaling and percolation in the small-world network model. Physical Review E 60: 7332–7342.

Nieto, M. A., Bradley, L. C., and Wilkinson, D. G. 1991. Conserved segmental expression of *Krox-20* in the vertebrate hindbrain and its relationship to lineage restriction. Development Suppl 2: 59–62.

Nieuwenhuys, R. 1994a. Comparative neuroanatomy: place, principles, practice and programme. Europ J Morphol 32: 142–155.

Nieuwenhuys, R. 1994b. The neocortex: an overview of its evolutionary development, structural organization and synaptology. Anat Embryol 190: 307–337.

Nieuwenhuys, R. 1998. Morphogenesis and general structure. In: Nieuwenhuys, R., Ten Donkelaar, H. J., Nicholson, C., editors. The central nervous system of vertebrates. Berlin: Springer Verlag, pp. 159–228.

Nieuwenhuys, R., and Bodenheimer, T. S. 1966. The diencephalon of the primitive bony fish

Polypterus in the light of the problem of homology. J Morphol 118: 415–450.

Nieuwenhuys, R., and Nicholson, C. 1969. A survey of the general morphology, the fiber connections, and the possible functional significance of the gigantocerebellum of mormyrid fishes. In: Llinás, R., editor. Neurobiology of cerebellar evolution and development. Chicago: American Medical Association/Education Research Foundation, pp. 107–134.

Nieuwenhuys, R., Ten Donkelaar, H. J., Nicholson, C. 1998. The central nervous system of vertebrates. Berlin: Springer Verlag.

Nilsson, G. E. 1996. Brain and body oxygen requirements of *Gnathonemus petersii*, a fish with an exceptionally large brain. J Exp Biol 199: 603–607.

Nilsson, G. E., Routley, M. H., and Renshaw, G. M. C. 2000. Low mass-specific brain Na+/K+-ATPase activity in elasmobranch compared to teleost fishes: Implications for the large brain size of elasmobranchs. Proc R Soc Bio Sci B 267: 1335–1339.

Nolfi, S. 1997. Using emergent modularity to develop control systems for mobile robots. Adaptive Behavior 5: 343–363.

Northcutt, R. G. 1978. Forebrain and midbrain organization in lizards and its phylogenetic significance. In: Greenberg, N., MacLean, P. D., editors. Behavior and neurology of lizards. Rockville, MD: National Institutes of Mental Health, pp. 11–64.

Northcutt, R. G. 1981. Evolution of the telencephalon in non-mammals. Ann Rev Neurosci 4: 301–350.

Northcutt, R. G. 1984. Evolution of the vertebrate central nervous system: patterns and processes. Am Zool 24: 701–716.

Northcutt, R. G. 1985a. Brain phylogeny: Speculations on pattern and cause. In: Cohen M. J., Strumwasser F., editors. Comparative Neurobiology. New York: Wiley, pp. 351–378.

Northcutt, R. G. 1985b. The brain and sense organs of the earliest craniates: reconstruction of a morphotype. In: Foreman, R. E., Gorbman, A., Dodds, J. M., Olson, R., editors. Evolutionary biology of primitive fishes. New York: Plenum Press, pp. 81–112.

Northcutt, R. G. 1986. Lungfish neural characters and their bearing on sarcopterygian phylogeny. J Morphol Suppl 1: 277–297.

Northcutt, R. G. 1991. Visual pathways in elasmobranchs: organization and phylogenetic implications. J Exp Zool Suppl 5: 97–107.

Northcutt, R. G. 1992. The phylogeny of octavolateralis ontogenies: a reaffirmation of Garstang's phylogenetic hypothesis. In: Webster, D. B., Fay, R. R., Popper, A. N., editors. The evolutionary biology of hearing. New York: Springer-Verlag, pp. 21–47.

Northcutt, R. G. 1995. The forebrain of gnathostomes: in search of a morphotype. Brain Behav Evol 46: 275–318.

Northcutt, R. G. 1998. Brains trust. Nature 392: 670–671.

Northcutt, R. G. 1999. Field homology: a meaningless concept. Eur J Morphol 37: 95–99.

Northcutt, R. G. 2001. Changing views of brain evolution. Brain Res Bull 55: 663–674.

Northcutt, R. G., and Braford, M. R. 1980. New observations on the organization and evolution of the telencephalon of actinopterygian fishes. In: Ebbesson, S. O. E., editor. Comparative neurology of the telencephalon. New York: Plenum, pp. 41–98.

Northcutt, R. G., and Kaas, J. H. 1995. The emergence and evolution of mammalian neocortex. Trends Neurosci 18: 373–379.

Northcutt, R. G., and Kicliter, E. 1980. Organization of the amphibian telencephalon. Comparative neurology of the telencephalon. New York: Plenum, pp. 203–255.

Northcutt, R. G., and Puzdrowski, R. L. 1988. Projections of the olfactory bulb and nervus terminalis in the silver lamprey. Brain Behav Evol 32: 96–107.

Northcutt, R. G., and Royce, G. J. 1975. Olfactory bulb projections in the bullfrog *Rana catesbeiana*. J Morphol 145: 251–268.

Northcutt, R. G., and Striedter, G. F. 2002. An explanation for telencephalic eversion in ray-finned fishes. Brain Behav Evol 60: 63.

Northcutt, R. G., and Wullimann, M. F. 1988. The visual system in teleost fishes: morphological patterns and trends. In: Atema, J., Fay, R. R., Popper, A. N., Tavolga, W. N., editors. Sensory biology of aquatic animals. New York: Springer, pp. 515–552.

Northmore, D. P. M. 1984. Visual and saccadic activity in the goldfish torus longitudinalis. J Comp Physiol A 155: 333–340.

Nudo, R. J., and Masterton, R. B. 1990a. Descending pathways to the spinal cord, III:

Sites of origin of the corticospinal tract. J Comp Neurol 296: 559–583.

Nudo, R. J., and Masterton, R. B. 1990b. Descending pathways to the spinal cord, IV: Some factors related to the amount of cortex devoted to the corticospinal tract. J Comp Neurol 296: 584–597.

O'Connell, J. F., Hawkes, K., and Blurton Jones, N. G. 1999. Grandmothering and the evolution of *Homo erectus*. J Human Evol 36: 461–485.

O'Connor, D. H., Fukui, M. M., Pinsk, M. A., and Kastner, S. 2002. Attention modulates responses in the human lateral geniculate nucleus. Nat Neurosci 5: 1203–1209.

O'Donohue, W., and Buchanan, J. A. 2001. The weaknesses of strong inference. Behavior and Philosophy 29: 1–20.

Ogawa, T. 1935. Über Nucleus ellipticus und den Nucleus ruber beim Delphin. Arb Anat Inst Sendaia 17: 55–61.

Ohama, T., Suzuki, T., Mori, M., Osawa, S., Ueda, T., Watanabe, K., and Nakase, T. 1993. Non-universal decoding of the leucine codon CUG in several Candida species. Nucl Acids Res 21: 4039–4045.

Ohno, S. 1970. Evolution by gene duplication. Berlin: Springer Verlag.

O'Leary, D. D. M. 1992. Development of connectional diversity and specificity in the mammalian brain by the pruning of collateral projections. Curr Opin Neurobio 2: 70–77.

O'Leary, D. D. M., and Nakagawa, Y. 2002. Patterning centers, regulatory genes and extrinsic mechanisms controling arealization of the neocortex. Curr Opin Neurobio 12: 14–25.

Olivares, R., Michalland, S., and Aboitiz, F. 2000. Cross-species and intraspecies morphometric analysis of the corpus callosum. Brain Behav Evol 55: 37–43.

Onodera, S., and Hicks, T. P. 1999. Evolution of the motor system: why the elephant's trunk works like a human's hand. Neuroscientist 5: 217–226.

Oppenheim, R. W. 1985. Naturally occurring cell death during neural development. Trends Neurosci 8: 487–493.

Osorio, D., and Vorobyev, M. 1996. Colour vision as an adaptation to frugivory in primates. Proc R Soc Bio Sci B 263: 593–599.

Oster, H. 1998. Non-Mendelian genetics in humans. Oxford: Oxford University Press.

Owen, A. M., Herrod, N. J., Menon, D. K., Clark, J. C., Downey, S. P. M. J., Carpenter, T. A., Minhas, P. S., Turkheimer, F. E., Williams, E. J., Robbins, T. W., Sahakian, B. J., Petrides, M., and Pickard, J. D. 1999. Redefining the functional organization of working memory processes within human lateral prefrontal cortex. Eur J Neurosci 11: 567–574.

Owen, R. 1843. Lectures on the comparative anatomy and physiology of the invertebrate animals. London: Longmans Green.

Owen, R. 1846. Report on the archetype and homologies of the vertebrate skeleton. Report of the British association for the advancement of science (Southampton meeting), pp. 169–340.

Owen, R. 1857. On the characters, principles of division, and primary groups of the class MAMMALIA. J Proc Linn Soc 2: 1–37.

Owen, R. 1859. Contributions to the natural history of the anthropoid apes. No VIII. On the external characters of the Gorilla (*Troglodytes Gorilla*, Sav.). Transact Zool Soc 5: 243–283.

Oyama, S. 1985. The ontogeny of information. New York: Cambridge University Press.

Pagel, M. D., and Harvey, P. H. 1989. Taxonomic differences in the scaling of brain on body weight among mammals. Science 244: 1589–1593.

Pakkenberg, B., and Gundersen, H. J. G. 1997. Neocortical neuron number in humans: effect of sex and age. J Comp Neurol 384: 312–320.

Pandya, D. N. 1995. Anatomy of the auditory cortex. Rev Neurol 151: 486–494.

Panger, M. A., Brooks, A. S., Richmond, B. G., and Wood, B. 2002. Older than the Oldowan? Rethinking the emergence of hominin tool use. Evol Anthrop 11: 235–245.

Papadimitriou, A., and Wynne, C. D. L. 1999. Preserved negative patterning and impaired spatial learning in pigeons (*Columba livia*) with lesions of the hippocampus. Behav Neurosci 113: 683–690.

Papez, J. W. 1929. Comparative neurology. New York: Thomas Y. Crowell Co.

Parent, A. 1986. Comparative neurology of the basal ganglia. New York: Wiley.

Parent, A., and Hazrati, L.-N. 1995a. Functional anatomy of the basal ganglia. I. The corticobasal ganglia-thalamo-cortical loop. Brain Res Rev 20: 91–127.

Parent, A., and Hazrati, L.-N. 1995b. Functional anatomy of the basal ganglia. II. The place of subthalamic nucleus and external pallidum in basal ganglia circuitry. Brain Res Rev 20: 128–154.

Parker, S. T., and Gibson, K. R. 1977. Object manipulation, tool use, and sensorimotor intelligence as feeding adaptations in cebus monkeys and great apes. J Human Evol 6: 623–641.

Parkins, E. J. 1997. Cerebellum and cerebrum in adaptive control and cognition: a review. Biol Cybern 77: 79–87.

Passingham, R. E. 1973. Anatomical differences between the neocortex of man and other primates. Brain Behav Evol 7: 337–359.

Passingham, R. E. 1975. Changes in the size and organisation of the brain in man and his ancestors. Brain Behav Evol 11: 73–90.

Passingham, R. E. 2002. The frontal cortex: does size matter? Nat Neurosci 5: 190–192.

Passingham, R. E., Stephan, K. E., and Kötter, R. 2002. The anatomical basis of functional localization in the cortex. Nat Rev Neurosci 3: 606–616.

Pavlidis, P., Madison, D. V. 1999. Synaptic transmission in pair recordings from CA3 pyramidal cells in organotypic culture. J Neurophysiol 81: 2787–2797.

Pawlowski, B. 1998. Why are human newborns so big and fat. Human Evol 13: 65–72.

Payne, B. R. 1993. Evidence for visual cortical area homologs in cat and macaque monkey. Cerebral cortex 3: 1–25.

Payne, R. S., and McVay, S. 1971. Songs of humpback whales. Science 173: 585–597.

Pearce, S. L., Miles, T. S., Thompson, P. D., and Nordstrom, M. A. 2003. Responses of single motor units in human masseter to transcranial magnetic stimulation of either hemisphere. J Physiol 549: 583–596.

Pearson, R. 1972. The avian brain. London: Academic Press.

Peccei, J. S. 2001. A critique of the grandmother hypotheses: old and new. Am J Hum Biol 13: 434–452.

Peichl, L., Behrmann, G., and Kröger, R. H. H. 2001. For whales and seals the ocean is not blue: a visual pigment loss in marine mammals. Eur J Neurosci 13: 1520–1528.

Penn, A. A., Riquelme, P. A., Feller, M. B., and Shatz, C. J. 1998. Competition in retinogeniculate patterning driven by spontaneous activity. Science 279: 2108–2112.

Pepperberg, I. M. 1981. Functional vocalizations by an African gray parrot (*Psittacus erithacus*). Z Tierpsychol 55: 139–160.

Percy, W. 1975. The message in the bottle: how queer man is, how queer language is, and what one has to do with the other. New York: Farrar, Straus and Giroux.

Pérez-Clausell, J. 1988. Organization of zinc-containing terminal fields in the brain of the lizard *Podarcis hispanica*: a histochemical study. J Comp Neurol 267: 153–171.

Petersen, S. E., Fox, P. T., Posner, M. I., Mintun, M., Raichle, M. E. 1988. Positron emission tomographic studies of the cortical anatomy of single-word processing. Nature 331: 585–589.

Peterson, I. 1999. The honeycomb conjecture. Science News 156: 60.

Petrides, M., and Pandya, D. N. 1999. Dorsolateral prefrontal cortex: comparative cytoarchitectonic analysis in the human and the macaque brain and corticocortical connection patterns. Eur J Neurosci 11: 1011–1036.

Petrides, M., and Pandya, D. N. 2001. Comparative cytoarchitectonic analysis of the human and the macaque ventrolateral prefrontal cortex and corticocortical connection patterns in the monkey. Eur J Neurosci 16: 291–310.

Pettigrew, J. 1989. Phylogenetic relations between microbats, megabats and primates (Mammalia, Chiroptera and primates). Phil Trans R Soc Lond B 325: 489–559.

Pettigrew, J. D. 1979. Binocular visual processing in the owl's telencephalon. Proc R Soc Lond B 204: 435–454.

Pettigrew, J. D. 1986. Flying primates? Megabats have the advanced pathway from eye to midbrain. Science 231: 1304–1306.

Pettigrew, J. D., and Konishi, M. 1976. Neurons selective for orientation and binocular disparity in the visual Wulst of the barn owl (*Tyto alba*). Science 193: 675–678.

Pierrot-Deseilligny, E. 2002. Propriospinal transmission of part of the corticospinal excitation in humans. Muscle and Nerve 26: 155–172.

Pirlot, P., Kamiya, T. 1982. Relative size of brain and brain components in three gliding placentals (Dermoptera: Rodentia). Can J Zool 60: 565–572.

Platel, R. 1979. Brain weight - body weight relationships. In: Gans, C., Northcutt, R. G., Ulinski, P., editors. Biology of the reptilia. London: Academic Press, pp. 147–171.

Platel, R., and Vesselkin, N. P. 1989. Etude comparé de l'encéphalisation chez 3 esp´ces de Pétromyzontidae (Agnatha): *Petromyzon marinus, Lampetra fluviatilis* et *Lampetra planeri*. J Hirnforsch 30: 23–32.

Platel, R., Ridet, J. M., Bauchot, R., and Diagne, M. 1977. L'organisation encéphalique chez Amia, Lepisosteus et Polypterus: Morphologie et analyse quantitative compareés. J Hirnforsch 18: 69–73.

Platt, J. R. 1964. Strong inference. Science 146: 347–353.

Poincaré, J. H. 1902. La Science et l'Hypothese. Paris: Flammarion.

Polyak, S. L. 1957. The vertebrate visual system. Chicago: University of Chicago Press.

Pombal, M. A., and Puelles L. 1999. Prosomeric map of the lamprey forebrain based on calretinin immunocytochemistry, Nissl stain, and ancillary markers. J Comp Neurol 414: 391–422.

Portmann, A. 1947. Étude sur la cérébralisation chez les oiseaux. II. Les indices intracérébraux. Alauda 15: 1–15.

Portmann, A. 1962. Cerebralisation und Ontogenese. Medizinische Grundlagenforschung 4: 1–62.

Preuss, T. M. 1995a. Do rats have prefrontal cortex? The Rose-Woolsey-Akert program reconsidered. J Cognitive Neurosci 7: 1–24.

Preuss, T. M. 1995b. The argument from animals to humans in cognitive neuroscience. In: Gazzaniga M. S., editor. The cognitive neurosciences. Cambridge, MA: MIT Press, pp. 1227–1241.

Preuss, T. M. 2001. The discovery of cerebral diversity: an unwelcome scientific revolution. In: Falk D., Gibson K., editors. Evolutionary anatomy of the primate cerebral cortex. Cambridge: Cambridge University Press, pp. 138-164.

Preuss, T. M. In press. Evolutionary specializations of primate brain systems. In: Ravoso, M. J., Dagosto, M., editors. Primate origins

and adaptations. New York: Kluwer Academic/Plenum Press.

Preuss, T. M., and Coleman, G. Q. 2002. Human-specific organization of primary visual cortex: alternating compartments of dense Cat-301 and calbindin immunoreactivity in layer 4A. Cereb Cortex 12: 671–691.

Preuss, T. M., and Goldman-Rakic, P. S. 1991a. Architectonics of the parietal and temporal association cortex in the strepsirhine primate *Galago* compared to the anthropoid primate *Macaca*. J Comp Neurol 310: 475–506.

Preuss, T. M., and Goldman-Rakic, P. S. 1991b. Ipsilateral cortical connections of granular frontal cortex in the strepsirhine primate *Galago*, with comparative comments on anthropoid primates. J Comp Neurol 310: 507–549.

Preuss, T. M., and Goldman-Rakic, P. S. 1991c. Myelo- and cytoarchitecture of the granular frontal cortex and surrounding regions in the strepsirhine primate *Galago* and the anthropoid primate *Macaca*. J Comp Neurol 310: 429–474.

Preuss, T. M., and Kaas, J. H. 1999. Human brain evolution. In: Zigmond, M. J., Bloom, F. E., Landis, S. C., Roberts, J. L., Squire, L. R., editors. Fundamental Neuroscience. San Diego, CA: Academic Press, pp. 1283–1308.

Preuss, T. M., Stepniewska, I., and Kaas, J. H. 1996. Movement representation in the dorsal and ventral premotor areas of owl monkeys: a microstimulation study. J Comp Neurol 371: 649–676.

Pritz, M. B. 1974. Ascending connections of a thalamic auditory area in a crocodile, *Caiman crocodilus*. J Comp Neurol 153: 199–214.

Pritz, M. B., and Stritzel, M. E. 1992. A second auditory area in the non-cortical telencephalon of a reptile. Brain Res 569: 146–151.

Prothero, J. 1997. Scaling of cortical neuron density and white matter volume in mammals. J Brain Res 4: 513–524.

Puelles, L. 2001. Thoughts on the development, structure and evolution of the mammalian and avian telencephalic pallium. Phil Trans R Soc London B 356: 1583–1589.

Puelles, L., and Medina, L. 1994. Development of neurons expressing tyrosine hydroxylase and dopamine in the chicken brain: a comparative segmental analysis. In: Smeets, W. J. A. J., Reiner, A., editors. Phylogeny and development of catecholamine systems in the CNS of vertebrates. Cambridge: Cambridge University Press, pp. 381–404.

Puelles, L., and Medina, L. 2002. Field homology as a way to reconcile genetic and developmental variability with adult homology. Brain Res Bull 57: 243–255.

Puelles, L., and Rubenstein, J. L. R. 1993. Expression patterns of homeobox and other putative regulatory genes in the embryonic mouse forebrain suggest a neuromeric organization. Trends Neurosci 16: 472–479.

Puelles, L., and Rubenstein, J. L. R. 2003. Forebrain gene expression domains and the evolving prosomeric model. Trends Neurosci 26: 469–476.

Puelles, L., Amat, J. A., and Martinez-de-la Torre, M. 1987a. Segment-related, mosaic neurogenetic pattern in the forebrain and mesencephalon of early chick embryos: I. Topography of AChE-positive neuroblasts up to stage HH18. J Comp Neurol 266: 247–268.

Puelles, L., Doménech-Ratto, G., and Martínez-de-la-Torre, M. 1987b. Location of the rostral end of the longitudinal brain axis: Review of an old topic in the light of marking experiments on the closing rostral neuropore. J Morphol 194: 163–171.

Puelles, L., Kuwana, E., Puelles, E., Bulfone, A., Shimamura, K., Keleher, J., Smiga, S., and Rubenstein, J. L. R. 2000. Pallial and subpallial derivatives in the embryonic chick and mouse telencephalon, traced by the expression of the genes *Dlx-2, Emx-1, Nkx-2.1, Pax-6* and *Tbr-1*. J Comp Neurol 424: 409–438.

Pugesek, B. H., and Tomer, A. 1996. The Bumpus house sparrow data: a reanalysis using structural equation models. Evol Ecol 10: 387–404.

Purves, D. 1988. Body and brain: a trophic theory of neural connections. Cambridge, MA: Harvard University Press.

Purves, D., Riddle, D. R., and LaMantia, A.-S. 1992. Iterated patterns of brain circuitry (or how the cortex got its spots). Trends Neurosci 15: 362–368.

Purvis, A. 1995. A composite estimate of primate phylogeny. Phil Trans R Soc London B 348: 405–421.

Raff, R. A. 1996. The shape of life: genes, development, and the evolution of animal form. Chicago: University of Chicago Press.

Raff, R. A., and Kaufman, T. C. 1983. Embryos, genes, and evolution. New York: MacMillan Co.

Rahn, H., Paganelli, C. V., Ar, R. 1975. Relation of avian egg weight to body weight. The Auk 92: 750–765.

Raizada, R. D. S., and Grossberg, S. 2003. Towards a theory of the laminar architecture of cerebral cortex: computational clues from the visual system. Cereb Cortex 13: 100–113.

Rakic, P. 1972. Mode of cell migration to the superficial layers of fetal monkey neocortex. J Comp Neurol 145: 61–84.

Rakic, P. 1977. Prenatal development of the visual system in rhesus monkey. Phil Trans R Soc London B 278: 245–260.

Rakic, P. 1981. Development of visual centers in the primate brain depends on binocular competition before birth. Science 214: 928–931.

Rakic, P. 1986. Mechanisms of ocular dominance segregation in the lateral geniculate nucleus: competitive elimination hypothesis. Trends Neurosci 9: 11–15.

Rakic, P., Suñer, I., and Williams, R. W. 1991. A novel cytoarchitectonic area induced experimentally within the primate visual cortex. Proc Natl Acad Sci (USA) 88: 2083–2087.

Ramón-Moliner, E., and Nauta, W. J. H. 1966. The isodendritic core of the brain stem. J Comp Neurol 126: 311–336.

Ramón y Cajal, S. 1909. Histologie du Systéme Nerveux de l'Homme et des Vertébrés. Azoulay, translator. Paris: A. Maloine. Translated from Spanish by L. Azoulay.

Reader, S. M., and Laland, K. N. 2002. Social intelligence, innovation, and enhanced brain size in primates. Proc Natl Acad Sci (USA) 99: 4436–4441.

Redies, C., and Takeichi, M. 1996. Cadherins in the developing nervous system: an adhesive code for segmental and functional subdivisions. Dev Biol 180: 413–423.

Redies, C., Ast, M., Nakagawa, S., Takeichi, M., Martínez-de-la-Torre, M., and Puelles, L. 2000. Morphologic fate of diencephalic prosomeres and their subdivisions revealed by mapping cadherin expression. J Comp Neurol 421: 481–514.

Redies, C., Kovjanic, D., Heyers, D., Medina, L., Hirano, S., Suzuki, S. T., and Puelles, L. 2002. Patch/matrix patterns of gray matter differentiation in the telencephalon of chicken and mouse. Brain Res Bull 57: 489–493.

Regan, B. C., Julliot, C., Simmen, B., Viénot, F., Charles-Dominique, P., and Mollon, J. D. 2001. Fruits, foliage and the evolution of primate colour vision. Phil Trans R Soc Lond B 356: 229–283.

Rehkämper, G. 1984. Remarks upon Ebbesson's presentation of a parcellation theory of brain development. Z Zool Syst Evolutionsforsch 22: 321–327.

Reiner, A. 1993. Neurotransmitter organization and connections of turtle cortex: implications for the evolution of mammalian isocortex. Comp Biochem Physiol 104A:735–748.

Reiner, A. 2000. A hypothesis as to the organization of cerebral cortex in the common amniote ancestor of modern reptiles and mammals. In: Bock, G. R., Carew, G., editors. Evolutionary developmental biology of the cerebral cortex. Chichester: Wiley, pp. 83–113.

Reiner, A., Medina, L., and Veenman, C. L. 1998. Structural and functional evolution of the basal ganglia in vertebrates. Brain Res Rev 28: 235–285.

Reiner, A., Perkel, D. J., Bruce, L. L., Butler, A. B., Csillag, A., Kuenzel, W., Medina, L., Paxinos, G., Shimizu, T., Striedter, G. F., Wild, M., Ball, G., Durand, S., Güntürkün, O., Lee, D. W., Mello, C. V., Powers, A., White, S. A., Hough, G., Kubikova, L., Smulders, T. V., Wada, K., Douglas-Ford, J., Husband, S., Yamamoto, K., Yu, J., Siang, C., and Jarvis, E. D. 2004. Revised nomenclature for avian telencephalon and some related brainstem nuclei. J Comp Neurol 473: 377–414.

Rendell, L., and Whitehead, H. 2001. Culture in whales and dolphins. Behav Brain Sci 24: 309–382.

Rensch, B. 1960. Evolution above the species level. New York: Columbia University Press.

Retzius, G. 1893. Ependym und Neuroglia bei den Cyclostomen. Biol Untersuch (Stockh) 5: 15–18.

Richardson, M. K. 1999. Vertebrate evolution: the developmental origins of adult variation. BioEssays 21: 604–613.

Richardson, M. K., Hanken, J., Gooneratne, M. L., Pieau, C., Raynaud, A., Selwood, L., and Wright, G. M. 1997. There is no highly conserved embryonic stage in vertebrates: implications for current theories of evolution and development. Anat Embryol 196: 91–106.

Ridgway, S. H. 1986. Physiological observations on dolphin brains. In: Schusterman, R. J., Thomes, J. A., Wood, F. G., editors. Dolphin cognition and behavior: a comparative approach. Hillsdale, N.J.: L. Erlbaum Associates, pp. 31–59.

Ridley, M. 1983. The explanation of organic diversity: the comparative method and adaptations for mating. Oxford: Oxford University Press.

Ridley, M. 1986. Evolution and classification: the reformation of cladism. Essex: Longman Group.

Rilling, J. K., and Insel, T. R. 1998. Evolution of the cerebellum in primates: differences in relative volume among monkeys, apes and humans. Brain Behav Evol 52: 308–314.

Rilling, J. K., and Insel, T. R. 1999a. Differential expansion of neural projection systems in primate brain evolution. Neuroreport 10: 1453–1459.

Rilling, J. K., and Insel, T. R. 1999b. The primate neocortex in comparative perspective using magnetic resonance imaging. J Human Evol 37: 191–223.

Rilling, J. K., and Seligman, R. A. 2002. A quantitative morphometric comparative analysis of the primate temporal lobe. J Human Evol 42: 505–533.

Ringo, J. L. 1991. Neuronal interconnection as a function of brain size. Brain Behav Evol 38: 1–6.

Ringo, J. L., Doty, R. W., Demeter, S., and Simard, P. Y. 1994. Time is of the essence: a conjecture that hemispheric specialization arises from interhemispheric conduction delay. Cereb Cortex 4: 331–343.

Rink, E., and Wullimann, M. F. 2002. Connections of the ventral telencephalon and tyrosine hydroxylase distribution in the zebrafish brain (Danio rerio) lead to identification of an ascending dopaminergic system in a teleost. Brain Res Bull 57: 385–387.

Rizzolatti, G., and Arbib, M. A. 1998. Language within our grasp. Trends Neurosci 21: 188–194.

Roberts, B. L., and Ryan, K. P. 1975. Cytological features of the giant neurons controlling electric discharge in the ray Torpedo. J Mar Biol Assoc UK 55: 123–131.

Robson, J. A. 1983. The morphology of corticofugal axons to the dorsal lateral geniculate nucleus in the cat. J Comp Neurol 216: 89–103.

Rockel, A. J., Hiorns, R. W., and Powell, T. P. S. 1980. The basic uniformity in structure of the neocortex. Brain 103: 221–244.

Rödel, R. M.W., Laskawi, R., and Markus, H. 2003. Tongue representation in the lateral cortical motor region of the human brain as assessed by transcranial magnetic stimulation. Annal Otol Rhinol Laryngol 112: 71–76.

Rodman, H. R., and Moore, T. 1997. Development and plasticity of extrastriate visual cortex in monkeys. In: Rockland, K. S., Kaas, J. H., Peters, A., editors. Cerebral Cortex, Vol. 12: Extrastriate cortex in primates. New York: Plenum Press, pp. 639–672.

Rogers, L. J., and Andrews, R. J., editors. 2002. Comparative vertebrate lateralization. Cambridge: Cambridge University Press.

Rogers, S. W. 1999. *Allosaurus*, crocodiles, and birds: evolutionary clues from spiral computed tomography of an endocast. Anat Rec (New Anat) 257: 162–173.

Roland, P. E. 1993. Partition of the cerebellum in sensory motor activities, learning and cognition. Can J Neurol Sci 20: 75–77.

Romanski, L. M., Giguere, M., Bates, J. F., and Goldman-Rakic, P. S. 1997. Topographic organization of medial pulvinar connections with the prefrontal cortex in the rhesus monkey. J Comp Neurol 379: 313–332.

Romer, A. S. 1955. The vertebrate body. Philadelphia: W. B. Saunders Co.

Romer, A. S., and Parsons, T. S. 1977. The vertebrate body. Philadelphia: Saunders College Publishing.

Rosa, M. P. G. 1999. Topographic organisation of extrastriate areas in the flying fox: implications for the evolution of mammalian visual cortex. J Comp Neurol 411: 503–523.

Rosa, M. P. G. 2002. Visual maps in the adult primate cerebral cortex: some implications for brain development and evolution. Braz J Med Biol Res 35: 1485–1498.

Rosa, M. P. G., and Krubitzer, L. A. 1999. The evolution of visual cortex: where is V2? Trends Neurosci 22: 242–248.

Rosa, M. P. G., and Schmid, L. M. 1994. Topography and extent of visual-field representation in the superior colliculus of the megachiropteran *Pteropus*. Visual Neurosci 11: 1037–1057.

Rose, G., and Heiligenberg, W. 1985. Structure and function of electrosensory neurons in the torus semicircularis of *Eigenmannia*: morphological correlates of phase and amplitude sensitivity. J Neurosci 5: 2269–2280.

Rose, J. E. 1942. The ontogenetic development of the rabbit's diencephalon. J Comp Neurol 77: 61–129.

Rose, J. E., and Woolsey, C. N. 1948. The orbitofrontal cortex and its connections with the mediodorsal nucleus in rabbit, sheep and cat. Research Publications - Association for Research in Nervous Mental Disease 27: 210–232.

Rose, M. R. 1980. The mental arms race amplifier. Human Ecol 8: 2.

Rosen, D. E., Forey, P. L., Gardiner, B. G., and Patterson, C. 1981. Lungfishes, tetrapods, paleontology, and plesiomorphology. Bull Am Mus Nat Hist 167: 163–275.

Rosenberg, A. R. 2001. How is biological explanation possible? Brit J Phil Sci 52: 735–760.

Rosenberg, K. R. 1992. The evolution of modern human childbirth. Yearbook Phys Anthropol 35: 89–124.

Ross, C. 1996. Adaptive explanation for the origin of the anthropoidea (primates). Am J Primatol 40: 205–230.

Roth, G., Blanke, J., and Ohle, M. 1995. Brain size and morphology in miniaturized plethodontid salamanders. Brain Behav Evol 45: 84–95.

Rouquier, S., Blancher, A., and Giorgi, D. 2000. The olfactory receptor gene repertoire in primates and mouse: evidence for reduction of the functional fraction in primates. Proc Natl Acad Sci (USA) 97: 2870–2874.

Rowe, M. 1990. Organization of the cerebral cortex in monotremes and marsupials. In: Jones, E. G., Peters, A., editors. Cerebral cortex. New York: Plenum, pp. 263–334.

Rowe, T. 1996. Coevolution of the mammalian middle ear and neocortex. Science 273: 651–654.

Rowe, T. 1997. Comparative rates of development in Monodelphis and Didelphis. Science 275: 684.

Rowlett, R. M. 2000. Fire control by *Homo erectus* in East Africa and Asia. Acta Anthropol Sin 19: 198–208.

Ruben, J. A., Jones, T. D., and Geist, N. R. 1998. Respiratory physiology of the dinosaurs. Bioessays 20: 852–859.

Rubenstein, J. L. R., and Puelles, L. 1994. Homeobox gene expression during development of the vertebrate brain. Curr Topics Dev Biol 29: 1–65.

Rubenstein, J. L. R., Martinez, S., Shimamura, K., and Puelles, L. 1994. The embryonic vertebrate forebrain: the prosomeric model. Science 266: 578–580.

Rubidge, B. S., and Sidor, C. A. 2001. Evolutionary patterns among Permo-Triassic therapsids. Ann Rev Ecol Syst 32: 449–480.

Ruff, C. B., Trinkaus, E., and Holliday, T. W. 1997. Body mass and encephalization in Pleistocene *Homo*. Nature 387: 173–176.

Rumelhart, D. E., McClelland, J. L., P.DP Research Group, editors. 1986. A general framework for parallel distributed processing. Cambridge, MA: MIT Press.

Rupp, B., and Northcutt, R. G. 1998. The diencephalon and pretectum of the white sturgeon (*Acipenser transmontanus*): a cytoarchitectonic study. Brain Behav Evol 51: 239–262.

Ruse, M. 1996. Monad to man: the concept of progress in evolutionary biology. Cambridge, MA: Harvard University Press.

Russel, E. S. 1916. Form and function: a contribution to the history of animal morphology. Chicago: University of Chicago Press.

Sacher, G. A. 1970. Allometric and factorial analysis of brain structure in insectivores and primates. In: Noback, C. R., Montagna, W., editors. The primate brain. New York: Appleton-Century-Crofts, pp. 109–135.

Sacher, G. A. 1973. Maturation and longevity in relation to cranial capacity in hominid evolution. In: Tuttle, R. H., editor. Primate functional morphology and evolution. The Hague: Mouton, pp. 417–441.

Sagan, C. 1977. The dragons of Eden. New York: Ballantine Books.

Salas, C., Broglio, C. V., J.P., and Rodríguez, F. 2003. Evolution of forebrain and spatial cognition in vertebrates: conservation across diversity. Brain Behav Evol 62: 72–82.

Salkoff, L., Butler, A., Fawcett, G., Kunkel, M., Mcardle, C., Paz-Y-Mino, G., Nonet, M., Walton, N., Wang, Z-W., Yuan, A., and Wei, A. 2001. Evolution tunes the excitability of individual neurons. Neuroscience 103: 853–859.

Sanderson, K. J. 1974. Lamination of the dorsal lateral geniculate nucleus of carnivores of the weasel (Mustelidae), racoon (Procyonidae) and fox (Canidae) families. J Comp Neurol 153: 239–266.

Sandmann, D., Engelmann, R., and Peichl, L. 1997. Starburst cholinergic amacrine cells in the tree shrew retina. J Comp Neurol 389:161-176.

Sanes, J. R., and Lichtman, J. W. 1999. Can molecules explain long-term potentiation? Nat Neurosci 2: 597–604.

Sanides, F. 1967. Comparative architectonics of the neocortex of mammals and their evolutionary interpretation. Ann NY Acad Sci 167: 404–423.

Santos, L. R., Ericson, B. N., and Hauser, M. D. 1999. Constraints on problem solving and inhibition: object retrieval in cotton-top tamarins (*Sanguinus oedipus oedipus*). J Comp Psychol 113: 186–193.

Scalia, F., and Ebbesson, S. O. E. 1971. The central projections of the olfactory bulb in a teleost, *Gymnothorax funebris*. Brain Behav Evol 4: 376–399.

Scalia, F., Halpern, M., Knapp, H., and Riss, W. 1968. The efferent connections of the olfactory bulb in the frog: a study of degenerating unmyelinated fibers. J Anat 103: 245–262.

Scannell, J. W., and Young, M. P. 1993. The connectional organization of neural systems in the cat cerebral cortex. Current Biol 3: 191–200.

Scannell, J. W., Blakemore, C., and Young, M. P. 1995. Analysis of connectivity in the cat cerebral cortex. J Neurosci 15: 1463–1483.

Schall, J. D., and Bichot, N. P. 1991. Neural correlates of visual and motor decision processes. Curr Opin Neurobio 8: 211–217.

Schlosser, G., Kintner, C., and Northcutt, R. G. 1999. Loss of ectodermal competence for lateral line placode formation in the direct developing frog *Eleutherodactylus coqui*. Dev Biol 213: 354–369.

Schmidt-Nielsen, K. 1984. Scaling: why animal size is so important. Cambridge: Cambridge University Press.

Schneider, G. E. 1969. Two visual systems. Science 163: 895–902.

Schneider, G. E. 1973. Early lesions of superior colliculus: factors affecting the formation of abnormal retinal connections. Brain Behav Evol 8: 73–109.

Schober, W., and Brauer, K. 1974. Makromorphologie des Zentralnervensystems. II. Teil. Das Gehirn. In: Helmcke J. G., Starck, D., Wermuth, H., editors. Handbuch der Zoologie. Berlin: De Gruyter, pp. 1–26.

Schoen, J. H. R. 1964. Comparative aspects of the descending fibre systems in the spinal cord. Prog Brain Res 11: 203–222.

Schoenbaum, G., and Setlow, B. 2001. Integrating orbitofrontal cortex into prefrontal theory: common processing themes across species and subdivisions. Learning and Memory 8: 134–147.

Schultz, A. H. 1969. The life of primates. New York: Universe Books.

Schüz, A., and Deminanenko, G. P. 1995. Constancy and variability in cortical structure. A study on synapses and dendritic spines in hedgehog and monkey. J Hirnforsch 36: 113–122.

Schüz, A., and Palm, G. 1989. Density of neurons and synapses in the cerebral cortex of the mouse. J Comp Neurol 286: 442–455.

Scudder, S. H. 1874. In the laboratory with Agassiz, by a former pupil. Every Saturday XVI (April 4): 369–370.

Sedgwick, A. 1894. On the law of development commonly known as von Baer's law; and on the significance of ancestral rudiments in embryonic development. Quart J Microscop Sci 36: 35–52.

Semendeferi, K. 2002. Humans and great apes share a large frontal cortex. Nat Neurosci 5: 272–276.

Semendeferi, K., Damasio, H., Frank, R., and van Hoesen, G. W. 1997. The evolution of the frontal lobes: a volumetric analysis based on three-dimensional reconstructions of magnetic resonance scans of human and ape brains. J Human Evol 32: 375–388.

Semendeferi, K., Armstrong, E., Schleicher, A., Zilles, K., and van Hoesen, G. W. 1998. Limbic frontal cortex in hominoids; a comparative study of area 13. Am J Phys Anthrop 106: 129–155.

Semendeferi, K., Armstrong, E., Schleicher, A., Zilles, K., and van Hoesen, G. W. 2001. Prefrontal cortex in humans and apes: a comparative study of area 10. Am J Phys Anthrop 114: 224–241.

Serluca, F. C., and Fishman, M. C. 2001. Pre-pattern in the pronephric kidney field of zebrafish. Development 128: 2233–2241.

Shakespeare, W. c. 1600. Hamlet, Act II, sc. Ii: 303-308

Sharpe, F. A., and Dill, L. M. 1997. The behavior of Pacific herring schools in response to artificial humpback whale bubbles. Can J Zool 75: 725–730.

Shea, J. J. 2003. Neandertals, competition, and the origin of modern human behavior in the Levant. Evol Anthrop 12: 173–187.

Sheehan, W. 1996. The planet Mars: a history of observation and discovery. Tuscon: University of Arizona Press.

Shepherd, G. M. 1991. Foundations of the neuron doctrine. New York: Oxford University Press.

Sherman, S. M., and Koch, C. 1986. The control of retinogeniculate transmission in the mammalian lateral geniculate nucleus. Exp Brain Res 63: 1–20.

Sibbing, F. A., Osse, J. W. M., and Terlouw, A. 1986. Food handling in the carp (*Cyprinus carpio*): its movement patterns, mechanisms, and limitations. J Zool Lond (A) 210: 161–203.

Sibley, C. G., Ahlquist, J. E., and Monroe, B. L., Jr. 1988. A classification of the living birds of the world based on DNA-DNA hybridization studies. Auk 105: 409–423.

Silberberg, A., Fujita, K. 1996. Pointing at smaller food amounts in an analogue of Boysen and Berntson's procedure. J Exp Anal Behav 66: 143–147.

Silver, R. A., Lübke, J., Sakman, B., and Feldmeyer, D. 2003. High-probability uniquantal transmission at excitatory synapses in barrel cortex. Science 302: 19811984.

Simek, J. 1992. Neanderthal cognition and the Middle to Upper Paleolithic transition. In: Bräuer, G., Smith, F. H., editors. Continuity or replacement? Controversies in *Homo sapiens* evolution. Rotterdam: Balkema, pp. 231–245.

Simmons, R. E., and Scheepers, L. 1996. Winning by a neck: sexual selection in the evolution of giraffe. Am Nat 148: 771–786.

Simonyan, K., and Jürgens, U. 2003. Efferent subcortical projections of the laryngeal motor cortex in the rhesus monkey. Brain Res 974: 43–59.

Simpson, G. G. 1958. The study of evolution: methods and present status of theory. In: Roe, A., Simpson, G. G., editors. Behavior and evolution. New Haven: Yale University Press, pp. 7–26.

Singer, C. 1952. Vesalius on the human brain. Oxford: Oxford University Press.

Singer, W. 1977. Control of thalamic transmission by corticofugal and ascending reticular pathways in the visual system. Physiol Rev 57: 386–420.

Singer, W., and Gray, C. M. 1995. Visual feature integration and the temporal correlation hypothesis. Ann Rev Neurosci 18: 555–586

Slack, J. M. W., Holland, P. W. H., and Graham, C. F. 1993. The zootype and the phylotypic stage. Nature 361: 490–492.

Smirnov, Y. 1989. Intentional human burial: Middle Paleolithic (last glaciation) beginnings. J World Prehist 3: 199–233.

Smith, B. H., Tompkins, R. L. 1995. Toward a life-history of the hominidae. Ann Rev Anthropol 24: 257–279.

Smith, C. U. M. 1987. "Clever beasts who invented knowing": Nietzsche's evolutionary biology of knowledge. Biol Philosophy 2: 65–91.

Smith, G. E. 1910. Some problems relating to the evolution of the brain. Lancet, Jan 15: 147–153.

Smith, H. M. 1967. Biological similarities and homologies. Syst Zool 16: 101–102.

Smith, R. J. 1980. Rethinking allometry. J Theor Biol 87: 97–111.

Smith, R. J. 1994. Degrees of freedom in interspecific allometry: an adjustment for the effects of phylogenetic constraint. Am J Physic Anthrop 93: 95–107.

Smith, R. J. 2002. Estimation of body mass in paleontology. J Human Evol 42: 271–287.

Smith-Fernandez, A., Pieau, C., Reperant, J., Boncinelli, E., and Wassef, M. 1998. Expression of the *Emx-1* and *Dlx-1* homeobox genes define three molecularly distinct domains in the telencephalon of mouse, chick, turtle, and frog embryos; implications for the evolution of telencephalic subdivisions in amniotes. Development 125: 2099–2111.

Snell, O. 1891. Die Abhängigkeit des Hirngewichtes von dem Körpergewicht und den geistigen Fähigkeiten. Arch Psychiat Nervenkrankh 23: 436–446.

Sober, E. 2000. Philosophy of biology. Boulder, CO: Westview Press.

Sol, D., Timmermans, S., and Lefebvre, L. 2002. Behavioural flexibility and invasion success in birds. Animal Behaviour 63: 495–502.

Spemann, H. 1915. Zur Geschichte und Kritik des Begriffs der Homologie. Kultur der Gegenwart, Part 3 4(1): 63–86.

Spemann, H. 1927. Croonian Lectures. Organizers in animal development. Proc R Soc Bio Sci B 102: 177–187.

Sperry, R. W. 1963. Chemoaffinity in the orderly growth of nerve fiber patterns and connections. Proc Natl Acad Sci (USA) 50: 703–710.

Sporns, O., Tononi, G., and Edelman, G. M. 2000. Theoretical neuroanatomy: relating anatomical and functional connectivity in graphs and cortical connection matrices. Cereb Cortex 10: 127–141.

Stamos, D. N. 1996. Popper, falsifiability, and evolutionary biology. Biol Philosophy 11: 161–191.

Stanford, C. B. 1999. The hunting apes. Princeton, N.J.: Princeton University Press.

Stanley, S. M. 1973. An explanation for Cope's rule. Evolution 27: 1–26.

Stedman, H. H., Kozyak, B. W., Nelson, A., Thesier, D. M., Su, L. T., Low, D. W., Bridges, C. R., Shrager, J. B., Minugh-Purvis, N., and Mitchell, M. A. 2004. Myosin gene mutation correlates with anatomical changes in the human lineage. Nature 428: 415–418.

Stephan, H., and Andy, O. J. 1970. The allocortex in primates. In: Noback, C. R., Montagna, W., editors. The primate brain. New York: Appleton-Century-Crofts, pp. 109–135.

Stephan, H., Bauchot, R., and Andy, O. J. 1970. Data on the size of the brain and of various brain parts in insectivores and primates. In: Noback, C. R., Montagna, W., editors. The primate brain. New York: Appleton-Century-Crofts, pp. 289–297.

Stephan, H., Frahm, H., and Baron, G. 1981a. New and revised data on volumes of brain structures in insectivores and primates. Folia Primatol 35: 1–29.

Stephan, H., Nelson, J. E., and Frahm, H. D. 1981b. Brain size comparisons in Chiroptera. Z Zool Syst Evolutionsforsch 19: 195–222.

Stephan, H., Baron, G., and Frahm, H. D. 1988. Comparative size of brains and brain components. In: Steklis, H. D., Erwin, J., editors.

Comparative primate biology. New York: Liss, pp. 1–38.

Steriade, M. 2003. The corticothalamic system in sleep. Frontiers in Biosci 8:D878-D899.

Stern, D. L., and Emlen, D. J. 1999. The developmental basis for allometry in insects. Development 126: 1091–1101.

Sternberger, L. A., Hardy, P. H., Cuculis, J. J., and Meyer, H. G. 1970. The unlabeled antibody-enzyme method of immunohistochemistry. J Histochem Cytochem 18: 315–333.

Stevens, C. F. 1989. How cortical interconnectedness varies with network size. Neural Computation 1: 473–479.

Stevens, C. F. 2001. An evolutionary scaling law for the primate visual system and its basis in cortical function. Nature 411: 193–195.

Stingelin, W. 1958. Vergleichend morphologische Untersuchungen am Vorderhirn der Vögel ouf cytologischer und cytoarchitektonischer Grundlage. Basel: Helbing & Lichtenhahn.

Strausfeld, N. J., Homburg, U., and Kloppenberg, P. 2000. Parallel organization in honey bee mushroom bodies by peptidergic Kenyon cells. J Comp Neurol 424: 179–195.

Stretavan, D. W., and Shatz, C. J. 1986. Prenatal development of retinal ganglion cell axons: segregation into eye-specific layers within the cat's lateral geniculate nucleus. J Neurosci 6: 234–251.

Striedter, G. F. 1990a. The diencephalon of the channel catfish, *Ictalurus punctatus*. I. Nuclear organization. Brain Behav Evol 36: 329–354.

Striedter, G. F. 1990b. The diencephalon of the channel catfish, *Ictalurus punctatus*. II. Retinal, tectal, cerebellar and telencephalic connections. Brain Behav Evol 36: 355–377.

Striedter, G. F. 1991. Auditory, electrosensory, and mechanosensory lateral line pathways through the forebrain of channel catfishes. J Comp Neurol 312: 311–331.

Striedter, G. F. 1992. Phylogenetic changes in the connections of the lateral preglomerular nucleus in ostariophysan teleosts: a pluralistic view of brain evolution. Brain Behav Evol 39: 329–357.

Striedter, G. F. 1994. The vocal control pathways in budgerigars differ from those of songbirds. J Comp Neurol 343: 35–56.

Striedter, G. F. 1997. The telencephalon of tetrapods in evolution. Brain Behav Evol 49: 179–213.

Striedter, G. F. 1998a. Progress in the study of brain evolution: from speculative theories to testable hypotheses. Anat Rec (New Anat) 253: 105–112.

Striedter, G. F. 1998b. Stepping into the same river twice: homologues as recurring attractors in epigenetic landscapes. Brain Behav Evol 52: 218–231.

Striedter, G. F. 1999. Homology in the nervous system: of characters, embryology and levels of analysis. Homology, pp. 158–172.

Striedter, G. F. 2002. Brain homology and function: an uneasy alliance. Brain Res Bull 57: 239–242.

Striedter, G. F. 2003. Epigenesis and evolution of brains: from embryonic divisions to functional systems. In: Müller G. B., Newman S. A., editors. Origination of organismal form: beyond the gene in developmental and evolutionary biology. Cambridge, MA: MIT Press, pp. 287–303.

Striedter, G. F. 2004. Brain evolution. In: Paxinos G., May J. K., editors. The human nervous system (Second Edition). San Diego: Elsevier, pp. 3–21.

Striedter, G. F., and Beydler, S. 1997. Distribution of radial glia in the developing telencephalon of chicks. J Comp Neurol 387: 399–420.

Striedter, G. F., and Northcutt, R. G. 1989. Two distinct visual pathways through the superficial pretectum in a percomorph teleost. J Comp Neurol 283: 342–354.

Striedter, G. F., and Northcutt, R. G. 1991. Biological hierarchies and the concept of homology. Brain Behav Evol 38: 177–189.

Striedter, G. F., Marchant, T. A., and Beydler, S. 1998. The "neostriatum" develops as part of the lateral pallium in birds. J Neurosci 18: 5839–5849.

Stryker, M. P., and Zahs, K. R. 1983. On and Off sublaminae in the lateral geniculate nucleus of the ferret. J Neurosci 3: 1943–1951.

Stuart, G., and Spruston, N. 1998. Determinants of voltage attenuation in neocortical pyramidal neuron dendrites. J Neurosci 18: 3501–3510.

Sues, H.-D., and Reisz, R. R. 1998. Origins and early evolution of herbivory in tetrapods. Trends Ecol Evol 13: 141–145.

Suga, N., Xiao, Z., Ma, X., and Ji, W. 2002. Plasticity and corticofugal modulation for hearing in adult animals. Neuron 36: 9–18.

Supèr, H., and Uylings, H. B. M. 2001. The early differentiation of the neocortex: a hypothesis on neocortical evolution. Cereb Cortex 11: 1101–1109.

Sussman, R. W. 1991. Primate origins and the evolution of angiosperms. Am J Primatol 23: 209–223.

Suzuki, D. T., Griffiths, A. J. F., Miller, J. H., and Lewontin, R. C. 1989. An introduction to genetic analysis. New York: W. H. Freeman.

Svensson, P., Romaniello, A., Arendt-Nilsen, L., and Sessle, B. J. 2003. Plasticity in corticomotor control of the human tongue musculature induced by tongue-task training. Exp Brain Res 152: 42–51.

Swanson, L. W., and Petrovich, G. D. 1998. What is the amygdala? Trends Neurosci 21: 323–330.

Székely, A. D. 1999. The avian hippocampal formation: subdivisions and connectivity. Behav Brain Res 98: 219–225.

Szekely, T., Catchpole, C. K., deVoogd, A., Marchl, Z., and deVoogd, T. J. 1996. Evolutionary changes in a song control area of the brain (HVc) are associated with evolutionary changes in song repertoire among European warblers (Sylviidae). Proc R Soc Bio Sci B 263: 607–610.

Tague, R. G., and Lovejoy, C. O. 1986. The obstetric pelvis of A.L. 288–1 (Lucy). J Human Evol 15: 237–255.

Talwar, S., Musial, P. G., and Gerstein, G. L. 2001. Role of mammalian auditory cortex in the perception of elementary sound properties. J Neurophysiol 85: 2350–2358.

Tanaka, H., and Landmesser, T. 1986. Cell death of lumbosacral motoneurons in chick, quail, and chick-quail chimera embryos: a test of the quantitative matching hypothesis of neuronal cell death. J Neurosci 6: 2889–2899.

Tattersall, I., Schwartz, J. H. 1999. Hominids and hybrids: the place of Neanderthals in human evolution. Proc Natl Acad Sci (USA) 96: 7117–7119.

Tavaré, S., Marshall, C. R., Will, O., Soligo, C., and Martin, R. D. 2002. Using the fossil record to estimate the age of the last common ancestor of extant primates. Nature 416: 726–729.

Taylor, C. R., and Weibel, E. R. 1981. Design of the mammalian respiratory system. I. Problem and strategy. Respir Physiol 44: 1–10.

Taylor, G. M., Nol, E., and Boire, D. 1995. Brain regions and encephalization in anurans: adaptation or stability? Brain Behav Evol 45: 96–109.

Templeton, A. R. 2002. Out of Africa again and again. Nature 416: 45–51.

Terrace, H. S. 1979. Nim: a chimpanzee who learned sign language. New York: Washington Square Press.

Thanos, S., and Mey, J. 2001. Development of the visual system of the chick. II. Mechanisms of axonal guidance. Brain Res Rev 35: 205–245.

Thelen, K. 2003. How institutions evolve: insights from comparative historical analysis. In: Mahoney, J., Rueschemeyer, D., editors. Comparative historical analysis in the social sciences. Cambridge: Cambridge University Press, pp. 208–240.

Thewissen, J. G. M., Williams, T. M., Roe, L. J., and Hussain, S. T. 2001. Skeletons of terrestrial cetaceans and the relationship of whales to artiodactyls. Nature 413: 277–281.

Thireau, M. 1975. L'allométrie pondérales encéphalo-somatique chez les Urodéles. II Relations interspécifiques. Bull Mus natn Hist nat, Paris 279, Zool 207: 483–501.

Thompson, D. W. 1959. On growth and form. Cambridge: Cambridge University Press.

Tilney, F. 1927. The brain stem of Tarsius. A critical comparison with other primates. J Comp Neurol 43: 371–432.

Tinbergen, N. 1951. The study of instinct. New York: Oxford University Press.

Tinklepaugh, O. L. 1932. Maze learning of a turtle. J Comp Psychol 13: 201–206.

Tobias, P. V. 1973. Brain evolution in the Hominoidea. In: Tuttle, R. H., editor. Primate functional morphology and evolution, pp. 353–392.

Tobias, P. V. 1996. The dating of linguistic beginnings. Behav Brain Sci 19: 789.

Tobias, P. V. 2001. Re-creating ancient hominid virtual endocasts by CT-scanning. Clinical Anat 14: 134–141.

Tomasello, M., Hare, B., and Agnetta, B. 1999. Chimpanzees, *Pan troglodytes*, follow gaze direction geometrically. Anim Behav 58: 769–777.

Tomback, D. F. 1980. How nutcrackers find their seed stores. Condor 82: 10–19.

Tower, D. B. 1954. Structural and functional organization of mammalian cerebral cortex: the correlation of neurone density with brain size. J Comp Neurol 101: 19–51.

Tower, D. B., and Elliott, K. A. C. 1952. Activity of acetylcholine system in cerebral cortex of various unanesthetized mammals. Am J Physiol 168: 747–759.

Tower, S. S. 1940. Pyramidal lesion in the monkey. Brain 63: 36–90.

Tremblay, L., and Schultz, W. 1999. Relative reward preference in primate orbitofrontal cortex. Nature 398: 704–708.

Tulving, E. 1985. Memory and consciousness. Can Psychol 26: 1–12.

Turlejski, K. 1996. Evolutionarily ancient roles of serotonin: long-lasting regulation of activity and development. Acta Neurobiologiae Experimentalis 56: 619–636.

Tyner, C. F. 1975. The naming of neurons: applications of taxonomic theory to the study of cellular populations. Brain Behav Evol 12: 75–96.

Ulfig, N., and Chan, W. Y. 2001. Differential expression of calcium-binding proteins in the red nucleus of the developing and adult human brain. Anat Embryol 203: 95–108.

Ulinski, P. 1983. Dorsal ventricular ridge: a treatise on forebrain organization in reptiles and birds. New York: Wiley.

Ulinski, P. 1990. The cerebral cortex of reptiles. In: Jones, E. G., Peters, A., editors. Cerebral cortex, Vol. 8A., Part I. New York: Plenum, pp. 139–215.

Uylings, H. B. M., and van Eden, C. G. 1990. Qualitative and quantitative comparison of the prefrontal cortex in rat and in primates, including humans. Prog Brain Res 85: 31–62.

Valverde, F. 1986. Intrinsic neocortical organization: some comparative aspects. Neuroscience 18: 1–23.

Valverde, F., de Carlos, J. A., López-Mascaraque, L., and Donate-Oliver, F. 1986. Neocortical layers I and II of the hedgehog (*Erinaceus europaeus*). Anat Embryol 175: 167–179.

van Dongen, B. A. M. 1998. Brain size in vertebrates. In: Nieuwenhuys R., Ten Donkelaar H. J., Nicholson C., editors. The central nervous system of vertebrates. Berlin: Springer, pp. 2099–2134.

van Essen, D. C. 1997. A tension-based theory of morphogenesis and compact wiring in the central nervous system. Nature 385: 313–318.

van Essen, D. C., Anderson, C. H., and Felleman, D. J. 1992. Information processing in the primate visual system: an integrated systems perspective. Science 255: 419–423.

Vanduffel, W., Fize, D., Peuskens, H., Denys, K., Sunaert, S., Todd, J. T., and Orban, G. A. 2002. Extracting 3D from motion: differences in human and monkey intraparietal cortex. Science 298: 413–415.

Varela, F. J. 1995. Resonant cell assemblies: a new approach to cognitive functions and neural synchrony. Biol Res 28: 81–95.

Veenman, C. L., and Gottschaldt, K.-M. 1986. The nucleus basalis-neostriatum complex in the goose (*Anser anser* L.). Adv Anat Embryol Cell Biol 96: 1–85.

Velasco, A., Jimeno, D., Lillo, C., Caminos, E., Lara, J. M., and Aijón, J. 1999. Enzyme histochemical identification of microglial cells in the retina of a fish (*Tinca tinca*). Neurosci Lett 263: 101–104.

Verhaart, W. J. C. 1970. The pyramidal tract in primates. In: Noback, C. R., Montagna, W., editors. The primate brain. New York: Appleton-Century-Crofts, pp. 83–108.

Vernadakis, A., and Roots, B. I., editors. 1995. Neuron-glia interrelations during phylogeny. Totowa, N.J: Humana Press.

Verney, C., Zecevic, N., and Puelles, L. 2001. Structure of longitudinal brain zones that provide the origin for the substantia nigra and ventral tegmental area in human embryos, as revealed by cytoarchitecture and typrosine hydroxylase, calretinin, calbindin, and GABA immunoreactions. J Comp Neurol 429: 22–44.

Vesalius, A. 1543. De humani corporis fabrica libri septem. Brussels: Basileae.

Vesalius, A. 2002. On the fabric of the human body, Books III and IV. Novato, CA: Norman Publishing.

Vigne, J.-D., Guilaine, J., Debue, K., Haye, L., and Gérard, P. 2004. Early taming of the cat in Cyprus. Science 304: 259.

Volkert, L. G., and Conrad, M. 1998. The role of weak interactions in biological systems: the dual dynamics model. J theor Biol 193: 287–306.

Volman, S. F., Grubb., T. C., and Schuett, K. C. 1997. Relative hippocampal volume in relation to food-storing behavior in four species of woodpeckers. Brain Behav Evol 49: 110–120.

von Baer, K. E. 1828. Über Entwicklungsgeschichte der Thiere: Beobachtung und Reflexion. Königsberg: Bornträger.

von Bonin, G. 1941. Sidelights on cerebral evolution: brain size of lower vertebrates and degree of cortical folding. J Gen Psychol 25: 273–282.

von Economo, C. 1929. The cytoarchitectonics of the human cortex. Oxford: Oxford University Press.

von Haller, A. 1762. Elementa physiologiae corporis humani. Lausanne: Sumptibus M.-M. Bousquet et Sociorum

von Staden, H. 1992. The discovery of the body: human dissection and its cultural contexts in ancient Greece. Yale J Biol and Med 65: 223–241.

Vonderschen, K., Bleckmann, H., and Hofmann, M. H. 2002. A direct projection from the cerebellum to the telencephalon in the goldfish, *Carassius auratus*. Neurosci Lett 320: 37–40.

Voogd, J. 2003. Cerebellum and precerebellar nuclei. In: Paxinos G., May J. K., editors. The human nervous system (Second Edition). San Diego: Elsevier, pp. 321-392.

Voogd, J., Nieuwenhuys, R., van Dongen, P. A. M., and Ten Donkelaar, H. J. 1998. Mammals. In: Nieuwenhuys, R., Ten Donkelaar, H. J., Nicholson, C., editors. The central nervous system of vertebrates. Berlin: Springer, pp. 1637–2097.

Vrba, E. S. 1995. The fossil record of African antelopes relative to human evolution. In: Vrba, E. S., Denton, G. H., Partridge, T. C., Burkle, L. H., editors. Paleoclimate and evolution, with emphasis on human evolution. New Haven: Yale University Press, pp. 385–424.

Waddell, P. J., Kishino, H., and Ota, R. 2001. A phylogenetic foundation for comparative mammalian genomics. Genome Informatics 12: 141–154.

Wagner, G. P., and Altenberg, L. 1996. Complex adaptations and the evolution of evolvability. Evolution 50: 967–976.

Wagner, H., and Frost, B. 1994. Binocular responses of neurons in the barn owl's visual Wulst. J Comp Physiol A 174: 661–670.

Wallace, M. N., Johnston, P. W., and Palmer, A. R. 2002. Histochemical identification of cortical areas in the auditory region of the human brain. Exp Brain Res 143: 499–508.

Wallis, J. D., and Miller, E. K. 2003. Neuronal activity in primate dorsolateral and orbital prefrontal cortex during performance of a reward preference task. Eur J Neurosci 18: 2069–2081.

Wallis, J. D., Dias, R., Robbins, T. W., and Roberts, A. C. 2001. Dissociable contributions of the orbitofrontal and lateral prefrontal cortex of the marmoset to performance on a detour reaching task. Eur J Neurosci 13: 1797–1808.

Walls, G. L. 1942. The vertebrate eye and its adaptive radiation. Bloomfield Hills, MI: Cranbrook Institute of Science.

Walls, G. L. 1953. The lateral geniculate nucleus and visual histophysiology. Berkeley: University of California Press.

Wang, S., Bickford, M. E., Van Horn, S. C., Erisir, A., Godwin, D. W., and Sherman, S. M. 2001. Synaptic targets of thalamic reticular nucleus terminals in the visual thalamus of the cat. J Comp Neurol 440: 321–341.

Wang, Y., Hu, Y., Meng, J., and Li, C. 2001. An ossified Meckel's cartilage in two Cretaceous mammals and origin of the mammalian middle ear. Science 294: 357–361.

Ward, J. P., and Masterton, B. 1970. Encephalization and visual cortex in the tree shrew (*Tupaia glis*). Brain Behav Evol 3: 421–469.

Wasowicz, M., Pierre, J., Repérant, J., Ward, R., Vesselkin, N. P., and Versaux-Botteri, C. 1994. Immunoreactivity to Glial Fibrillary Acidic Protein (GFAP) in the brain and spinal cord of the lamprey (*Lampetra fluviatilis*). J Brain Res 35: 834–837.

Wasowicz, M., Ward, R., and Repérant, J. 1999. An investigation of astroglial morphology in *Torpedo* and *Scyliorhinus*. J Neurocytol 28: 639–653.

Watanabe, M., Inoue, Y., Sakimura, K., Mishina, M. 1992. Developmental changes in distribution of NMDA receptor channel subunit messenger RNAs. Neuroreport 3: 1138–1140.

Watts, D. J., Strogatz, S. H. 1998. Collective dynamics of 'small-world' networks. Nature 393: 440–442.

Webster, D. B., Fay, R. R., and Popper, A. N., editors. 1992. The evolutionary biology of hearing. New York: Springer-Verlag.

Webster, K. E. 1979. Some aspects of the comparative study of the corpus striatum. In: Divac, I., Öberg, R. G. E., editors. The neostriatum. New York: Elsevier/Pergamon, pp. 107–126.

Weibel, E. R. 2000. Symmorphosis: on form and function in shaping life. Cambridge, MA: Harvard University Press.

Weir, A. A. S., Chappell, J., and Kacelnik, A. 2002. Shaping of hooks in New Caledonian crows. Science 297: 981.

Weiskrantz, L. 1982. Blindsight revisited. Curr Opin Neurobio 6: 215–220.

Welker, W. 1990. Why does cerebral cortex fissure and fold? A review of determinants of gyri and sulci. In: Jones, E. G., Peters, A., editors. Cerebral cortex, Vol. 8B. New York: Plenum, pp. 3–136.

Welker, W. I., and Campos, G. B. 1963. Physiological significance of sulci in somatic sensory cerebral cortex in mammals of the family Procyonidae. J Comp Neurol 120: 19–36.

Whewell, W. 1837. History of the inductive sciences. London: J.W. Parker.

White, T. 2003. Early hominids - diversity or distortion? Science 299: 1994–1997.

White, T., and Kimbel, B. 1988. A revised reconstruction of the adult skull of *Australopithecus afarensis*. J Human Evol 17: 545–550.

Whiten, A., and Byrne, R. W., editors. 1997. Machiavellian intelligence II. Cambridge: Cambridge University Press.

Whiten, A., Custance, D. M., Gomez, J.-C., Teixidor, P., and Bard, K. 1996. Imitative learning of artificial fruit processing in children (*Homo sapiens*) and chimpanzees (*Pan troglodytes*). J Comp Psychol 110: 3–14.

Wicht, H., and Nieuwenhuys, R. 1998. Hagfishes (Myxinoidea). In: Nieuwenhuys, R., Ten Donkelaar, H. J., Nicholson, C., editors. The central nervous system of vertebrates. Berlin: Springer Verlag, pp. 497–549.

Wicht, H., and Northcutt, R. G. 1992. The forebrain of the Pacific hagfish: a cladistic reconstruction of the ancestral craniate forebrain. Brain Behav Evol 40: 25–64.

Wicht H., and Northcutt, R. G. 1993. Secondary olfactory projections and pallial topography in the Pacific hagfish, *Eptatretus stouti*. J Comp Neurol 337: 529–542.

Wicht, H., Derouiche, A., and Korf, H.-W. 1994. An immunocytochemical investigation of glial morphology in the Pacific hagfish: radial and astrocyte-like glia have the same phylogenetic age. J Neurocytol 23: 565–576.

Wikler, K. C., and Rakic, P. 1991. Distribution of photoreceptor subtypes in the retina of diurnal and nocturnal primates. J Neurosci 10: 3390–3401.

Wilczynski, W. 1984. Central nervous systems subserving a homoplasous periphery. Amer Zool 24: 755–763.

Wild, J. M. 1987. The avian somatosensory system: connections of regions of body representation in the forebrain of the pigeon. Brain Res 412: 205–223.

Wild, J. M. 1989. Pretectal and tectal projections to the homolog of the dorsal lateral geniculate nucleus in the pigeon - an anterograde and retrograde tracing study with choleratoxin conjugated to horseradish-peroxidase. Brain Res 479: 130–137.

Wild, J. M., and Farabaugh, S. M. 1996. Organization of afferent and efferent projections of the nucleus basalis prosencephali in a passering, *Taeniopygia guttata*. J Comp Neurol 365: 306–328.

Wild, J. M., and Williams, M. N. 2000. Rostral Wulst in passerine birds. I. Origin, course, and terminations of an avian pyramidal tract. J Comp Neurol 416: 429–450.

Wild, J. M., Cabot, J. B., Cohen, D. H., and Karten, H. J. 1979. Origin, course and terminations of the rubrospinal tract in the pigeon (*Columba livia*). J Comp Neurol 187: 639–654.

Wild, J. M., Arends, J. J. A., and Zeigler, H. P. 1985. Telencephalic connections of the trigeminal system in the pigeon (*Columba livia*): a trigeminal sensorimotor circuit. J Comp Neurol 234: 441–464.

Wild, J. M., Arends, J. J. A., and Zeigler, H. P. 1990. Projections of the parabrachial nucleus in the pigeon (*Columba livia*). J Comp Neurol 293: 499–523.

Wild, J. M., Kubke, M. F., and Carr, C. E. 2001. Tonotopic and somatotopic representation in the nucleus basalis of the barn owl, *Tyto alba*. Brain Behav Evol 57: 39–62.

Wilkinson, D. G., and Krumlauf, R. 1990. Molecular approaches to the segmentation of the hindbrain. Trends Neurosci 13: 335–339.

Wilkinson, D. G., Bhatt, S., Chavrier, P., Bravo, R., and Charnay, P. 1989. Segment-specific expression of a zinc finger gene in the developing nervous system of the mouse. Nature 337: 461–465.

Williams, R. W. 2000. Mapping genes that modulate mouse brain development: a quantitative genetic approach. In: Goffinet, A. F., Rakic, P., editors. Mouse brain development. New York: Springer Verlag, pp. 21–49.

Williams, R. W., and Herrup, K. 1988. The control of neuron number. Ann Rev Neurosci 11: 423–453.

Willshaw, D. J., and von der Malsburg, C. 1976. How patterned neural connections can be set up by self-organization. Proc R Soc Bio Sci B 194: 431–445.

Wise, S. P., and Murray, E. A. 2000. Arbitrary associations between antecedents and actions. Trends Neurosci 23: 271–276.

Wolpoff, M. H. 1999. Paleoanthropology. Boston: McGraw-Hill.

Wolpoff, M. H., Senut, B., Pickford, M., and Hawks, J. 2002. Sahelanthropus or 'Sahelpithecus'? Nature 419: 581–582.

Wong-Riley, M. T. T. 1972. Neuronal and synaptic organization of the normal dorsal lateral geniculate nucleus of the squirrel monkey, *Saimiri sciureus*. J Comp Neurol 144: 25–60.

Wrangham, R., and Conklin-Brittain, N. 2003. Cooking as a biological trait. Comp Biochem and Physiol A 136: 35–46.

Wright, R. D. 1934. Some mechanical factors in the evolution of the central nervous system. J Anat 69: 86–88.

Wright, S. 1986. Evolution and the genetics of populations. Chicago: University of Chicago Press.

Wu, C. W.-H., Bichot, N. P., and Kaas, J. H. 2000. Converging evidence from microstimulation, architecture, and connections for multiple motor areas in the frontal and cingulate cortex of prosimian primates. J Comp Neurol 423: 140–177.

Wullimann, M. F. 1997. The central nervous system. In: Evans, D. H., editor. The physiology of fishes. New York: CRC Press, pp. 245–282.

Wullimann, M. F., and Puelles, L. 1999. Postembryonic neural proliferation in the zebrafish forebrain and its relationship to prosomeric domains. Anat Embryol 329: 329–348.

Wullimann, M. F., and Rooney, D. J. 1990. A direct cerebellotelencephalic projection in an electrosensory mormyrid fish. Brain Res 520: 354–357.

Wurst, W., and Bally-Cuif, L. 2001. Neural plate patterning: upstream and downstream of the isthmic organizer. Nat Rev Neurosci 2: 99–108.

Yang, H.-W., and Lemon, R. N. 2003. An electron microsopic examination of the corticospinal projection to the cervical spinal cord in the rat: lack of evidence for cortico-motoneuronal synapses. Exp Brain Res 149: 458–469.

Young, L. J., Wang, Z., and Insel, T. R. 1998. Neuroendocrine bases of monogamy. Trends Neurosci 21: 71–75.

Young, M. P. 1992. Objective analysis of the topological organization of the primate cortical visual system. Nature 358: 152–155.

Young, M. P., Scannell, J. W., and Burns, G. 1995. The analysis of cortical connectivity. Berlin: Springer-Verlag.

Young, R. M. 1990. Mind, brain and adaptation in the nineteenth century. Oxford: Clarendon Press.

Zardoya, R., and Meyer, A. 2001. The evolutionary position of turtles revised. Naturwissenschaften 88: 193–200.

Zeki, S. 2003. Improbable areas in the visual brain. Trends Neurosci 26: 23–26.

Zhang, K., and Sejnowski, T. J. 2000. A universal scaling law between gray matter and white matter of cerebral cortex. Proc Natl Acad Sci (USA) 97: 5621–5626.

Zilles, K., and Wree, A. 1995. Cortex: areal and laminar structure. In: Paxinos, G., editor. The rat nervous system. San Diego: Academic Press, pp. 649–685.

Zilles, K., Armstrong, E., Moser, K. H., Schleicher, A., and Stephan, H. 1989. Gyrification in the cerebral cortex of primates. Brain Behav Evol 34: 143–150.

Index

About the Book

Editor: Graig Donini
Project Editor: Chelsea D. Holabird
Copy Editor: Kerry Falvey
Index: Kerry Falvey
Production Manager: Christopher Small
Book Production: Janice Holabird
Illustration Program: Georg Striedter
Book Design: Jefferson Johnson
Cover Concept: Georg Striedter
Cover and Book Manufacture: Courier Companies, Inc.